TITANIUM

Rapid Solidification Technology

COVER PHOTO
Solidification substructures in Ti-40W produced by splat quenching.

TITANIUM

Rapid Solidification Technology

Proceedings for the four session symposium on "Titanium, Rapid Solidification Technology," sponsored by the Titanium Committee of The Metallurgical Society held at the 1986 TMS-AIME Annual Meeting, New Orleans, Louisiana, March 2-6, 1986.

Edited by

F.H. Froes
Air Force Wright Aeronautical Laboratories
Materials Laboratories
Wright-Patterson Air Force Base, Ohio

D. Eylon
Metcut-Materials Research Group
Now With The University of Dayton
Dayton, Ohio

A Publication of The Metallurgical Society, Inc.

Library of Congress Cataloging - in - Publication Data

Titanium rapid solidification technology.

Includes index.
1. Titanium--Rapid solidification processing--Congresses. **I.** Froes, F.H. **II.** Eylon, Daniel. **III.** Metallurgical Society of AIME. Titanium Committee. **IV.** Metallurgical Society of AIME. Meeting.
TS247.T58 1986 673'.73223 86-19176
ISBN 0-87339-050-4

A Publication of The Metallurgical Society, Inc.
420 Commonwealth Drive
Warrendale, Pennsylvania 15086
(412) 776-9000

Printed in the United States of America.
Library of Congress Catalogue Number 86-19176
ISBN NUMBER 0-87339-050-4

Preface

The contents of this book come from the four session symposium on "Titanium, Rapid Solidification Technology," held 2-6 March 1986 during the 115th Annual TMS Meeting, New Orleans, Louisiana, and a number of additional related written contributions. The symposium was sponsored by the Titanium Committee of The Metallurgical Society.

Rapid solidification has emerged over the past three decades, since the pioneering work of Duwez in the United States and Salli in the Soviet Union, to the status of the most exciting and promising technology in metals research. Amorphous materials are being studied from a scientific point of view and are seeing applications as improved magnetic, electric, superconducting, corrosion resistant, high strength, and brazing alloys. Crystalline materials such as aluminum alloys, superalloys, tool steels and steels produced by rapid solidification have shown enhanced behavior over conventional products, particularly mechanical property combinations.

The study of rapid solidification of titanium has only recently been initiated mainly because of the inherent reactivity of titanium in the molten state. However, the high level of interest in this area can be gauged by the large number of attendees at the symposium and by the almost thirty contributions to the present volume. These papers deal with alloy design, production methods, processing and fabrication, microstructures, and mechanical properties. Little mention is made of applications, a reflection of the "early days" of this technology. The ability of rapid solidification to circumvent many of the restrictions placed by equilibrium phase diagrams combined with the microstructural refinement inherent to this technology omens well for development of new titanium alloys which can be used in higher stressed components, at increased temperatures, in conjunction with reduced density. Who would not want a titanium alloy with the strength of steel, the temperature capability of the superalloys, the density of aluminum, coupled with the inherent corrosion and fracture resistance of titanium.

The organizers of the symposium are also the editors of this proceedings. We would like to acknowledge the support of the Titanium Committee of The Metallurgical Society for sponsoring this symposium and also thank the many verbal and written contributors who have made this project such a success. We are especially grateful to Dr. S. M. L. Sastry who helped in the early organization of the symposium and served as a session chairman, and Dr. J. A. Hall who supported us both as chairman of the Titanium Committee and as a chairman of one of the symposium sessions. A special "thank you" is also due to Ms. Karen A. Sitzman who helped us so much with the final preparation of the manuscripts.

Dr. F. H. Froes
Air Force Wright Aeronautical Laboratories
Materials Laboratory, AFWAL/MLLS
Wright-Patterson AFB, Ohio

Dr. D. Eylon
Metcut-Materials Research Group
now with the University of Dayton
Graduate Materials Engineering
Dayton, Ohio

June 1986

Preface

Table of Contents

RAPIDLY SOLIDIFIED TITANIUM - A REVIEW*

F. H. Froes[a] and R. G. Rowe[b]

[a]Air Force Wright Aeronautical Laboratories
Materials Laboratory, AFWAL/MLLS
Wright-Patterson AFB, OH 45433

[b]General Electric Company
Corporate Research and Development Center
P.O. Box 8, Bldg. K-1, 263MB
Schenectady, NY 12301

Abstract

The science and technology of rapid solidification (RS) is discussed and related to the characteristics of the titanium system, particularly the reactivity in the molten state. The various titanium alloy classes which have been produced by the RS approach are presented: conventional alloys such as Ti-6Al-4V, rare earths, metalloids, eutectoid formers, beta alloys, amorphous materials, high aluminum alloys, aluminides, and other minor classes. Finally some thoughts on the future of this technology are discussed.

*This is a condensed version of two recent review papers written by the authors: "Rapidly Solidified Titanium," in Rapidly Solidified Alloys and Their Mechanical and Magnetic Properties, vol. 58, ed. B. C. Giessen, D. Polk, and A. I. Taub (New York: North Holland, 1986), in print; and "Titanium Alloy Rapid Solidification," in Rapid Solidification of Crystalline Alloys, ed. J. L. Wood (London: The Institute of Metals, 1986), in print.

Introduction

Scope

Powder metallurgy (PM) can conveniently be separated into two general areas, the traditional press-and-sinter approach, where alloy elements are blended together, and the newer prealloyed technique [1]. For the titanium system a second sub-division of the prealloyed method can be made [2-4]. In the first category the PM approach is used as a low cost method for production of conventional alloy parts at essentially the same mechanical behavior levels as cast and wrought ingot metallurgy (IM) material. The second involves RS leading to production of unique alloys with enhanced mechanical behavior levels or unique microstructures in conventional alloys again with the same effect. It is this subject of unique alloys/microstructures, particularly with structural applications in mind, which will be presented in the present paper.

Rapid Solidification

Attributes. The advantages which can be gained by RS can most conveniently be divided into constitutional and microstructural effects [5]. RS allows large departures from equilibrium constitution (the identity and compositional ranges of phases formed) resulting in large extensions of solid solubility, formation of non-equilibrium or metastable crystalline phases and metallic glasses, and retention of disordered crystalline structures in normally ordered materials such as intermetallic compounds. Additionally it leads to changes in the morphology and dimensions of the microstructure (the size and location of the phases present) to a more uniform and finer microstructure with a large reduction in solute segregation effects.

Requirements. RS usually refers to rapid solidification of the molten metal to the solid state, involving rapid cooling through the solidification temperature range. This is achieved with two basic requirements, a small cross-sectional dimension of molten metal which must be in good contact with an effective heat extraction medium [5, 6].

Titanium Characteristics. The extreme reactivity of titanium in the molten state sets it apart from most other metals when considering RS techniques. Thus containment must be of major concern in developing RS methods for titanium. Contamination can take place from the mold, guide tubes (tundish) or nozzle. Thus the RS of titanium alloys has generally involved the use of containerless processes (e.g., local melting and centrifugal atomization) which however can lead to problems because of the absence of a large mass of molten metal to give chemical homogeneity and the lack of super-heat.

Total Cooling Path. Although the technology being discussed is referred to as rapid "solidification" it is also critical that subsequent solid state cooling must be rapid, especially in the high solid state temperature range where diffusion is most rapid [6]. This requirement for a fast solid state quench is particularly necessary because the metastability of extended solid solutions or new phases, and the fine microstructure make them extremely vulnerable to solid state decomposition [5].

Titanium Phase Diagrams and Potential for RS

Titanium Phase Diagrams. A number of authors have attempted to categorize titanium alloy phase diagrams [7, 8], all agreeing that there are two major divisions -- alpha stabilized and beta stabilized systems. Of these perhaps the most convenient is that developed by Molchanova [8], and this is discussed in detail in a companion paper in these proceedings by Bomberger and Froes.

Solubility Limits. A summary of the solubility extensions determined in titanium alloy systems to date is shown in Table I. Here the maximum equilibrium solubility is the solubility in the high temperature beta phase. The extended solubility level indicated is the value judged by the present authors to be the level to which the matrix in question did not exhibit second phase particles as solidify, but not necessarily after cool down to room temperature. Estimates of the extension of metastable alpha phase solubility of the beta isomorphous and eutectoid forming elements in titanium, using a somewhat different criterion, have been reported elsewhere [9].

Table I. Solubility Limits of Some Alloying Elements Under Equilibrium and Rapid Solidification Conditions

Element	Matrix	Maximum Equilibrium Solubility[a], at%	Extended Solubility[b], at%	Technique	Reference
B	Binary	0.5	6.0	Splat	10, 11
C	Binary	3.1	10.0	Splat	10, 11
Si	-10at%Zr	5.0	6.0	Splat	10, 12, 13
Ge	-5Al-2.5Sn	8.2	>5.0	EBSQ[c]	14
Nd	Binary	∼0.3	>1.0	EBSQ	15
Y	-8Al	∼0.5	>2.0	Laser	16
Gd	Binary	∼0.3	>0.5	EBSQ	15
La	-5Al, -5Sn	∼1.5	≧1.5	Splat	11, 17
Ce	Binary	∼1.5	>1.0	Splat	11, 18
Er	-5Al	∼0.3	≧1.5	Splat	11
Dy	Binary	∼0.3	>0.6	EBSQ	15

[a] In beta titanium.

[b] As-rapidly solidified product.

[c] Electron beam splat quench.

RS Potential for Titanium Alloys. The classes of elements for which the alloy potential in the titanium system using RS appears greatest, can be summarized as:

(a) the rare earths (including Y) which normally exhibit low solubility

(b) the metalloids (B, C, Si) which exhibit low solubility

(c) the eutectoid formers, which although they exhibit quite high solubility, are very segregation prone.

Titanium Rapid Solidification

Powder Production

The production of RS titanium alloys is complicated by the fact that metal containment is difficult. Even the most chemically stable ceramic compounds are dissolved in liquid titanium [19], therefore most RS processes for titanium alloys rely upon either cold hearth [20] or local melting schemes [2-4] for melt containment.

Titanium powder is produced on a semi-commercial basis by the plasma rotating electrode process (PREP) [2-4] with an estimated cooling rate of 10^3°C/sec to 10^5°C/sec [21]. Heterogeneity of the powder due to local fluctuation of the ingot is occasionally observed [22]. A process similar to PREP in concept is laser spin atomization (LSA) [23, 24] with a cooling rate for 100 micron diameter powder produced by this technique of 10^5°C/sec. Other local melting/centrifugal atomization techniques have been developed for titanium alloys [2, 3, 6, 26].

Gas atomization of RS titanium alloys has been reported by two entirely different techniques: ultrasonic gas atomization (USGA) [27] and gas atomization utilizing cold hearth melting and bottom ejection through a refractory metal-skull nozzle [28, 29].

Titanium RS processes based upon substrate quenching have also been developed such as pendant drop melt extraction (PDME) [6, 30-32]. Another localized melting technique which has been developed is the electron beam splat quench (EBSQ) process [6, 27] which produces horseshoe-shaped splat foils. Prototype scale cold hearth arc melting systems have also been developed in order to melt spin titanium alloys [6, 30, 33, 34]. Ribbon is produced which is somewhat thinner (30-50 μm) than that which is achieved in melt extraction processes because of the higher wheel speeds that can be tolerated in melt spinning [35].

Titanium alloys have also been rapidly solidified for microstructural studies by the process of laser or electron beam surface melting [6, 36] and direct titanium powder production [37].

Compaction and Processing

Thermomechanical Processing (TMP). To date mainly conventional compaction techniques such as extrusion, vacuum hot pressing, and hot isostatic pressing have been used with RS titanium alloys [2-4]. A small amount of work using high strain rate compaction techniques such as gun compaction [38] and explosive methods [39-41] has also been carried out.

Thermochemical Processing (TCP). Hydrogen can easily be introduced into titanium powder in a hydrogen rich atmosphere, and equally easily removed by a vacuum anneal leading to development of refined microstructures of various types in the Ti-6Al-4V and similar alloys [2-4, 42]. Recently, TCP was used to modify the microstructure of eutectoid forming elements [43] and dispersion strengthened high temperature RS titanium alloys [44] with some success.

Alloy Systems

Matrix Microstructure Development. The matrix microstructure of as-produced RS titanium alloys can be significantly more complicated than alloys such as aluminum because of the allotropic change and the occurrence of martensite. In richly beta stabilized alloys RS material quenches to a beta matrix which is relatively uncomplicated [45]. In leaner alloys however the RS material passes through the beta region and is unstable enough that on quenching to room temperature a hexagonal martensite structure occurs [21]. On subsequent heating, the matrix can then evolve in two ways depending on the beta grain size. If the beta grain size is relatively large (≧5 μm) lenticular alpha forms from the martensite lathes [21] in a manner analagous to that which occurs in ingot metallurgy material. If, however, the beta grain size is small (≦5 μm), a quite different equiaxed alpha morphology develops. This alpha grows to consume the martensitic structure in a similar manner to that of the normal lenticular alpha. Examples of this alpha morphology are shown in Figure 1 [14].

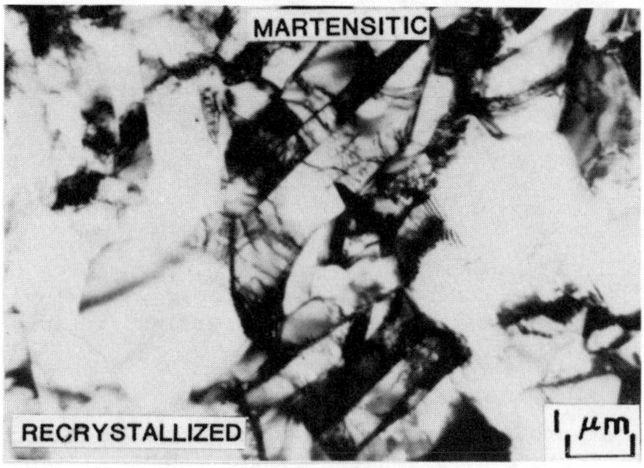

Figure 1 - Equiaxed alpha growing from martensite plates in a rapidly solidified Ti-5Al-2.5Sn-0.5Si alloy aged for 5 hr at 535°C (995°F) [14].

Conventional Alloys. The most comprehensive study done to date on RS conventional titanium alloys has been on the Ti-6Al-4V alloy [2-4, 21] using cooling rates varying from 10^4-10^7°C/sec. From this work two significant effects were noted: the beta grain size decreased as the cooling rate was increased (Figure 2), and at the higher cooling rate studied, equiaxed alpha was produced on subsequent high temperature annealing (Figure 3).

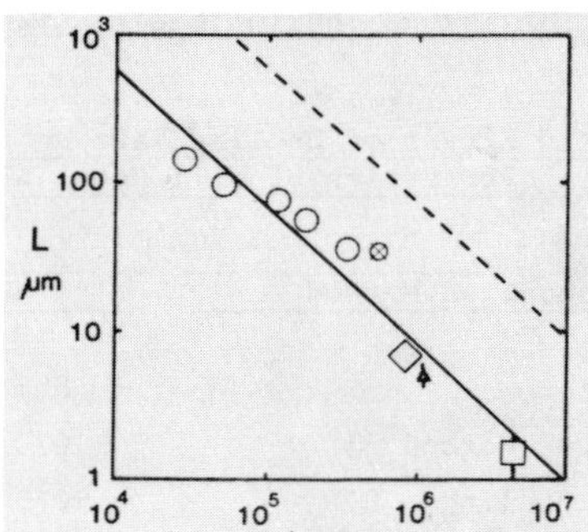

Figure 2 - Variation of beta grain size in Ti-6Al-4V with cooling rate [21]. Dashed line is predicted for aluminum [46].

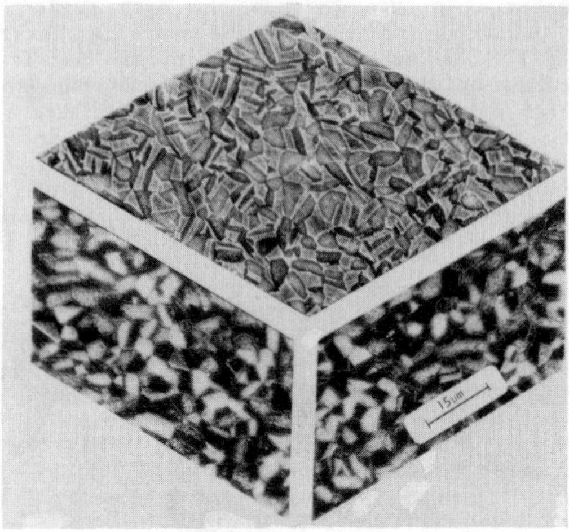

Figure 3 - Equiaxed alpha morphology in Ti-6Al-4V produced after annealing fine grained rapidly solidified material [21].

The result of plotting beta grain size (L) as a function of estimated cooling rate (T) (Figure 4) gives the single power relation:

$$L = A_2 \dot{T}^{-n_2}$$

where $n_2 = 0.93 \pm 0.2$ and $A_2 = 3.1 \times 10^6$ µm $(K/s)^{n_2}$

The exponent n_2 in particular does not differ significantly from that predicted for aluminum [46].

Other work on conventional alloys has shown the advantage to be gained from RS processing. In an alloy similar to Ti-6Al-4V, but containing Mo instead of V, both strength and ductility were increased as the cooling rate was increased from $10^3/10^4$ to $10^6/10^7$ K/s (Table II) [47].

Table II. Effect of Cooling Rate on Tensile Properties of a Ti-Mo-Al Alloy [47]

Production Method	UTS, MPa	RA, %
Casting	895	12
Cast + Hot Worked	1070	22
$10^3/10^4$ Cooling + HIP	1070	22
Flake $10^6/10^7$ Cooling + HIP	1135	30

A study of the Beta III alloy (Ti-11.5Mo-6Zr-4.5Sn) indicated that the beta grain size decreased as the cooling rate increased (∿40 μm in PREP, 2-4 μm in EBSQ) [45] and for the Ti-15V-3Al-3Sn-3Cr alloy a relationship was developed between the calculated cooling rate ($\dot{T}$, K/s) and the secondary dendrite arm spacing (d, mm) using a heat transfer model [36] of:

$$d = 80\dot{T}^{-0.34}$$

<u>Rare Earth Additions</u>. The RS of titanium alloys containing a rare earth element produced an ultrafine dispersion of stable rare earth oxides with particle sizes much finer than that produced in earlier work on IM material [48-59]. Aging at various temperatures produced extremely fine dispersoids, with aging at 500°C (930°F) producing the finest dispersoid [55]. Fabrication of components requires exposure to consolidation cycles at temperatures up to 950°C (1740°F). The stability of the dispersoid is a critical issue to temperatures considerably higher than the service temperature. The stability of the Er_2O_3 dispersoid has been shown to be very good in alpha titanium-aluminum alloys as long as the alloy remains in the alpha titanium phase field. The alloy Ti-6Al-2Sn-4Zr-2Er (wt%), rapidly solidified by melt extraction, was annealed for 1 hr at 950°C [30, 31]. A dispersoid 200-400Å diameter, with an interparticle spacing of ∿0.3-0.4 micron, was maintained in the grain interiors of this alloy.

The rare earth sulfides and oxysulfides form a fine dispersion in alpha titanium alloys and exhibit good resistance to coarsening even after exposure to temperatures as high as 1000°C (1830°F) [31, 32, 60].

Dispersoid coarsening has been observed to be significantly higher near grain boundaries for both Er_2O_3 and CeS dispersoids after elevated temperature exposure [30-32, 60, 61]. Figure 4 [61] shows the depletion and coarsening in Ti_3Al-2Er which had been consolidated by HIP at a temperature of 850°C (1560°F) [61]. The Er_2O_3 dispersoid was relatively unaffected in grain interiors, yet heavily depleted and coarsened near grain boundaries.

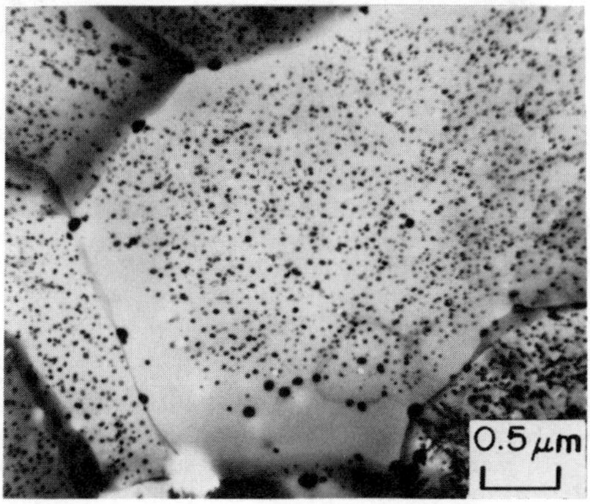

Figure 4 - Depleted grain boundary region and particle coarsening in a Ti_3Al-2Er alloy HIP'd at 850°C (1560°F) [61].

The rare earth dispersoid in titanium alloys appears to effect the behavior of the material in a number of ways. First, the dispersoid produces hardening due to the Orowan effect [62]. There are also indirect effects due to a grain size reduction [15, 31, 32, 60, 63] and scavenging of interstitial elements thereby reducing interstitial strengthening.

Preliminary work on a Ti-6Al-2Sn-4Zr-2Mo-0.1Si-2Er alloy produced using the PREP process [64] indicates that even after compaction, by extrusion at 760°C (1400°F), a relatively fine dispersion of rare earth oxide is present (maximum size 500Å, majority of the particles <100Å) (Figure 5). This suggests that the PREP process may be useful for production of alloys containing rare earth oxides and that in fact a cooling rate of $\sim 10^4$ K/s [21] may be more realistic for this production method than the often quoted 10^2 K/s.

Figure 5 - Fine dispersion of Er_2O_3 particles in PREP Ti-6Al-2Sn-4Zr-2Mo-0.1Si-2Er alloy as-extruded at 760°C (1400°F). Note majority of particles are <100Å [64].

Metalloid Additions (C, B, Si, Ge). The metalloid elements C, B, Si, and Ge form stable titanium compounds whose chemical stability is exceeded only by the rare earth oxides and sulfides. However, the stability of these elements is only moderate in titanium alloys, and their useful temperature range is below that of rare earth oxide dispersion strengthened alloys [57], i.e., below 600°C (1110°F).

RS titanium alloys containing boron have been produced by laser surface melting, splat quenching, melt extraction, and melt spinning [30, 48, 57, 65, 66]. In virtually all studies with the possible exception of hammer and anvil splat quenching, boron decorates the solidification interfaces. Because of this segregation, boride-containing alloys exhibit stronger resistance to grain coarsening of annealed ribbon than alloys containing rare earth additions [30]. Fine, needle-like TiB [30] precipitates develop at temperatures above 500°C (930°F) (Figure 6) [11, 57, 65, 66] and coarsen dramatically at 800°C (1470°F) [11, 30, 66].

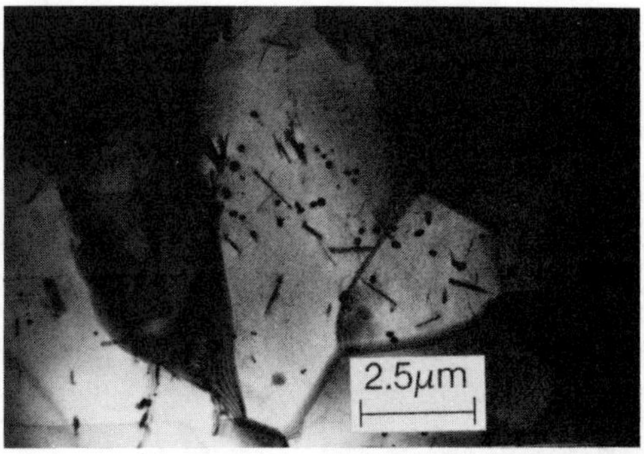

Figure 6 - Fine needle-like TiB precipitates formed by exposing material at 925°C (1700°F) [30].

Silicon and germanium do not segregate as strongly as boron in RS alloys [57, 67-69]. Annealing these RS alloys at intermediate temperatures (550-700°C [1020-1290°F]) produced fine generally spherical precipitates [13, 57, 59, 67, 70] although at the higher temperatures, the particles were relatively coarse [59, 68].

Rapidly solidified titanium alloys containing carbon have been shown to have good strength and ductility as-quenched. However, they lose both strength and ductility after aging at 700°C (1290°F) due to precipitation and rapid coarsening of titanium carbide [66]. Annealing carbon-bearing alloys at 900°C (1650°F) produced ½ micron diameter carbides.

The inherent instability of the metalloid compounds in titanium alloys limits the upper temperature bound of their use to temperatures on the order of 500-600°C (930-1110°F). However RS is an effective means of introducing relatively high volume fractions of fine titanium-metalloid compounds. The potential for alloys based upon metalloid additions is yet to be fully identified.

Eutectoid Formers. The eutectoid formers (e.g., Ni, Cu, Co, Fe, Cr, Si, and W) have been studied in detail using ingot metallurgy [71]. It is normally possible to obtain relatively high levels of solute in the beta phase and lower though still significant levels in solution in the alpha phase. However the eutectoid formers normally show a high partitioning coefficient between the liquid and the solid phase and a large temperature difference between the liquidus and solidus meaning that they are very segregation prone. Thus though there have been a number of attempts to develop titanium alloys containing eutectoid formers using conventional ingot metallurgy only limited success has been achieved due to ingot segregation and also poor workability [72-75]. The great advantage in avoiding segregation which can be achieved using RS is shown in Figure 7 for a Ti-22Fe alloy [76].

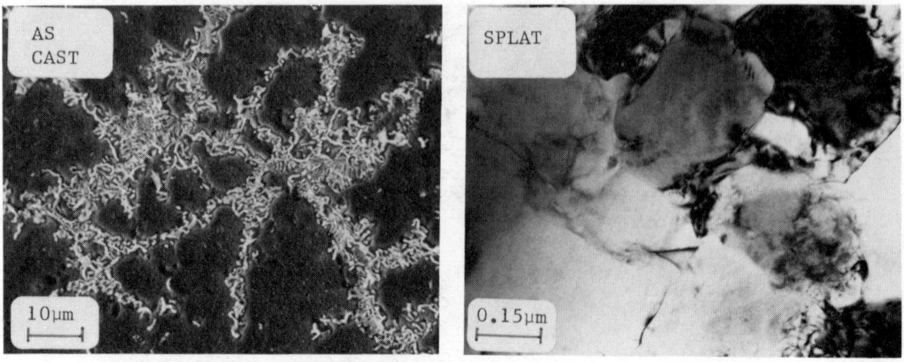

Figure 7 - Ti-22Fe alloy showing difference in segregated eutectic between ingot material and RS product [76].

Studies of RS eutectoid formers [77-89] have indicated that alloys such as Ti-3Ni-6Al and Ti-15Cr-4Al appeared to have the best combination of mechanical properties. In addition, preliminary results on a Ti-8V-5Fe-1Al PREP powder product indicate that strengths in excess of 1380 MPa (200 ksi) can be obtained with ductility levels close to 10% elongation [90].

Thus it can be concluded that RS processing does offer an advantage for the Ti-eutectoid system. Specifically a more homogeneous, more thermally stable precipitation reaction occurs in the RS material. However little has yet been done to optimize compositions to obtain the best mechanical property combination.

Beta Alloys. Except for the conventional beta alloys such as Beta III and Ti-15V-3Al-3Sn-3Cr discussed above there has been little done on beta alloys using the RS technique. Three exceptions are a study of the microstructure and creep behavior of a Ti-6Al-15V-2Er alloy [91] and microstructural studies of a Ti-25V-4Ce-0.6S [92] and a heavily stabilized Ti-24V-10Cr alloy [93, 94]. The Ti-25V-4Ce-0.6S alloy showed a very fine dispersoid leading to a predicted strength in excess of 1725 MPa (250 ksi). The Ti-6Al-15V-2Er alloy showed no dispersoid coarsening to temperatures as high as 760°C (1400°F).

Amorphous Alloys. Metallic glasses or amorphous alloys have been classified in two major groups, metal-metal and metal-metalloid systems [95]. The major theme of the present paper is towards the use of RS titanium alloys for structural applications, however, the interested reader is referred to a number of other articles for further details of titanium base glassy materials for enhanced physical properties such as electrical and magnetic behavior [96-101] and general discussions on metallic glasses in this system [57, 102-108].

The strength of Ti-25Zr-10Be (Metglas 2204, $Ti_{50}Be_{40}Zr_{10}$) was estimated from hardness values to be 2345 MPa (340 ksi) UTS [77]. Even higher values have been obtained in a series of amorphous Ti-M-Si (M = Mn, Fe, Co, Ni, Cu) alloys where strengths close to 2750 MPa (400 ksi) were recorded [102]. In other work [103] Ti-Zr-Ni-B and Ti-Zr-Ni-Si glasses were formed using the hammer and anvil splat quenching technique, and subsequently devitrified by aging. Hardness values corresponding to strengths in excess of 2070 MPa (300 ksi) UTS were obtained with grain sizes less than 1 μm even after exposures as long as 2 hr at 950°C (1740°F). High hardness were also found in Ti-Si, Ti-M-Si, and Ti-M-B alloys [107].

Much further work is needed in this area, particularly on the use of amorphous precursors for usable crystalline structural alloys [108].

High Aluminum, Alpha Titanium Alloys. High aluminum alloys offer the potential for good high temperature behavior due to a dispersion of Ti_3Al precipitates, however ductility is generally low. Work on highly alloyed alpha titanium alloys indicated that consolidated alloys containing a dispersoid are significantly finer in grain size than the same alloy without a dispersoid [31, 32]. It appears that the coarse dispersoid reduces the tendency to planar slip and allows higher aluminum contents which, without dispersoid modification, would have had unacceptable post-creep brittleness [31].

Titanium Aluminides. The titanium aluminides because of their ordered structure are potentially useful alloys for high temperature applications. However the planarity of slip results in low ductility at room temperature. Rapid solidification offers the potential for improved ductility in these intermetallics by disordering [5] grain refinement, and development of fine dispersoid particles. When these fine dispersoids are oxides an additional gain in ductility can result from deoxidation of the matrix.

Early work on RS of this alloy class utilized TiAl-W alloys atomized by the PREP process. This material exhibited good creep resistance due to a fine precipitate of tungsten rich beta phase in the alpha-2 + gamma lath structure of the TiAl composition [110].

Rapid solidification of alloys with a base composition of Ti_3Al with the addition of 0.4 at%Er has been shown to produce a fine dispersoid of Er_2O_3, Figure 8 [61, 111]. This dispersoid does not coarsen in grain interiors to temperatures as high as 1000°C (1830°F) [111]. Under HIP conditions Er_2O_3 dispersoid coarsening in a Ti_3Al-Nb alloy was comparable to that observed during an anneal at the same temperature. However extrusion caused rapid coarsening; which was contrasted with cerium sulfide or cerium oxysulfide containing alloys where virtually no coarsening occurred [60, 61]. The presence of a dispersoid in Ti_3Al-Nb alloys also produced a refinement in grain size. The ductility and fracture toughness of dispersoid containing alloys was dependent on the test used. Hardness indentation ductility was improved by grain refinement but fracture toughness was not [60]. Much further work is required in this generic system to optimize the microstructure and hence mechanical properties.

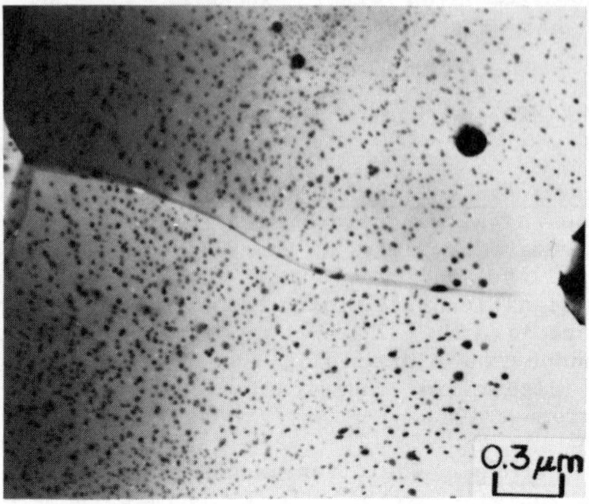

Figure 8 - Fine stable dispersion of Er_2O_3 particles in Ti_3Al aged 750°C (1380°F)/1 hr [60].

Other Systems. Other alloy elements which could be added to titanium to give attractive characteristics are the low density elements Li, Mg (Al and Si have a similar effect). The challenge here is to successfully add these elements which have a boiling point below the melting point of titanium. However work is in progress in a number of laboratories to achieve these alloying possibilities by innovative melting techniques [112, 113], use of the ALTi powder making method [37], blended elemental approaches [114], mechanical alloying [115], and direct production from the vapor phase [116].

Conclusions and Future Thoughts

The study of RS titanium alloys has got off to a later start than in some other systems such as aluminum primarily because of problems such as the reactivity of molten titanium. However there is obviously much to be gained from this approach. Conventional alloys can be produced with refined and more desirable microstructures. The addition of the generally catastropic rare earths and metalloids can be greatly refined and made useful via RS. Segregation in alloys containing eutectoid formers can be controlled much better using the RS technique making this alloy class particularly attractive for high strength applications. Preliminary work suggests that new and improved beta alloys can be synthesized by RS. The surface has only just been scratched in using amorphous glassy materials as a precursor to production of useful crystalline alloys. Some of the inherent problems associated with the ordered aluminides may be circumvented using much faster cooling rates. Finally, development of lower density titanium alloys may be possible with RS.

The technology of RS titanium has effectively been explored for only about a half decade. Yet already a myriad of very attractive avenues for exploration have been defined. In concert with coating developments for high temperature applications, and useful fiber reinforcement, this technology should see many applications within the next decade.

Acknowledgments

The authors would like to acknowledge useful discussions and comments from their colleagues: H. I. Aaronson, H. B. Bomberger, R. R. Boyer, T. F. Broderick, D. Eylon, H. L. Fraser, S. Fujishiro, M. F. X. Gigliotti, F. J. Hehmann, J. P. Hirth, A. G. Jackson, H. Jones, Y-W. Kim, E. F. Koch, D. G. Konitzer, S. Krishnamurthy, H. A. Lipsitt, Y. R. Mahajan, I. A. Martorell, R. E. Omlor, S. M. L. Sastry, S. J. Savage, P. R. Smith, D. B. Snow, J. A. Sutliff, K. R. Teal, R. G. Vogt, I. Weiss, S. H. Whang, J. C. Williams, and C. F. Yolton. In addition we would like to recognize the high level of efficiency of Ms. Karen A. Sitzman in typing the text.

References

1. R. M. German, Powder Metallurgy Science (MPIF, Princeton, NJ, 1984).

2. F. H. Froes and J. R. Pickens, J. of Metals 36 (1), 14-28 (1984).

3. F. H. Froes and D. Eylon, in Titanium, Science and Technology, Vol. 1, edited by G. Lutjering, U. Zwicker, and W. Bunk (DGM, Oberursel, West Germany, 1985), pp. 267-286.

4. Titanium Technology: Present Status and Future Trends, edited by F. H. Froes, D. Eylon, and H. B. Bomberger (The Titanium Development Association, Dayton, OH, 1985).

5. H. Jones, Rapid Solidification of Metals and Alloys, Monograph No. 8, (Institution of Metallurgists, London, 1982).

6. S. J. Savage and F. H. Froes, J. of Metals 36 (4), 20-33 (1984).

7. H. Margolin and J. P. Nielson, in Modern Materials, Advances in Development and Application, Vol. 2, edited by H. H. Hauser (Academic Press, 1960), pp. 225-325.

8. E. K. Molchanova, Phase Diagrams of Titanium Alloys, (Israel Program for Scientific Translations, Jerusalem, 1965).

9. S. H. Whang and C. S. Chi, in Rapidly Solidified Alloys and Their Mechanical and Magnetic Properties, ed. B. C. Giessen, D. Polk, and A. I. Taub, vol. 58 (New York: North Holland, 1986), in print.

10. C. S. Chi and S. H. Whang (private communication, 1985).

11. S. H. Whang (private communication, 1985).

12. J. E. O'Neal, S. M. L. Sastry, T. C. Peng, and J. F. Tesson, Micro. Sci. 11, 143 (1983).

13. S. H. Whang, Y. Z. Lu, and Y-W. Kim, J. Mat. Sci. Letts. 4, 883 (1985).

14. A. G. Jackson, PhD thesis, University of Cincinnati, 1983.

15. S. M. L. Sastry, P. J. Meschter, and J. E. O'Neal, Met. Trans. A 15A, 1451 (1984).

16. D. G. Konitzer, B. C. Muddle, and H. L. Fraser, Met. Trans. A 14A, 1979 (1983).

17. Y. Z. Lu, C. S. Chi, and S. H. Whang, in Rapidly Quenched Metals, Vol. 1, edited by S. Steeb and H. Warlimont (Elsevier, Amsterdam, 1985), pp. 949-952.

18. S. H. Whang, U.S. Patent No. 4 512 826 (23 April 1985).

19. C. E. Holcombe and T. R. Serandos, Met. Trans. B 14B, 497 (1983).

20. S. J. Noesen, in International Transactions Vacuum Metallurgy Conference 1967, edited by E. L. Foster (Am. Vac. Soc., New York, 1967), p. 503.

21. T. F. Broderick, A. G. Jackson, H. Jones, and F. H. Froes, Met. Trans. A, 16A, 1951 (1985).

22. C. H. Smith, General Electric Company (Schenectady, NY) Report No. 84CRD071, 1984.

23. D. G. Konitzer, K. W. Walters, E. L. Heiser, and H. L. Fraser, Met. Trans. B 15B, 149 (1984).

24. T. C. Peng, S. M. L. Sastry, and J. E. O'Neal, Met. Trans. A 16A, 1897 (1985).

25. D. G. Konitzer and H. L. Fraser, Proceedings of the Powder Metallurgy Conference, Toronto, Ontario, Canada, 1984 (to be published).

26. N. F. Anoshkin (private communication, 1980).

27. S. M. L. Sastry, T. C. Peng, P. J. Meschter, and J. E. O'Neal, J. of Metals 35 (9), 21-28 (1983).

28. C. F. Yolton and J. H. Moll, U.S. Patent No. 4 544 404 (1 October 1985).

29. C. F. Yolton and J. H. Moll, "Gas Atomized Titanium Powder," in Rapidly Solidified Materials ed. P. W. Lee and R. S. Carbonara (Metals Park, OH: American Society for Metals, 1985), 97-100.

30. R. G. Rowe, T. F. Broderick, E. F. Koch, and F. H. Froes, "Microstructural Study of Rapidly Solidified Titanium Alloys Containing Erbium and Boron," in Rapidly Solidified Materials ed. P. W. Lee and R. S. Carbonara (Metals Park, OH: American Society for Metals, 1985), 107-114.

31. M. F. X. Gigliotti, R. G. Rowe, G. E. Wasielewski, G. K. Scarr, and J. C. Williams, "Tensile and Creep Properties of Rapidly Solidified Titanium Alloys Containing Complex Matrices and Fine Dispersions," in Rapidly Solidified Alloys and Their Mechanical and Magnetic Properties, vol. 58, ed. B. C. Giessen, D. Polk, and A. I. Taub (New York: North Holland, 1986), in print.

32. R. G. Rowe and E. F. Koch, "Rapidly Solidified Titanium Alloys Containing Cerium Sulfide and Oxysulfide Dispersions," in Rapidly Solidified Materials ed. P. W. Lee and R. S. Carbonara (Metals Park, OH: American Society for Metals, 1985), 115-120.

33. R. Ray, U.S. Patent No. 4 471 831 (18 September 1984).

34. S. H. Whang, C. S. Chi, and Y. Z. Lu, in Rapidly Quenched Metals, vol. 1, edited by S. Steeb and H. Warlimont (Elsevier, Amsterdam, 1985), pp. 115-118.

35. R. B. Pond and J. M. Winter, Mat. Sci. Engr. 23, 87 (1976).

36. T. C. Peng, S. M. L. Sastry, J. E. O'Neal, and J. F. Tesson, in Lasers in Materials Processing, edited by E. A. Metzbower (ASM, Metals Park, OH, 1983), p. 241.

37. F. H. Hayes, H. B. Bomberger, F. H. Froes, L. Kaufman, and H. M. Burte, J. of Metals 36 (6), 70-76 (1984).

38. D. Raybould, in Progress in Powder Metallurgy 1982, Vol. 38, edited by J. G. Bewley and S. W. McGee (MPIF, Princeton, NJ, 1983), pp. 575-592.

39. O. V. Roman and V. G. Gorobtsov, in Shock Waves and High-Strain-Rate Phenomena in Metals, edited by M. A. Meyers and L. E. Murr (Plenum Press, New York, 1980).

40. V. D. Linse, in Sagamore Army Materials Research Conference Proceedings, edited by Gordon Bruggeman and Volker Weiss (1983), pp. 381-404.

41. O. V. Roman, Y. V. Boiko, V. G. Gorobtsov, and V. I. Shelegov, in Titanium, Science and Technology, Vol. 1, edited by G. Lutjering, U. Zwicker, and W. Bunk (DGM, Oberursel, West Germany, 1985), pp. 333-337.

42. B. A. Kolachev and V. K. Nosov, in Titanium, Science and Technology, Vol. 1, edited by G. Lutjering, U. Zwicker, and W. Bunk (DGM, Oberursel, West Germany, 1985), pp. 625-631.

43. K. R. Teal, A. G. Jackson, and F. H. Froes (work in progress, 1984-1986).

44. M. F. X. Gigliotti (private communication, 1985).

45. T. F. Broderick, F. H. Froes, and A. G. Jackson, in Rapidly Solidified Metastable Materials, Vol. 28, edited by B. H. Kear and B. C. Giessen (Elsevier Science Publishers, New York, 1984), p. 345.

46. P. G. Boswell and G. A. Chadwick, Scripta Met., 2, 459 (1977).

47. A. F. Belov and I. S. Polkin, "Modern Trends in Titanium Production and Processing," Germany Metallurgical Society Workshop, University of Nuremberg, Erlangen, July 1982.

48. T. C. Peng, S. M. L. Sastry, and J. E. O'Neal, in Lasers in Metallurgy, edited by K. Mukherjee and J. Mazumder (TMS Publications, New York, 1981), pp. 279-292.

49. J. E. O'Neal, T. C. Peng, and S. M. L. Sastry, in Proceedings of 39th Annual EMSA Meeting, edited by G. W. Bailey (Claitor's Publishing Division, Baton Rouge, LA, 1981), pp. 66-67.

50. B. C. Muddle, D. G. Konitzer, and H. L. Fraser, in Strength of Metals and Alloys, Vol. 1, edited by R. C. Gifkins (Pergamon Press, Oxford, 1983), pp. 313-318.

51. I. G. Wright and B. A. Wilcox, Battelle Columbus Laboratories Report No. AD-781-133, 1974.

52. D. B. Snow, in Laser Processing of Materials, edited by K. Mukherjee and J. Mazumder (TMS Publications, Warrendale, PA, 1984), pp. 83-98.

53. D. G. Konitzer, B. C. Muddle, R. Kirchheim, and H. L. Fraser, in Rapidly Quenched Metals, Vol. 1, edited by S. Steeb and H. Warlimont (Elsevier, Amsterdam, 1985) pp. 953-956.

54. D. G. Konitzer, B. C. Muddle, and H. L. Fraser, Scripta Met. 17, 963 (1983).

55. D. G. Konitzer, B. C. Muddle, H. L. Fraser, and R. Kirchheim, in Titanium, Science and Technology, Vol. 1, edited by G. Lutjering, U. Zwicker, and W. Bunk (DGM, Oberursel, West Germany, 1985), pp. 405-410.

56. D. G. Konitzer (private communication, 1985).

57. S. H. Whang, J. of Metals 36 (4), 34-40 (1984).

58. H. L. Fraser (private communication, 1985).

59. S. H. Whang, "Rapidly Solidified Ti Alloys for High Temperature Applications" (to appear in J. Mat. Sci., July 1986).

60. R. G. Rowe, J. A. Sutliff, and E. F. Koch, "Dispersoid Modification of Ti-3Al-Nb Alloys," in Rapidly Solidified Alloys and Their Mechanical and Magnetic Properties, vol. 58, ed. B. C. Giessen, D. Polk, and A. I. Taub (New York: North Holland, 1986), in print.

61. J. A. Sutliff and R. G. Rowe, "Rare Earth Oxide Dispersoid Stability and Microstructural Effects in Rapidly Solidified Ti_3Al and Ti_3Al-Nb," in Rapidly Solidified Alloys and Their Mechanical and Magnetic Properties, vol. 58, ed. B. C. Giessen, D. Polk, and A. I. Taub (New York: North Holland, 1986), in print.

62. L. M. Brown and R. K. Ham, in Strengthening Methods in Crystals, edited by A. Kelly and R. B. Nicholson (Elsevier, New York, 1971), 9.

63. S. M. L. Sastry, T. C. Peng, and L. P. Beckerman, Met. Trans. A 15A, 1465 (1984).

64. R. G. Vogt, D. Eylon, and F. H. Froes (work in progress, 1983-1986).

65. S. M. L. Sastry, T. C. Peng, P. J. Meschter, and J. E. O'Neal, J. of Metals 35 (9), 21-28 (1983).

66. S. M. L. Sastry, T. C. Peng, and J. E. O'Neal, in Titanium, Science and Technology, Vol. 1, edited by G. Lutjering, U. Zwicker, and W. Bunk (DGM, Oberursel, West Germany, 1985), pp. 397-404.

67. A. G. Jackson and F. H. Froes, in Titanium, Science and Technology, Vol. 1, edited by G. Lutjering, U. Zwicker, and W. Bunk (DGM, Oberursel, West Germany, 1985), pp. 381-387.

68. C. S. Chi and S. H. Whang, in Rapidly Solidified Metastable Materials, Vol. 28, edited by B. H. Kear and B. C. Giessen (Elsevier Science Publishers, New York, 1984), pp. 353-360.

69. P. R. Frausto, A. G. Jackson, A. H. Clauer, and J. L. McCall, "Metallographic Studies of Consolidated Rapidly Solidified Titanium Alloys," in Microstructural Science, vol. 10, ed. W. E. White, J. H. Richardson, and J. L. McCall (New York: Elsevier Science Publishing, 1982), 103-112.

70. S. H. Whang, Y. Z. Lu, and B. C. Giessen, in Rapidly Solidified Metastable Materials, Vol. 28, edited by B. H. Kear and B. C. Giessen (Elsevier Science Publishers, New York, 1984), pp. 367-373.

71. G. W. Franti, J. C. Williams, and H. I. Aaronson, Met. Trans. A 9A (11), 1641 (1978).

72. H. B. Bomberger and F. H. Froes, AFWAL Report No. TR-84-4164, 1985.

73. J. Gross, Mallory-Sharon Titanium Corporation Internal Report No. 1000R164, 1956.

74. P. A. Farrar and H. Margolin, Army Materials and Mechanics Research Center Report No. WAL TR 401/303-13, 1967.

75. D. B. Hunter, AFWAL Report TR-405/2-15, Part II, 1966.

76. S. Krishnamurthy, R. G. Vogt, D. Eylon, and F. H. Froes, in Progress in Powder Metallurgy, Vol. 39, edited by H. S. Nayar, S. M. Kaufman, and K. E. Meiners (MPIF, Princeton, NJ, 1984), pp. 603-623.

77. E. W. Collings, R. E. Maringer, and C. E. Mobley, AFML Report No. TR-78-70, 1978.

78. S. M. L. Sastry, T. C. Peng, and J. E. O'Neal, presented at the 1984 MPIF/APMI International PM Conference, Toronto, Ontario, Canada (to be published).

79. T. C. Peng, S. M. L. Sastry, and J. E. O'Neal, in Rapid Solidification Technology III, edited by R. Mehrabian (NBS, Gaithersburg, MD, 1982), pp. 452-457.

80. W. A. Baeslack III, S. Krishnamurthy, and F. H. Froes, in Strength of Metals and Alloys, Vol. 2, edited by H. J. McQueen, J-P. Bailon, J. I. Dickson, J. J. Jonas, and M. G. Akben (Pergamon Press, Oxford, 1985), pp. 1633-1638.

81. S. Krishnamurthy, A. G. Jackson, H. Jones, and F. H. Froes, "Beta-Eutectoid Decomposition in Rapidly Solidified Titanium-Nickel Alloys," (submitted to Met. Trans., 1984).

82. S. Krishnamurthy, A. G. Jackson, D. Eylon, R. R. Boyer, and F. H. Froes, in Rapidly Quenched Metals, Vol. 1, edited by S. Steeb and H. Warlimont (Elsevier Science Publishers, Amsterdam, 1985), pp. 945-948.

83. S. Krishnamurthy, I. Weiss, D. Eylon, and F. H. Froes, in Strength of Metals and Alloys, Vol. 2, edited by H. J. McQueen, J-P. Bailon, J. I. Dickson, J. J. Jonas, and M. G. Akben (Pergamon Press, Oxford, 1985), pp. 1627-1632.

84. W. A. Baeslack III, L. Weeter, S. Krishnamurthy, P. R. Smith, and F. H. Froes, in Rapidly Solidified Metastable Materials, Vol. 28, edited by B. H. Kear and B. C. Giessen (Elsevier Science Publishers, NY, 1984), pp. 375-379.

85. S. Krishnamurthy, R. G. Vogt, D. Eylon, and F. H. Froes, in Rapidly Solidified Metastable Materials, Vol. 28, edited by B. H. Kear and B. C. Giessen (Elsevier Science Publishers, NY, 1984), pp. 361-366.

86. W. A. Baeslack III, S. Krishnamurthy, and F. H. Froes, "A Study of Rapidly Quenched Microstructures in a Hypereutectoid Ti-40 wt% W Alloy," J. of Materials Science Letters, 5 (1986), 315-318.

87. S. Krishnamurthy, A. G. Jackson, I. Weiss, and F. H. Froes, "Aging Response of Rapidly Solidified Titanium-Tungsten Alloys with Nickel and Silicon Additions," in Rapidly Solidified Materials ed. P. W. Lee and R. S. Carbonara (Metals Park, OH: American Society for Metals, 1985), 121-127.

88. S. Krishnamurthy, D. Eylon, I. Weiss, and F. H. Froes, "Age Hardening Behavior of a Rapidly Solidified Ti-15Cr-4Al Alloy," in Mechanical Behavior of Rapidly Solidified Materials, ed. S. M. L. Sastry and B. A. MacDonald (Warrendale, PA: TMS Publications, 1986), 247-256.

89. S. Krishnamurthy and F. H. Froes (work in progress, 1985).

90. R. G. Vogt, P. R. Smith, D. Eylon, and F. H. Froes (work in progress, 1983-1986).

91. T. F. Broderick, F. H. Froes, and J. A. Snide (work in progress, 1985-1986).

92. T. F. Broderick and F. H. Froes (work in progress, 1984-1986).

93. F. H. Froes and P. R. Smith (unpublished work, 1985).

94. M. Gutierrez, I. A. Martorell, and F. H. Froes (work in progress, 1986).

95. D. E. Polk and B. C. Giessen, Metallic Glasses, (ASM, Metals Park, OH, 1978), pp. 1-35.

96. R. S. Carbonara, T. A. Gaspar, R. V. Raman, R. E. Maringer, and J. L. McCall, MCIC Report No. 81-45, 1981.

97. Yu. K. Kovneristy, E. K. Osipov, E. A. Trofimova, A. A. Baikov, N. D. Tomashov, and I. B. Skvortsova, in Titanium, Science and Technology, Vol. 1, edited by G. Lutjering, U. Zwicker, and W. Bunk (DGM, Oberursel, West Germany, 1985), pp. 203-208.

98. L. E. Tanner and R. Ray, Scripta Met. 11, 783 (1977).

99. L. E. Tanner, Scripta Met. 12, 703 (1978).

100. L. E. Tanner and Ranjan Ray, Scripta Met. 14, 657 (1980).

101. L. E. Tanner and R. Ray, Acta Met. 27, 1727 (1979).

102. A. Inoue et al., in Proceedings of the Industrial Conference on Titanium, (1980), pp. 1936-1944.

103. S. H. Whang and Y. Z. Lu, in Proceedings of RSP, Principles and Technologies, III, edited by R. Mehrabian (NBS, Gaithersburg, MD, 1983), pp. 286-290.

104. S. H. Whang, Mat. Sci. Eng. 57, 87 (1983).

105. S. H. Whang, J. Non-Cryst. Sol. 61 and 62, 841 (1984).

106. S. H. Whang, Scripta Met. 18, 309 (1984).

107. A. Inoue, H. Kimura, S. Sakai, and T. Masumoto, in Titanium '80, Science and Technology, Vol. 2, edited by H. Kimura and O. Izumi (TMS Publications, Warrendale, PA, 1980), pp. 1137-1145.

108. C. Suryanarayana, A. Inoue, and T. Masumoto, in Titanium '80, Science and Technology, Vol. 1, edited by H. Kimura and O. Izumi (TMS Publications, Warrendale, PA, 1980), pp. 699-707.

109. Y. Q. Gao and S. H. Whang, J. Non-Cryst. Sol. 70, 85 (1985).

110. P. L. Martin, M. G. Mendiratta, and H. A. Lipsitt, Met. Trans. A 14A, 2170 (1983).

111. D. G. Konitzer and H. L. Fraser, "The Production and Thermal Stability of a Refined Dispersion of Er_2O_3 in Ti_3Al Using Rapid Solidification Processing," in High-Temperature Ordered Intermetallic Alloys, vol. 39, ed. C. C. Koch, C. T. Liu, and N. S. Stoloff (Pittsburgh, PA: Materials Research Society, 1985), 437-442.

112. H. B. Bomberger and F. H. Froes (work in progress, 1983-1986).

113. C. F. Yolton, P. R. Smith, H. B. Bomberger, and F. H. Froes (work in progress, 1985-1986).

114. S. M. L. Sastry (work in progress, 1985-1986).

115. Y. R. Mahajan, F. H. Froes, H. B. Bomberger (work in progress, 1984-1986).

116. C. Dekock, F. H. Froes, and R. G. Vogt (work in progress, 1985).

PROSPECTS FOR DEVELOPING NOVEL TITANIUM

ALLOYS USING RAPID SOLIDIFICATION

H. B. Bomberger[a] and F. H. Froes[b]

[a]Visiting Scientist
385 Bradford Drive
Canfield, OH 44406

[b]Air Force Wright Aeronautical Laboratories
Materials Laboratory, AFWAL/MLLS
Wright-Patterson AFB, OH 45433

Abstract

The attributes of rapid solidification have been used to advantage in the development of new and improved alloys in a number of metal systems including those of aluminum, iron, and nickel. Rapid solidification offers a number of advantages such as extended solubility, very fine microstructures, alloy homogeneity, and the formation of metastable crystalline and amorphous phases. Prospects for developing new titanium alloys using rapid solidification technology will be discussed in the context of phase diagrams, alloy theory, and state of the art. Specific systems likely to provide enhanced strength, greater temperature capability, and reduced density are presented.

Introduction

Conventional titanium alloying methods, using ingot metallurgy (IM), have resulted in a series of important and useful compositions [1]. Although these methods have not been exhausted, their potential for large new improvements is diminishing; no important new titanium alloys have been conceived and developed in the United States in the past decade. The newer IM alloys being promoted were developed in an earlier period. Availability of continuously improved and superior titanium products is needed by the aerospace industries and would benefit both the end users and producers of such materials.

The potential for major improvements in metal products has been renewed with the advent of rapid solidification (RS) technology in combination with powder metallurgy (PM) and composite technologies. RS is being used to achieve major improvements in physical, mechanical, and corrosion properties [2, 3]. The benefits of RS are well known and include extended alloy solubility, very fine and new metastable structures, and more homogeneous compositions.

Current structural applications for RS products are limited mainly to a few aluminum- and nickel-base alloys [4, 5], but the benefits from RS extend to titanium systems. Work on titanium RS began at a later date largely because of contamination problems related to the metals' high reactivity at elevated temperatures. These problems have now been largely overcome and a number of methods are available for producing limited quantities of good quality RS powder, flake, fibers, foils, and ribbons [6-8]. Figure 1 illustrates some of the methods used to produce RS metal products.

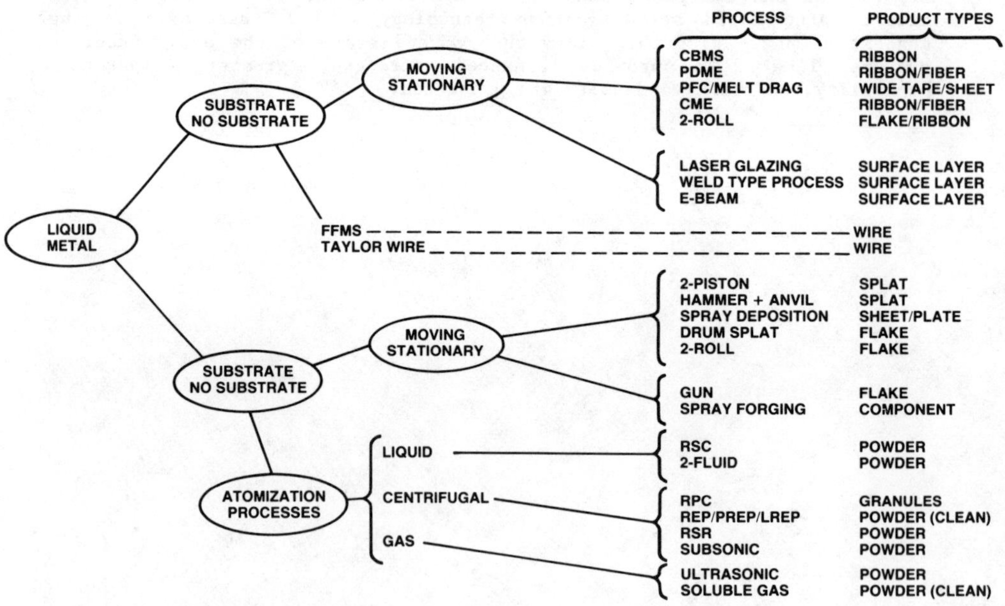

Figure 1 - Some production routes for RS products [8].

True RS products are limited initially to at least one small dimension to facilitate rapid heat extraction. Furthermore, attainable cooling rates are also limited by the production methods used, as noted in Table I. These variables, in turn, affect the maximum alloyability which can be achieved in most systems because extended solid solubility increases with increasing cooling rate from the liquid phase. Beta grain sizes also decrease with increasing cooling rates as noted in Figure 2. Thus it is well for cooling rates to be reported along with metastable solubility and other data.

Table I. Some Techniques of Interest for Rapidly Quenching Titanium Alloys [8]

RS Technique (abbreviation)	Cooling Rate, K/s	Product	
		Form	μm
Gas Atomization (GA)	10^2-10^3	Spherical powder	50-100 diameter
Ultrasonic Gas Atomization (USGA)	to 10^6	Spherical powder	10-50 diameter
Rotating Electrode Process (REP)	10^2	Spherical powder	125-200 diameter
Hammer and Anvil (HA)	10^4-10^6	Splat	5-300 thick
Electron Beam Splat Quenching (EBSQ)	10^4-10^7	Elongated splat	40-100 thick
CBMS Melt Spinning (MS)	10^5-10^7	Tape	10-100 thick to 3mm wide
Pendant Drop Melt Extraction (PDME)	10^5-10^6	Filament, fiber	20-100 thick
Surface Melting (SM)	10^5-10^8	Surface layer	10-1000 thick

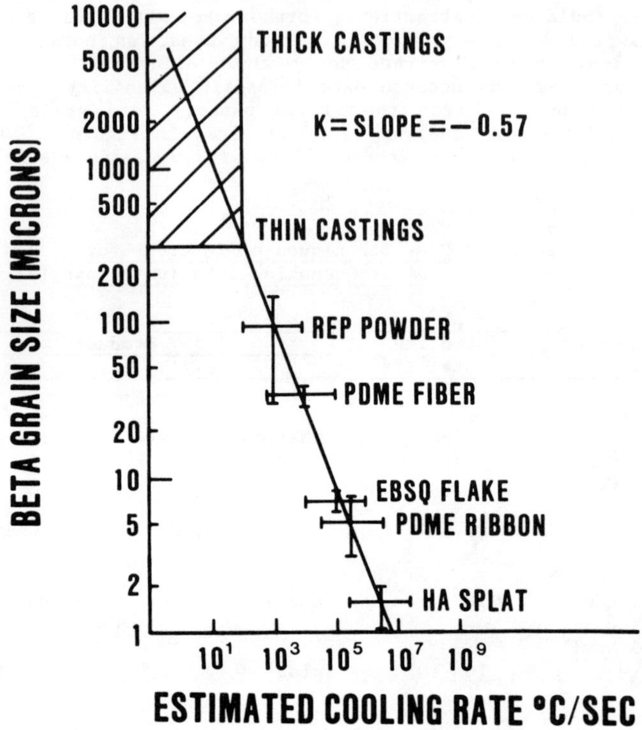

Figure 2 - Effect of cooling rates on prior beta grain size of Ti-6Al-4V alloy [10]. Abbreviations are identified in Table I.

Alloying Concepts

Conventional phase diagrams are of limited use for predicting solubility extension and metastable structures by RS. Nevertheless, such diagrams are of great value in developing conventional alloys and for explaining equilibrium structures. One useful concept is to classify alloying elements according to their influence on phase transformations. Figure 3 provides such a classification and suggests general diagrams for the four main categories. While such generalizations can be useful it is well to note that they have several limitations. For example, Zr and Hf are completely soluble in both the alpha and beta phases but do not decrease the beta transus temperature as much as other beta isomorphous elements Mo, Nb, Ta, and V. Cr, Fe, and Mn sharply lower the beta transus temperature but alloys from these elements can undergo slow eutectoid decomposition at aging temperatures forming corresponding intermetallic phases. Co, Cu, Ni, and Si have a similar effect but their eutectoid reactions are relatively rapid. Sn is usually classified as an alpha stabilizer (rather than beta) with a peritectoid reaction, although it has very little effect on the beta transus temperature; it is quite soluble, and is a useful strengthener, in both the alpha and beta phases. Except for N, O, and Al, most alpha stabilizers have little solubility in the alpha phase and, consequently, do not have much effect on the beta transus. Systems of these elements generally involve peritectoid reactions. The common alpha stabilizing elements are useful for solid solution strengthening and small to moderate additions of the beta stabilizer elements are used to impart heat treatment response, solid solution strengthening, precipitation hardening, and improved workability [12].

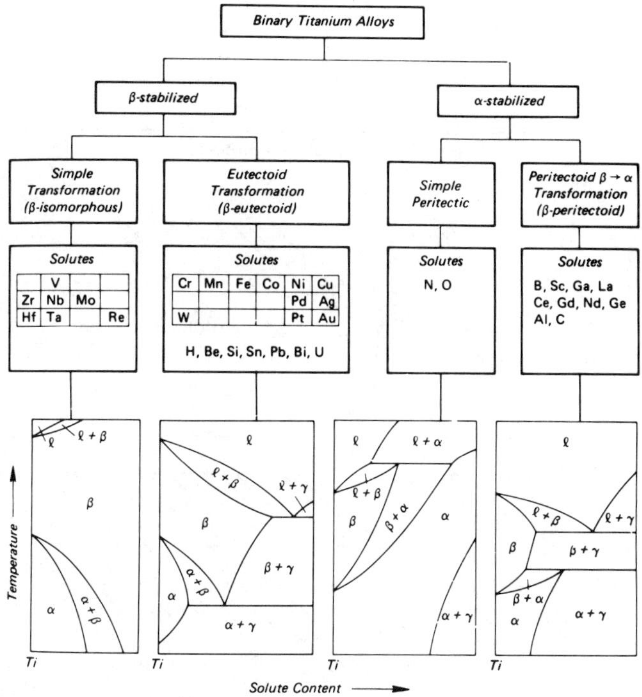

Figure 3 - General categories of titanium alloy phase diagrams [11].

Extensive substitutional solid solubility is generally obtained in alloy systems if the Hume-Rothery conditions are satisfied [13, 14]. Briefly stated, solid solubility favors solute and solvent elements with the same crystal structures, similar valences and electronegativity values and less than 15% difference in atomic radii. The very small atoms, H, N, and O dissolve interstitially and are exceptions with relatively high solubility, but B and C, being a little too large, have very little interstitial solubility in titanium.

As noted in Figure 4, a large number of elements meet the size requirements for titanium. However, in most cases one or more of the other properties, listed in Table II, are not met. All conditions for solid solubility are met with the homologous elements Hf and Zr and, as a consequence, they are fully soluble in both the alpha and beta phases. As various restrictions are imposed, solid solubility is decreased, intermetallic phases, eutectics, eutectoids, peritectics and peritectoids appear, and account for great diversity observed in phase diagrams.

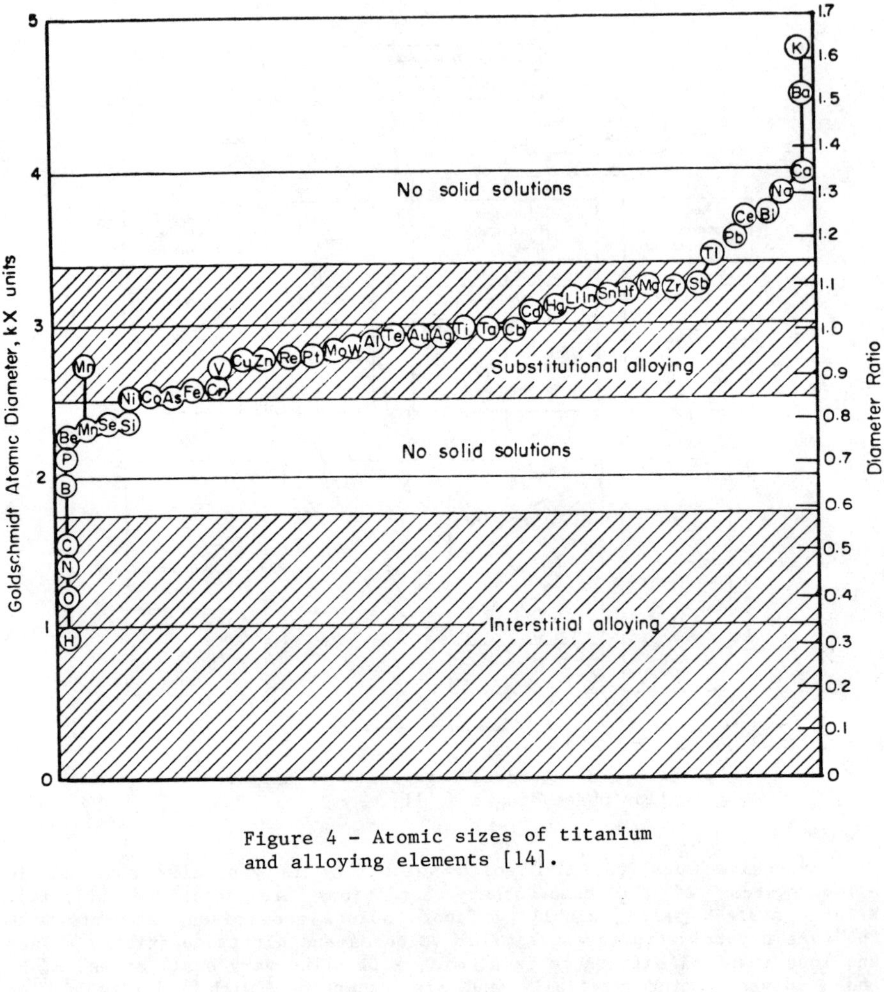

Figure 4 - Atomic sizes of titanium and alloying elements [14].

Table II. Properties of Selected Alloying Elements [12-14]

Element	Crystal Structure	Atomic diameter, Kx	Common Valence	Electro-negativity	Titanium Phase Stabilizer	Approximate Weight Percent[a]			
						Alpha Ti	Beta Ti	RT Beta Ti	Eutectoid
Al	fcc	2.86	3	1.48	A	31	35	-	-
Ag	fcc	2.89	1	1.8	B	6.5	27	-	22.6
B	ortho.	1.94	3	-	A	0.03	0.11	-	-
Be	hcp	2.22	2	1.41	B	<0.2	2	-	1
Bi	rhom.	3.11	3	1.78	B	1.5	33	28	29
C	h	1.42	4	-	A	0.5	0.8	-	-
Ce	fcc	3.64	3, 4	1.18	A	7	0.9	-	-
Co	hcp	2.50	3, 2	1.70	B	<1	17.5	7.0	9
Cr	bcc	2.50	3, 6, 2	1.64	B	<0.5	100	6.3	15
Cu	fcc	2.56	2, 1	1.80	B	<1	17	13	9
Fe	bcc/fcc	2.48	3, 2	1.75	B	<0.2	25	3.5	16
Ge	c	2.45	4	1.78	A	4	12	-	-
H	h	0.92	1	-	B	0.25	1.9	-	1.6
Hf	hcp/bcc	3.14	4	1.48	B	100	100	-	-
In	fct	3.25	3	1.38	B	24	>25	-	25
La	h	3.73	3	1.18	A	5	4	-	-
Mn	c	2.24	2, 3, 4	1.70	B	0.5	36	6.4	20
Mo	bcc	2.73	6, 3, 5	2.05	B	<0.8	100	10	-
Nb	bcc	2.86	5, 3	1.76	B	4	100	36	-
N	h	1.42	3, 5, 2	-	A	7	2	-	-
Nd	hcp	3.63	3	-	A	3	2	-	-
Ni	fcc	2.49	2, 3	1.80	B	<0.2	12.3	9	5.5
O	c	1.20	2	-	A	14	2	-	-
Pb	fcc	3.50	2, 4	1.73	B	16	45	35	26
Re	hcp	2.74	7, 4	-	B	<1	80	11	-
Sc	hcp	3.20	3	1.20	B	8	100	-	10
Si	c	2.35	4	1.82	B	0.4	3	-	8.5
Sn	tet	3.02	4, 2	1.87	A(?)	19	32	-	-

(continued)

Table II. Properties of Selected Alloying Elements (continued)

Element	Crystal Structure	Atomic diameter, Kx	Common Valence	Electro-negativity	Titanium Phase Stabilizer	Approximate Weight Percent[a]			
						Alpha Ti	Beta Ti	RT Beta Ti	Eutectoid
Ta	bcc	2.86	5	1.76	B	7.5	100	45	-
Ti	hcp/bcc	2.93	4, 3	1.90	-	-	-	-	-
V	bcc	2.63	5, 4, 2	1.85	B	2	100	14.9	-
W	bcc	2.73	6	1.07	B	0.8	50	22.5	28
Y	hcp	3.62	4	-	B	<0.1	0.3	-	0.1
Zr	hcp/bcc	3.17	4	1.48	B	100	100	20-80	-

[a]Values are for maximum solute solubility in alpha and beta titanium, for retaining the beta phase to room temperature (RT) and for the eutectoid composition.

This is not to suggest that extensive equilibrium solid solubility is required or is even desirable for structural alloys. However, a large decrease in solid solubility with temperature (as a result of a phase transformation or from a favorably small slope of the solvus curve) may be useful to precipitate a large volume fraction of another phase and thereby achieve considerable strengthening. Examples are small additions of C to Fe to produce the most widely used structural alloys, plain carbon steels, and the use of Al and Ti in some Fe- and Ni-base superalloys to obtain excellent high temperature strength from high volume fractions (to 65% or more) of finely dispersed thermally stable gamma prime [15]. A similar effect is achieved with certain aluminum alloys using RS to retain large supersaturation of transition elements and fine, stable dispersoids during powder consolidation [16].

RS Titanium Alloy Status and Prospects

Conventional Alloys

Most titanium alloy microstructures are considerably more complex than those of aluminum and some other alloys because of the allotropic change and, in some cases, the formation of martensite. Metastable beta alloys are relatively uncomplicated, but the leaner ones form martensitic structures when quenched from the beta region [17]. Structures formed on subsequent heating are determined by the beta grain size. Martensite from large beta grains (≧5μm) forms a lenticular alpha analogous to that found in IM materials [18]. If the beta grain size is very small (≦5μm) alpha grains grow to consume the martensite in a manner similar to that for normal lenticular alpha but the alpha structure formed is equiaxed, as shown in Figure 5. This is probably a result of the dominance of grain boundary alpha because the distances over which diffusion must occur are quite small. Such structures are not observed after annealing conventional PM and unworked IM materials [17].

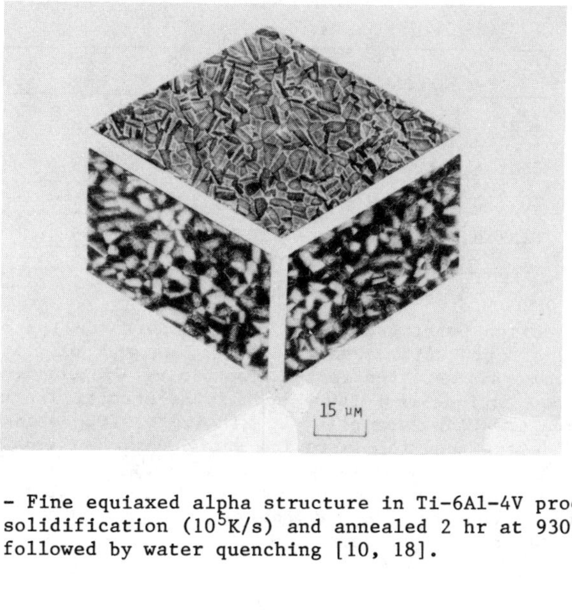

Figure 5 - Fine equiaxed alpha structure in Ti-6Al-4V produced by rapid solidification (10^5K/s) and annealed 2 hr at 930°C (1700°F) followed by water quenching [10, 18].

Laser glazing Ti-4Al-4Mo-2Sn-0.5Si [19], Ti-6Al-2Nb-1Ta-0.8Mo [20], Ti-8Al-1Mo-1V, Ti-6Al-4V, and Ti-6Al-2Sn-4Zr-6Mo [21-25] also resulted in martensitic surface structures. Subsequent alpha-beta annealing produced fine equiaxed alpha-beta microstructures.

The time (t) required for solidification, and cooling rates, can be determined from the secondary dendritic arm spacing (d) using the relationship:

$$d = Kt^{n}$$

With the aid of a heat transfer model, the relationship between d (in μm) and the cooling rate ($\dot{T}$, K/s) for the metastable beta alloy Ti-15V-3Al-3Sn-3Cr was found [26] to be:

$$d = 80\dot{T}^{-0.34}$$

The same general expression can also be used to relate grain diameters with cooling rates although other constants are required. For example, for the beta grain size of RS Ti-6Al-4V (in μm), n is about -0.93 and K is 3.1×10^{6} μm $(K/s)^{n}$ [18].

High cooling rates of an unidentified Ti-Al-Mo alloy were found to improve both strength and ductility of products as noted in Table III. The improvement in strength can be explained by the increase in dislocation densities and by structural refinements using Orowan and Hall-Petch relationships.

Table III. Effect of Cooling Rates on the Strength and Ductility of a Ti-Al-Mo Alpha-Beta Alloy [27]

Production Method	UTS, MPa	UTS, ksi	RA, %
Cast	900	130	12
Cast + Hot Work	1070	155	22
$10^{3}/10^{4}$ K/s + HIP	1070	155	22
Flakes, $10^{6}/10^{7}$ K/s + HIP	1135	165	30

Consolidation practices are of great importance for achieving optimum properties in RS products. Use of HIP'ing, as part of the consolidation of Ti-6Al-4V powders, is essential for obtaining wrought product properties including good fatigue resistance and little scatter in the data [10, 28]. Beta solution treating just prior to HIP'ing was reported to be useful for eliminating defects and for improving fatigue properties [29]. IMI-685 PSV powder was HIP'd to improve its microstructure over the beta forged material. Failure to attain the desired fatigue properties was attributed to insufficient bonding [30] but the problem appears to be correctable by changes in the process parameters.

RS shows the greatest promise for those alloys which cannot be produced satisfactorily with conventional ingot sizes and solidification rates. RS techniques are attractive where high solubility levels, homogeneity, and good workability are not attainable with IM [31]. Alloyability is especially enhanced for the segregation prone eutectoid formers, and other elements which have very little equilibrium solubility in solid titanium, including the metalloids B, C, and Si, and the rare earth elements.

Alpha-Beta Eutectoid Alloys

Comprehensive RS studies were made with a number of eutectoid-forming alloys including Ti-Co [32], Ti-Cr [33], Ti-Fe [33, 34], Ti-Ni [35-37], Ti-Si [33], Ti-W [33, 38], Ti-Cr-Al [39], Ti-W-Ni and Ti-W-Ni-Si [40]. Among these, Ti-9Co [32], Ti-Ni and Ti-3Ni-6Al [25, 35, 39, 41, 42] were the most interesting because of the high microhardness levels obtained. The Ti-3Ni-6Al alloy had the best combination of properties with a UTS of 145ksi and 6% elongation. Structures observed included predendritic, cellular, and equiaxed dendritic. A Ti-9Co alloy [32] had a finer and more uniform structure than most materials. Additional work needs to be done to determine how other important properties are affected and to develop compositions with a good combination of mechanical properties. Some of the benefits from eutectoid formers extend to metastable beta alloys.

Metastable Beta Alloys

Powder produced of the Ti-8V-5Fe-1Al metastable beta alloy by the plasma rotating electrode process (PREP) was found to have a strength in excess of 1380MPa (200ksi) with almost 10% elongation [43]. However, a few higher strengths are reported for other RS alloys in Table IV. The high strengths can be attributed largely to the fine metastable structures being saturated with eutectoid-forming elements. The structures in the first two alloys were saturated beta and the fourth was amorphous.

Table IV. Tensile Strengths of Titanium Alloy Filaments

Alloy, weight percent	UTS, MPa	UTS, ksi	UTS/ Density, 10^5cm	Ref. No.
Ti-5.5Al-3.6V-8Fe-3Cu	2346	340	50.9	44
Ti-5.4Al-3.6V-6Fe-5Cu	2387	346	51.9	44
Ti-1.5Al-1V-3Be-2B	1670	242	39.3	44
Ti-25Zr-10Be	2353	341	58.0	45

RS work on more conventional beta alloys has been quite limited although grain size, microstructures, and alloy homogeneity can be enhanced. The beta grain size of Ti-11.5Mo-6Zr-4Sn decreased from about 40μm in PREP to 2-4μm in EBSQ material [46]. Furthermore, these sizes are smaller than observed in corresponding Ti-6Al-4V product (about 135μm and 7μm, respectively) indicating high solute contents influence cooling rates [9]. Furthermore, precipitates occurred interdendritically on aging PREP material but at the grain boundaries of the more rapidly cooled EBSQ material. This feature suggests less solute redistribution occurs as the cooling rate increases. The addition of rare earth elements to beta alloys is discussed later.

Rare Earth Additions

Rare earth metals and yttrium are soluble in liquid titanium but, with few exceptions, they have very little room temperature equilibrium solubility in titanium [17, 31]. As noted in Table V, the terminal solubility of these and metalloid elements can be extended by RS. In most cases the extension is even in excess of the maximum elevated temperature solid solubility reported on phase diagrams and greatly in excess of apparent room temperature values. For example, there appears to be no room temperature solid solubility of Ce, Er, Gd, and Y in Ti and it is less than 2% for La and Nd. True solubilities of the metalloids B, C, and Si in Ti at low temperatures are well under 1%. Unfortunately, phase diagrams are not very accurate for small solubilities and such values are approximations. However, non-equilibrium terminal solubilities are extended as indicated by the fact that RS retains very significant quantities of these elements in solution which, on aging, can produce large amounts of fine dispersoids.

Table V. Terminal Solubility Extension by Rapid Solidification

Element	Matrix[a]	Maximum Equilibrium Solubility[b], at%	Extended Solubility[c], at%	Technique	Reference No.
B	Binary	0.5	6.0	Splat	47, 48, 49
C	Binary	3.1	10.0	Splat	47, 48
Ce	Binary	~0.25	>1.0	Splat	48, 50
Dy	Binary	~0.3	>0.6	EBSQ[d]	51
Er	Binary	~0.0	~1.0	Splat	51
Er	-5Al	~0.3	≧1.5	Splat	48
Gd	Binary	~0.3	>0.5	EBSQ	51
Ge	-5Al-2.5Sn	8.2	>5.0	EBSQ	52
La	Binary	~0.15	<0.3	Splat	51
La	-5Al, -5Sn	~0.15	≧1.5	Splat	48, 53
Nd	Binary	~0.6	>1.0	EBSQ	51
Si	Binary	5.0	~12.0	Splat	54
Si	-18Zr	5.0	6.0	Splat	23, 47, 54, 55
Y	Binary	~0.5	0.6	Splat	51
Y	-8Al	~0.5	>2.0	Laser	56

[a]Alloying elements shown are in weight percent.

[b]Solubility observed in the beta phase after RS.

[c]As-rapidly solidified.

[d]Electron beam splat quench.

Being very active, the RE elements combine with residual oxygen and precipitate from solution, in most cases, as fine, incoherent sesquioxide (RE_2O_3) dispersoids. The presence of other strongly electronegative elements can result in other, sometimes more complex, dispersoids including oxysulfides [51] and intermetallic phases.

A comprehensive study under U.S. Air Force sponsorship did much to define the potential for rare earth elements in high temperature titanium alloys [17, 26, 27, 42, 51, 56]. This work involved a series of alloys containing Ce, Dy, Er, Gd, La, Nd, and Y produced as flakes using the electron beam splat quench (EBSQ) process.

All these RS alloys developed very fine dispersoids on aging with the finest produced at 500°C (930°F). The oxide particles formed most rapidly in Ti-Ce and Ti-Nd and coarsened to >1μm diameter at 800°C (1470°F). Dispersoids in the Ti-Er, Ti-Gd, and Ti-Y alloys contained carbon as an impurity and coarsened relatively slowly. Particle sizes, spacings, and their calculated strengthening effects after annealing are listed in Table VI [51]. Dispersoids in the Ti-Er alloy grew at the lowest rate indicating the highest metallurgical stability. However, there is good evidence that the RE and oxygen contents of such dispersoids can vary over a wide range [51] and growth may be related to the RE and oxygen contents. There is also some evidence [57] that dispersoids in the Ti-Er alloy may be semi-coherent with the matrix and that the coarsening rate is influenced by the alloy content. These differences in dispersoid characteristics can be expected to influence their mechanical properties and usefulness but they should be controllable to some extent by optimizing the alloy composition and processing.

Table VI. Dispersoid Parameters and Theoretical Strengths of Some RS Ti-RE Alloys Annealed at 800°C (1470°F) for 1000 Minutes [51]

Alloy, weight percent	Mean Particle Diameter, nm	Mean Interparticle Spacing, nm	Orowan Strengthening, MPa
Ti-1.0Ce	1580	12,500	4
Ti-2.0Dy	98	411	73
Ti-2.0Er	65	332	85
Ti-1.0Gd	252	967	38
Ti-2.0La	73	810	35
Ti-1.5Nd	173	910	37
Ti-3.0Nd	82	425	68
Ti-1.0Y	150	620	53

Mechanical tests on two of the most promising RE alloy systems, Ti-Er (1% and 2% Er) and Ti-Nd (1.5% and 3% Nd), showed the dispersoids increased the ultimate and yield strengths from 40% to 110% over the values for unalloyed titanium at 25°C to 700°C (77°F to 1292°F). Creep rates and stress rupture times at 482°C to 700°C (900°F to 1292°F) were improved an order of magnitude [58]. Strengthening resulted from dislocation dispersion interactions. However, the loss in strength of Ti-5Al-5.4Er and Ti-5Al-4.5La alloys by annealing above 700°C (1292°F), shown in Figure 6, or on holding above 650°C (1202°F), noted in Figure 7, seems to be due to matrix softening [54].

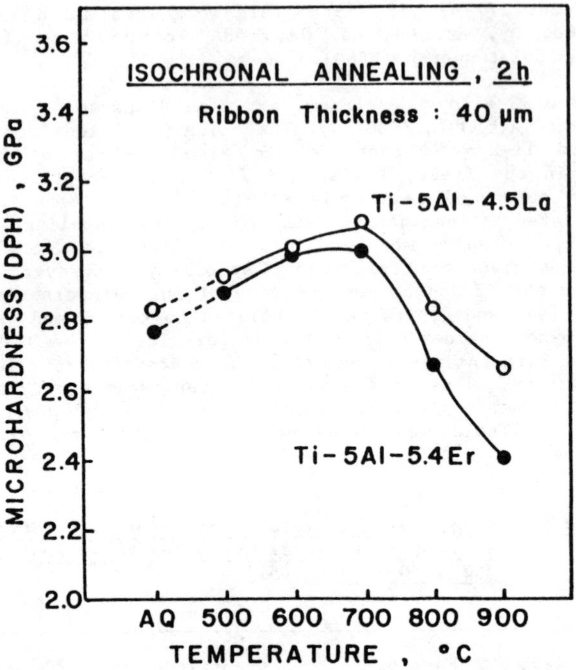

Figure 6 - Microhardness of Ti-5Al-4.5La and Ti-5Al-5.4Er after isochronal annealing [54].

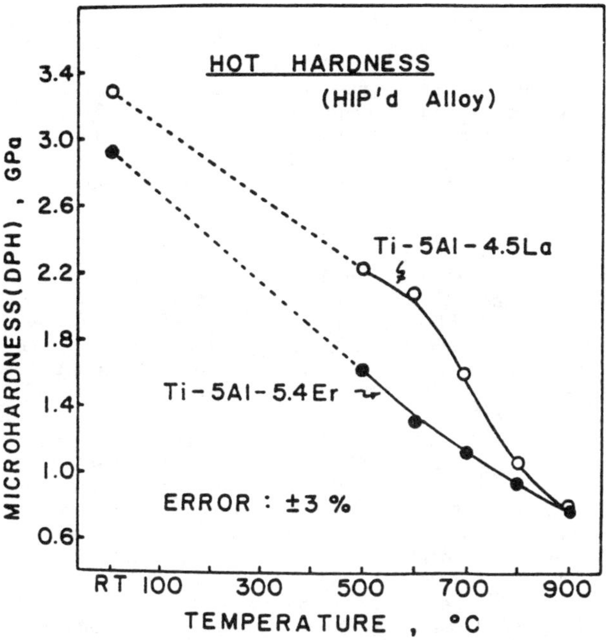

Figure 7 - Hot hardness values for HIP'd Ti-5Al-4.5La and Ti-5Al-5.4Er [54].

Dispersoids found in RS ($\sim 10^6$ K/s) Ti-5Al-3La and Ti-9.5Sn-5.3La annealed at 700°C (1292°F) consisted of uniform fine (~200Å diameter) precipitates of Al_3La and La_2Sn, respectively [59]. Particles in as-quenched ribbon were coarser than in splat-quenched foil of the same thickness reflecting a difference in cooling rate from the melt.

The stability of Er_2O_3 dispersoids was reported to be very good in alpha Ti-Al alloys as long as it remained in the alpha phase. Ti-6Al-2Sn-4Zr-2Er alloy made by melt extraction and annealed 1 hr at 950°C (1742°F) had a dispersoid diameter of 200-400Å and a spacing of about 0.3 to 0.4 microns within the grain interior [60, 61]. Ce and La dispersoids in RS Ti-Al and Ti-Sn do not have the same resistance to coarsening observed with Er_2O_3 [50]. On annealing such alloys at 800°C (1470°F) the particle diameters grew to about 1000 to 2000Å. Er_2O_3 particles in Ti-Al-Sn alloys were about 200 to 400Å after annealing at 950°C (1740°F) [53, 58, 61]. After extruding Ti-6Al-2Sn-4Zr-2Mo-0.1Si-2Er PREP material at 760°C (1400°F) the dispersoid diameters were mostly under 100Å with a few as large as 500Å [62]. The results indicate that the PREP process with a cooling rate of about 10^{-4} K/s may be useful for producing such alloys.

Laser surface melting ($\sim 5 \times 10^4$ K/s) has been used to produce homogeneous additions of ≤3% Er, Nd and Y to Ti-6Al-4V and 1% Er to Ti-6Al-2Sn-4Zr-6Mo [63]. The resulting fine oxide dispersions did not enhance tensile properties but an improvement in creep strength was anticipated. The solubility of Y in Ti-8Al-4Y was also studied by laser melting. Dispersoids of Y_2O_3 ranged mainly from 8 to 20nm [56]. A similar process was used to produce RS beta alloy Ti-6Al-15V-2Er [64] which experienced no dispersoid coarsening at temperatures as high as 760°C (1400°F). A second RS beta alloy, Ti-25V-4Ce-0.6S, had a predicted strength level in excess of 1725MPa (250ksi) [65]. RS Ti-24V-10Cr also showed potential for high strengths [66]. RE sulfides and oxysulfides form fine dispersoids in alpha titanium and have excellent resistance to coarsening at temperature as high as 1000°C (1832°F) [66]. Coarsening of CeS and Er_2O_3 dispersoids is more rapid near grain boundaries than within the grains apparently because of more rapid grain boundary diffusion.

Metalloid Additions

Dispersoids of metalloid elements B, C, Ge, and Si are less stable than RE dispersoids and this could limit their use to about 500 to 600°C (930 to 1110°F) [67]. Although Si is an important alloying element for several current IM high temperature alloys, the maximum useful temperature for the best of these alloys is about 600°C (1110°F). Hardening of as-quenched $Ti-Zr_{10}-Si_x$ foil (20μm) was found to be a direct function of the Si content up to at least 8 atomic percent. Zr is an important strengthening agent when Si is used by promoting the formation of a more stable silicide $(Ti, Zr)_5Si_3$. The martensite structure from RS provides numerous nucleation sites for silicide precipitation [55]. Work on Ti-5Al-2.5Sn modified with Si and Ge and rapidly solidified indicated that the eutectoid and eutectic reactions can be suppressed so the precipitate size, distribution, and morphology can be controlled [68].

RS Ti-0.5B forms a hexagonal alpha prime which, on annealing at 800 to 900°C (1470 to 1650°F) results in a fine grained alpha matrix and fine needle-shaped TiB particles. The effect of the boron was to increase the yield strength 100% and Young's modulus 30%. The increase in strength was said to persist at elevated temperatures [41]. B, however, has a greater tendency to segregate than Si and Ge, but B is said to provide greater resistance to grain growth than the rare earth elements. TiB, however, grows rapidly at 800°C (1470°F) [41, 61].

RS Ti-1C was found to have 7 to 8% ductility and strengths about 150% greater than unalloyed titanium [41], but such alloys lose strength and ductility on aging at 700°C (1290°F) with the precipitation and rapid growth of TiC [41].

Other Alloy Systems

High aluminum alloys are of great interest for elevated temperature applications. RS is being considered as a means to achieve needed improvements in ductility and other properties of these materials [17]. Work on highly alloyed alpha alloys indicated that RE additions result in fine grain sizes and significantly better post-creep ductility. However, the fine grain size also promotes high creep rates except when a beta anneal is employed to decrease planar slip [17]. Interesting work on TiAl-W alloys prepared by the PREP process had good creep resistance because of a W-rich beta phase in the alpha 2-gamma lath structure of TiAl [69].

Laser additions of 0.4 atomic percent Er to Ti_3Al produced a fine dispersion of Er_2O_3 which did not coarsen in 10 hr at 900°C (1650°F), but particle growth occurred at 1000°C (1830°F) [70]. During HIP'ing Er_2O_3 dispersoids coarsened noticeably in a Ti_3Al-Nb alloy and also on annealing at the same temperature. Extruding the alloy caused rapid particle coarsening in contrast with alloys containing cerium sulfide or cerium oxysulfide [71].

Interesting work on nickel aluminide (Ni_3Al) showed that boron additions within the solubility limit can increase grain boundary strength and improve ductility. A maximum strain of 22% was achieved with 0.75 atomic percent B [72]. This work shows that the ductility of an aluminide can be improved by critical alloying.

Although considerable work has been done on amorphous titanium alloys they have been of limited interest for structural applications. They are readily produced by quenching metal and metalloid compositions near deep eutectics. Certain other compositions can also be converted to glassy structures providing the quenching rates are sufficiently fast for the system. Most glassy titanium alloys involve near-eutectic amounts of B, Be, Co, Cu, Fe, Ni, and Si. Most are of interest for their physical rather than their mechanical properties [3, 73]. One of the most interesting alloys of this type has been Ti-25Zr-10Be [45]. Foils of this material had a calculated UTS of 2353MPa (341ksi) and a density of 4.1g/cm^3 (0.152 lb/in.^{3}) for an unusually high strength-to-weight ratio as noted in Table IV. However, strengths up to 2740MPa (398ksi) were reported for $Ti_{70}Mn_{20}Si_{10}$. Somewhat lower values were noted for other alloys in the same series of amorphous Ti-M-Si alloys (where M = Co, Cu, Fe, Mn or Ni) [74]. Splat-quenched Ti-Zr-Ni-B and Ti-Zr-Ni-Si glasses, when devitrified by aging, had strengths, calculated from hardnesses, higher than 2070MPa (300ksi). Grain sizes were less than 1μm even after 2 hr at 950°C (1740°F) [74]. Additional work is needed on these high strength materials to determine if they have other attractive properties and especially for structural uses. They are being evaluated for superconductivity applications [75].

Additional opportunities may include the possibility of adding other very light metals, Li and Mg (in addition to Al, Be, and Si) to achieve very significant density reductions and other useful properties [31]. These elements have almost no equilibrium solubility in titanium at low temperatures but appear to have at least some solubility in liquid titanium. However, practical additions of such relatively low boiling point metals may require new melting and RS techniques, co-reduction of powders or mechanical alloying.

Summary

Rapid solidification methods offer practical means for achieving finer grained, more homogeneous, and greater solute concentrations in titanium alloys than by any other means. A review of titanium alloy technology has helped to identify opportunities where the attributes of RS should be most useful. Special opportunities include the use of elements which have little equilibrium solid solubility or have a tendency to segregate considerably. Among these elements are the eutectoid formers, the metalloids, and rare earth elements to produce supersaturated metastable alloys which can be aged to high strength levels by the precipitation of large volume fractions of very fine dispersoids. The rare earth elements, and especially Er, are of special interest for high temperature strengthening because some of the oxides and other dispersoids have extraordinarily high stability at temperatures up to about 1000°C (1832°F). Amorphous alloys can also be readily produced by rapidly cooling near-eutectic compositions involving metals or metalloids, alloying elements. In the glassy conditions some display exceptional strength levels up to 2750MPa (400ksi) or can be devitrified by aging to achieve greater thermal stability but at somewhat lower strength levels. Although areas of opportunity are well defined, work remains to be done to develop specific alloys for practical applications.

Acknowledgments

The authors would like to acknowledge helpful discussions with R. R. Boyer, D. Eylon, S. Fujishiro, H. Jones, S. Krishnamurthy, P. Morton, R. G. Rowe, P. R. Smith, R. G. Vogt, and C. F. Yolton. In addition the assistance of Ms. Karen A. Sitzman in formating and typing the text is greatly appreciated.

References

1. F. H. Froes, D. Eylon, and H. B. Bomberger, ed., Titanium Technology: Present Status and Future Trends (Dayton, OH: Titanium Development Association, 1985).

2. R. S. Carbonara, T. A. Gaspar, R. V. Raman, R. E. Maringer, and J. L. McCall, "State-of-the-Art Review of Rapid Solidification Technology (RST)" (MCIC Report No. 81-45, Battelle Memorial Institute, Columbus, OH, 1981).

3. R. L. Ashbrook, ed., Rapid Solidification Technology Source Book (Metals Park, OH: American Society for Metals, 1983).

4. T. H. Sanders, Jr. and E. A. Starke, Jr., "Aluminum-Lithium Alloys: New Materials for Tomorrow's Technology," in Rapid Solidification Technology Source Book (Metals Park, OH: American Society for Metals, 1983), 396-402.

5. R. J. Patterson II, A. R. Cox, and E. C. van Reuth, "Rapid Solidification Rate Processing and Application to Turbine Engine Materials," in Rapid Solidification Technology Source Book (Metals Park, OH: American Society for Metals, 1983), 414-419.

6. S. Kavesh, "Principles of Fabrication," in Rapid Solidification Technology Source Book (Metals Park, OH: American Society for Metals, 1983), 73-110.

7. A. Lawley, "Fundamentals of Particulate Metallurgy," in Rapid Solidification Technology Source Book (Metals Park, OH: American Society for Metals, 1983), 47-65.

8. S. J. Savage and F. H. Froes, "Production of Rapidly Solidified Metals and Alloys," Journal of Metals, 36(4)(1984), 20-33.

9. H. Jones, Rapid Solidification of Metals and Alloys, Monograph No. 8 (London: The Institute of Metallurgists, 1982).

10. F. H. Froes and D. Eylon, "Powder Metallurgy of Titanium Alloys - A Review," in Titanium, Science and Technology, vol. 1, ed. G. Lutjering, U. Zwicker, and W. Bunk (Oberursel, West Germany: DGM, 1985), 267-286; also in Powder Metallurgy International (PMI), 17(4)(1985), 163-167 and continued in (5)(1985), 235-238.

11. E. K. Molchanova, Phase Diagrams of Titanium Alloys (Israel Program of Scientific Translations, Jerusalem, 1965).

12. Properties and Selection of Metals, Metals Handbook, vol. 1, 8th edition (Metals Park, OH: American Society for Metals, 1961), 42-51.

13. W. Hume-Rothery and G. V. Raynor, "The Structure of Metals and Alloys," in Monograph and Report Series No. 1 (London: The Institute of Metals, 1962).

14. D. J. Maykuth, H. R. Ogden, and R. I. Jaffee, "The Effects of Alloying Elements in Titanium" (DMIC Report No. 136A, Battelle Memorial Institute, Columbus, OH, 1960).

15. R. F. Decker, "Strengthening Mechanisms in Nickel-Base Superalloys," in Source Book on Materials for Elevated-Temperature Applications, ed. E. F. Bradley (Metals Park, OH: American Society for Metals, 1979).

16. W. E. Quist and C. H. Narayanan, "Powder Metallurgy (PM) Aluminum Alloys for Aerospace Use" (Paper presented at the 28th National SAMPE Meeting, Anaheim, CA, 1983).

17. F. H. Froes and R. G. Rowe, "Rapidly Solidified Titanium," in Rapidly Solidified Alloys and Their Mechanical and Magnetic Properties, ed. B. C. Giessen, D. Polk, and A. I. Taub (New York: North Holland, 1986), in print.

18. T. F. Broderick, A. G. Jackson, H. Jones, and F. H. Froes, "The Effect of Cooling Conditions on the Microstructure of Rapidly Solidified Ti-6Al-4V," Met. Trans. A, 16A (1985), 1951-1959.

19. K. Lipscomb, W. M. Steen, and D. R. F. West, "The Laser Glazing of a Cast Nickel Based Superalloy and an Alpha-Beta Titanium Alloy," in Rapid Solidification Processing, Principles and Technologies, ed. R. Mehrabian, B. H. Kear, and M. Cohen (Baton Rouge, LA: Claitor's Publishing Division, 1980), 189-194.

20. L. S. Weinman and C. Kim, "Rapid Solidification of Aluminum and Titanium Alloy Surfaces," in Rapid Solidification Processing, Principles and Technologies, ed. R. Mehrabian, B. H. Kear, and M. Cohen (Baton Rouge, LA: Claitor's Publishing Division, 1980), 117-122.

21. T. C. Peng, S. M. L. Sastry, J. E. O'Neal, and J. F. Tesson, "Microstructures and Mechanical Properties of Laser-Welded Titanium Alloys (Paper presented at the Fall Meeting of TMS-AIME, Louisville, KY, 11-16 October 1981).

22. T. C. Peng, S. M. L. Sastry, and J. E. O'Neal, "Exploratory Study of Laser Processing of Titanium Alloys" (Paper presented at the 110th Annual AIME Meeting, Chicago, IL, 22-26 February 1981).

23. J. E. O'Neal, S. M. L. Sastry, T. C. Peng, and J. F. Tesson, "Microstructures of Rapidly Solidified Titanium Alloys," in Microstructural Science, vol. 11, ed. R. T. DeHoff, J. D. Braun, and J. L. McCall (New York: Elsevier Science Publishing, 1983), 143-151.

24. S. M. L. Sastry, T. C. Peng, J. E. O'Neal, and L. P. Beckerman, "Consolidation, Thermomechanical Processing, and Mechanical Properties of Rapidly Solidified, Dispersion-Strengthened Titanium Alloys," in Rapid Solidification Processing, Principles and Technologies, III, ed. R. Mehrabian (Gaithersburg, MD: National Bureau of Standards, 1982), 579-584.

25. T. C. Peng, S. M. L. Sastry, and J. E. O'Neal, "Rapid-Solidification Processing of Titanium Alloys," in Rapid Solidification Processing, Principles and Technologies, III, ed. R. Mehrabian (Gaithersburg, MD: National Bureau of Standards, 1982), 452-457.

26. T. C. Peng, S. M. L. Sastry, J. E. O'Neal, and J. F. Tesson, "A Correlation Between Dendrite Arm Spacing and Cooling Rate for Laser-Melted Ti-15V-3Al-3Sn-3Cr," in Lasers in Materials Processing, ed. E. A. Metzbower (Metals Park, OH: American Society for Metals, 1983), 241-246.

27. A. F. Belov and I. S. Polkin, "Modern Trends in Titanium Production and Processing" (German Metallurgical Society Workshop, University of Nuremberg, Erlangen, July 1982).

28. J-P. Herteman, D. Eylon, and F. H. Froes, "Mechanical Properties of Advanced Titanium Powder Metallurgy Compacts," in Titanium, Science and Technology, vol. 1, ed. G. Lutjering, U. Zwicker, and W. Bunk (Oberursel, West Germany: DGM, 1985), 303-310.

29. J. Devillard, "Effect of Thermomechanical Treatments of Ti-6Al-4V PSV Powders on the Improvement of the Low Cycle Fatigue of Sintered Compacts," in Titanium, Science and Technology, vol. 1, ed. G. Lutjering, U. Zwicker, and W. Bunk (Oberursel, West Germany: DGM, 1985), 287-294.

30. B. Borchert, H. Schmid, and J. Wortmann, "Strength and Reliability of Powder Metallurgically Produced IMI 685," in Titanium, Science and Technology, vol. 1, ed. G. Lutjering, U. Zwicker, and W. Bunk (Oberursel, West Germany: DGM, 1985), 295-302.

31. H. B. Bomberger and F. H. Froes "Titanium Rapid Solidification Technology - Status and Potential" (Final Technical Report No. AFWAL-TR-84-4164, Wright-Patterson AFB, OH, 1984).

32. S. Krishnamurthy, I. Weiss, D. Eylon, and F. H. Froes, "Aging Response of a Rapidly Solidified Beta-Eutectoid Ti-9 wt% Co Alloy," in Strength of Metals and Alloys, vol. 2, ed. H. J. McQueen, J-P. Bailon, J. I. Dickson, J. J. Jonas, and M. G. Akben (Oxford: Pergamon Press, 1985), 1627-1632.

33. S. Krishnamurthy, R. G. Vogt, D. Eylon, and F. H. Froes, "Microstructures of Rapidly Solidified Titanium-Eutectoid Former Alloys," in Rapidly Solidified Metastable Materials, vol. 28, ed. B. H. Kear and B. C. Giessen (New York: Elsevier Science Publishing, 1984), 361-366.

34. W. A. Baeslack III, L. Weeter, S. Krishnamurthy, P. R. Smith, and F. H. Froes, "Rapidly Solidified Microstructures in Ti-22 wt% Fe Alloy," in Rapidly Solidified Metastable Materials, vol. 28, ed. B. H. Kear and B. C. Giessen (New York: Elsevier Science Publishing, 1984), 375-379.

35. S. Krishnamurthy, A. G. Jackson, D. Eylon, R. R. Boyer, and F. H. Froes, "Rapidly Solidified Microstructures and Precipitation in a Ti-5.5w/o Ni Alloy," in Rapidly Quenched Metals, vol. 1, ed. S. Steeb and H. Warlimont (Amsterdam: Elsevier Science Publishers B.V., 1985), 945-948.

36. W. A. Baeslack III, S. Krishnamurthy, and F. H. Froes, "Rapid Solidification and Aging of a Near-Eutectoid Titanium Nickel Alloys," in Strength of Metals and Alloys, vol. 2, ed. H. J. McQueen, J-P. Bailon, J. I. Dickson, J. J. Jonas, and M. G. Akben (Oxford: Pergamon Press, 1985), 1633-1638.

37. S. Krishnamurthy, A. G. Jackson, H. Jones, and F. H. Froes, "Beta-Eutectoid Decomposition in Rapidly Solidified Titanium-Nickel Alloys," (submitted to Met. Trans. for publication, 1984).

38. W. A. Baeslack III, S. Krishnamurthy, and F. H. Froes, "A Study of Rapidly Quenched Microstructures in a Hypereutectoid Ti-40 wt% W Alloy," J. of Materials Science Letters, 5 (1986), 315-318.

39. S. Krishnamurthy, D. Eylon, I. Weiss, and F. H. Froes, "Age Hardening Behavior of a Rapidly Solidified Ti-15Cr-4Al Alloy," in Mechanical Behavior of Rapidly Solidified Materials, ed. S. M. L. Sastry and B. A. MacDonald (Warrendale, PA: The Metallurgical Society, 1986), 247-256.

40. S. Krishnamurthy, A. G. Jackson, I. Weiss, and F. H. Froes, "Aging Response of Rapidly Solidified Titanium-Tungsten Alloys with Nickel and Silicon Additions," in Rapidly Solidified Materials, ed. P. W. Lee and R. S. Carbonara (Metals Park, OH: American Society for Metals, 1985), 121-127.

41. S. M. L. Sastry, T. C. Peng, and J. E. O'Neal, "Design and Development of Advanced Titanium Alloys by Rapid Solidification Technology," in Titanium, Science and Technology, vol. 1, ed. G. Lutjering, U. Zwicker, and W. Bunk (Oberursel, West Germany: DGM, 1985), 397-404.

42. S. M. L. Sastry, T. C. Peng, P. J. Meschter, and J. E. O'Neal, "Rapid Solidification Processing of Titanium Alloys," Journal of Metals, 35(9)(1983), 21-28.

43. R. G. Vogt, D. Eylon, R. R. Boyer, and F. H. Froes, "Production of High Strength Beta Titanium Alloys Through Powder Metallurgy," this proceedings.

44. E. W. Collings, C. E. Mobley, R. E. Maringer, and H. L. Gegel, "Selected Properties of Melt-Extracted Titanium-Base Polycrystalline Alloys," in Rapidly Quenched Metals III, ed. B. Cantor (London: Chameleon Press, 1978), 188-191.

45. L. E. Tanner and R. Ray, "Physical Properties of $Ti_{50}Be_{40}Zr_{10}$ Glass," Scripta Met., 11(9)(1977), 783-789.

46. T. F. Broderick, F. H. Froes, and A. G. Jackson, "Cooling Rate Effects on Ti-6Al-4V and Beta III Titanium Alloys," in Rapidly Solidified Metastable Materials, vol. 28, ed. B. H. Kear and B. C. Giessen (New York: Elsevier Science Publishing, 1984), 345-351.

47. C. S. Chi and S. H. Whang, private communication with author, Polytechnic University, 1985.

48. S. H. Whang, private communication with author, Polytechnic University, 1985.

49. S. M. L. Sastry, T. C. Peng, and J. E. O'Neal "Rapid Solidification and Powder Metallurgical Processing of Titanium Alloys," in Modern Developments in Powder Metallurgy, vol. 16, Ferrous and Nonferrous Materials, ed. E. N. Aqua and C. I. Whitman (Princeton, NJ: MPIF Publications, 1985), 577-606.

50. S. H. Whang, U.S. Patent No. 4,512,826, April 23, 1985.

51. S. M. L. Sastry, P. J. Meschter, and J. E. O'Neal, "Structure and Properties of Rapidly Solidified Dispersion-Strengthened Alloys: Part 1. Characterization of Dispersoid Distribution, Structure, and Chemistry," Met. Trans. A, 15A (1984), 1451-1463.

52. A. G. Jackson, "A Study of the Microstructure Developed in Rapidly Solidified Ti-5Al-2.5Sn with Additions of Si and Ge" (Ph.D. thesis, University of Cincinnati, 1983).

53. Y. Z. Lu, C. S. Chi, and S. H. Whang, "Second Phase Coarsening in Rapidly Solidified Ti-5Sn-4.5La System," in Rapidly Quenched Metals, vol. 1, ed. S. Steeb and H. Warlimont (Amsterdam: Elsevier Science Publishers B.V., 1985), 949-952.

54. S. H. Whang, "RS Titanium Alloys for High Temperature Applications," (submitted to Journal of Materials Science for publication, 1986).

55. S. H. Whang, Y. Z. Lu, and Y-W. Kim, "Microstructures and Age Hardening of Rapidly Quenched Ti-Zr-Si Alloys," J. of Materials Science Letters, (4)(7)(1985), 883-887.

56. D. G. Konitzer, B. C. Muddle, and H. L. Fraser, "A Comparison of the Microstructure of As-Cast and Laser Surface Melted Ti-8Al-4Y," Met. Trans. A, 14A(10)(1983), 1979-1988.

57. D. G. Konitzer, B. C. Muddle, R. Kirchheim, and H. L. Fraser, "The Production of Ultrafine Dispersions of Rare Earth Oxides in Ti Alloys Using Rapid Solidification," in Rapidly Quenched Metals, vol. 1, ed. S. Steeb and H. Warlimont (Amsterdam: Elsevier Science Publishers B.V., 1985), 953-956.

58. S. M. L. Sastry, T. C. Peng, and L. P. Beckerman, "Structure and Properties of Rapidly Solidified Dispersion-Strengthened Titanium Alloys: Part 2. Tensile and Creep Properties," Met. Trans. A, 15A (1984), 1465-1484.

59. C. S. Chi and S. H. Whang, "Microstructural Characteristics of Rapidly Quenched Alpha-Ti Alloys Containing La," (submitted to Met. Trans. for publication, 1986).

60. C. F. Yolton and J. H. Moll, "Gas Atomized Titanium Powder," in Rapidly Solidified Materials, ed. P. W. Lee and R. S. Carbonara (Metals Park, OH: American Society for Metals, 1985), 97-100.

61. R. G. Rowe, T. F. Broderick, E. F. Koch, and F. H. Froes, "Microstructural Study of Rapidly Solidified Titanium Alloys Containing Erbium and Boron," in Rapidly Solidified Materials, ed. P. W. Lee and R. S. Carbonara (Metals Park, OH: American Society for Metals, 1985), 107-114.

62. R. G. Vogt, D. Eylon, and F. H. Froes, "Effect of Er, Si, and W Additions on Powder Metallurgy High Temperature Titanium Alloys," this proceedings.

63. D. B. Snow, "Structure and Mechanical Properties of Laser-Consolidated Ti-6Al-4V and Ti-6Al-2Sn-4Zr-6Mo with Rare Earth Element Additions," in Laser Processing of Materials, ed. K. Mukherjee and J. Mazumder (Warrendale, PA: The Metallurgical Society, 1984), 83-98.

64. T. F. Broderick, J. Snide and F. H. Froes (work in progress, 1985).

65. T. F. Broderick and F. H. Froes (work in progress, 1985).

66. F. H. Froes and P. R. Smith (work in progress, 1985).

67. S. H. Whang, "Rapidly Solidified Ti Alloys Containing Novel Additions," Journal of Metals, 36(4)(1984), 34-40.

68. A. G. Jackson, T. F. Broderick, and F. H. Froes, "Microstructures of Rapidly Solidified Ti-5Al-2.5Sn with Si or Ge Additions," in Titanium, Science and Technology, vol. 1, ed. G. Lutjering, U. Zwicker, and W. Bunk (Oberursel, West Germany: DGM, 1985), 381-387.

69. P. L. Martin, M. G. Mendiratta, and H. A. Lipsitt, "Creep Deformation of TiAl and TiAl-W Alloys," Met. Trans. A, 14A (1983), 2170-2174.

70. D. G. Konitzer and H. L. Fraser, "The Production of Thermal Stability of a Refined Dispersion of Er_2O_3 in Ti_3Al Using Rapid Solidification Processing," in High-Temperature Ordered Intermetallic Alloys, vol. 39, ed. C. C. Koch, C. T. Liu, and N. S. Stoloff (Pittsburgh, PA: Materials Research Society, 1985), 437-442.

71. R. G. Rowe, J. A. Sutliff, and E. F. Koch, "Dispersoid Modification of Ti-3Al-Nb Alloys," in Rapidly Solidified Alloys and Their Mechanical and Magnetic Properties, vol. 58, ed. B. C. Giessen, D. Polk, and A. I. Taub (New York: North Holland, 1986), in print.

72. S. C. Huang, A. I. Taub, K. M. Chang, C. L. Briant, and E. L. Hall, "Boron Effects in Rapidly Solidified Ni_3Al," in Rapidly Quenched Metals, vol. 2, ed. S. Steeb and H. Warlimont (Amsterdam: Elsevier Science Publishers B.V., 1985), 1407-1410.

73. Metallic Glasses (Metals Park, OH: American Society for Metals, 1978).

74. A. Inoue, H. Kimura, S. Sakai, and T. Masumoto, "Mechanical Properties of Titanium-Based Amorphous Alloys with Metalloids," in Titanium '80, Science and Technology, vol. 2, ed. H. Kimura and O. Izumi (Warrendale, PA: TMS Publications, 1980), 1137-1145.

75. A. Inoue and T. Masumoto, "Melt Quenched Amorphous Superconducting Alloys," in Rapid Solidification Technology Source Book, ed. R. L. Ashbrook (Metals Park, OH: American Society for Metals, 1983), 292-327.

PRODUCTION AND CHARACTERIZATION OF RAPIDLY SOLIDIFIED TITANIUM AND OTHER ALLOY POWDERS MADE BY GAS ATOMIZATION

John H. Moll and C. F. Yolton

Crucible Materials Corporation
Crucible Research Center
P. O. Box 88
Pittsburgh, Pa. 15230

Abstract

Gas atomization has developed as a means for producing high quality rapidly solidified powders for aerospace and other high performance applications. The production and development techniques used to manufacture these powders are discussed. The unique application of gas atomization to titanium alloys is described. Physical characteristics of the titanium powder and the properties of consolidated product are presented.

Introduction

Gas atomization is becoming well established as a production process for producing high quality rapidly solidified powders of a wide variety of metallic materials. With the process, prealloyed powder can be produced which is virtually free of macroscopic segregation. Because of the fineness of the structure, any residual microscopic segregation can be readily homogenized. As a result, the powder can be used to produce fully dense products which have property uniformity unattainable by conventional ingot metallurgy. The process also provides a means for making new alloys that cannot be made by conventional melting, casting and hot working. The quality and characteristics of the powder produced by gas atomization is such that it can be used for the direct production of near-net shapes by hot-isostatic-pressing (1,2). Thus, hot-working operations are eliminated and machining operations can be substantially reduced.

In this paper, the production methods currently being used for commercial manufacture of tool steel and superalloy powder are presented. The physical and microstructural characteristics of gas atomized titanium alloys are also discussed.

The Gas Atomization Process

Gas atomization involves the impingement of a high pressure gas stream on a stream of molten metal which breaks the metal stream into droplets that subsequently solidify in flight. The process is usually used to produce prealloyed powder and normally involves the use of argon or nitrogen gas. Figure 1 shows the process schematically using two different nozzle designs. In closed stream atomization, the gas impinges on the metal stream immediately as it exits the nozzle. This configuration is highly energy efficient and produces the smallest powder particles. In open stream atomization, the metal stream freefalls some distance before being struck by the atomization gas. Compared to closed stream atomization, open stream atomization produces a coarser powder, but is less troublesome with regard to metal freeze-up at the nozzle. With either design, the gas may be delivered as an annulus concentric with the metal stream or as one or more discrete jets.

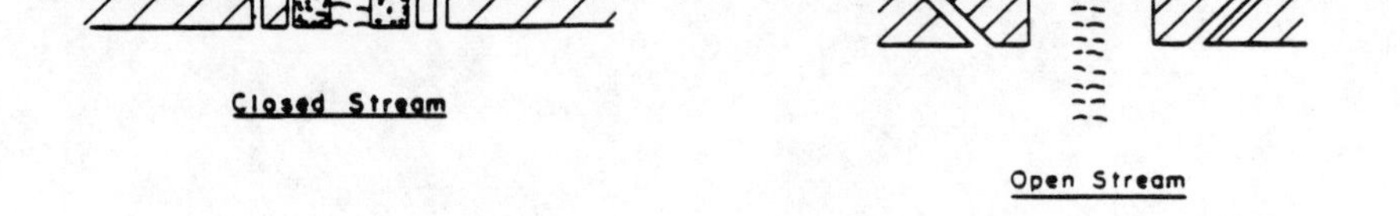

Figure 1. Nozzle configuration used for gas atomization.

Figure 2 shows a schematic drawing of a 5000 lb production atomizer. Figure 3 shows the actual unit which is capable of producing powder of a wide variety of alloys including tool steels, superalloys and corrosion resistant, nickel-base alloys. It features vacuum induction melting and inert gas atomization. The atomizer is of all stainless steel construction and all interior surfaces are polished to ease the job of cleaning between alloy changes.

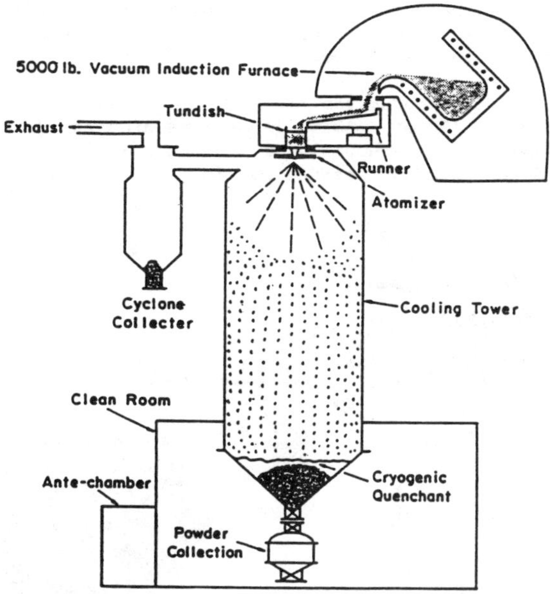

Figure 2. Schematic of production inert gas atomizer.

Figure 3. Production inert gas atomizer.

The base of the atomizer is housed in a positive-pressure clean room to provide a dust-free environment for cleaning operations. Powder is discharged directly into stainless steel transfer containers at the base of the atomizer without exposure to the atmosphere.

In contrast to the production sized atomizer described above, Figure 4 shows a schematic drawing of a small laboratory atomizer developed for gas atomizing titanium (3). The actual unit is shown in Figure 5. The activity of molten and powder titanium creates special problems for gas atomization.

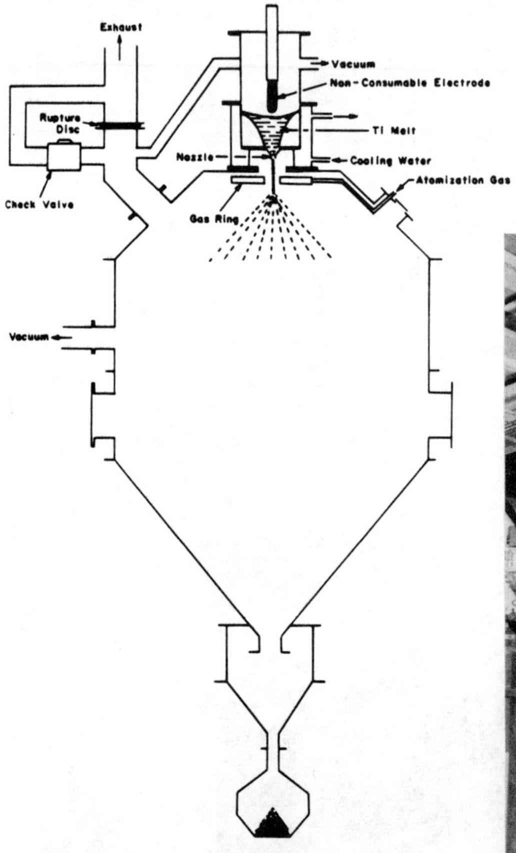

Figure 4. Schematic drawing of laboratory sized unit for gas atomizing titanium.

Figure 5. Laboratory unit for gas atomizing titanium.

The unit shown features skull melting using a non-consumable electrode. In skull melting, the molten metal contacts only metal of the same composition. A specially developed nonreactive nozzle is used to produce a controlled stream of molten titanium which is atomized with an annular gas ring. As with the production atomizer, the titanium atomizer is of all stainless steel construction to maintain powder cleanliness.

Figure 6 shows an example of gas atomized powder. Typically, the powder is spherical, has good flow characteristics and a tap density of about 65%. Large particles often have smaller satellite particles attached. Also, particles may have a partial coating of splat which results from the collision of a solid particle with a liquid particle. Experience has shown that the presence of satellites and occasional splat has no adverse effect on the quality of the powder. Since gas atomized powder is spherical, it is not suitable for cold pressing. However, the powder can be very effectively hot-isostatically-pressed (HIP) to complex shapes having full density and outstanding mechanical properties (1,2).

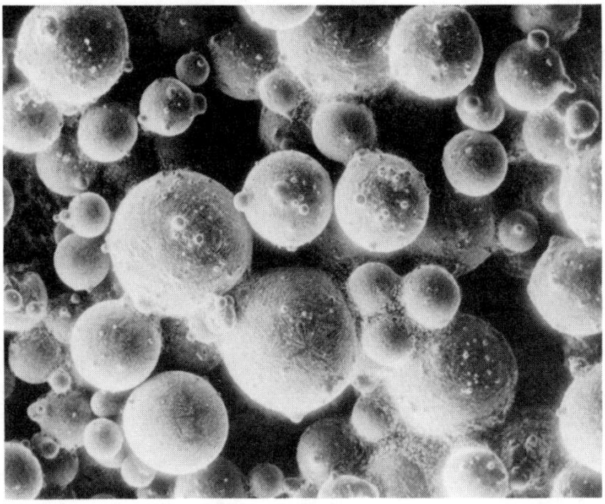

Figure 6. Argon gas atomized superalloy powder.

Gas Atomized Tool Steel and Superalloy Powder

Gas atomized tool steel powder has been in production for over 15 years. It is used to produce high quality tooling with unique performance advantages. Table I lists some alloys currently made by gas atomization. Tool steels are highly alloyed and subject to severe segregation during the slow solidification of ingots. The problem intensifies with increasing alloying content. This segregation results in poor hot-workability, low product yields and non-uniform product. With rapidly solidified gas atomized powder, the problem is minimized resulting in improved products and alloying flexibility not attainable with ingot technology (4).

Table I. Compositions of Several Tool Steels Produced from Gas Atomized Powder

Alloy	AISI Type Designation	C	Mn	S	Si	Cr	V	W	Mo	Co	Fe
CPM REX 15	T15	1.55	0.30	-	0.30	4.00	5.00	12.25	-	5.00	Bal.
CPM REX M4	M4	1.35	0.30	-	0.30	4.25	4.00	5.75	4.50	-	Bal.
CPM REX M2HSHC	M2	1.00	0.80	0.27	0.30	4.15	1.95	6.40	5.00	-	Bal.
CPM REX M3HS-2	M3	1.20	0.80	0.27	0.30	4.00	3.00	6.25	6.00	-	Bal.
CPM REX 76	-	1.50	0.30	-	0.30	3.75	3.10	10.0	5.25	9.00	Bal.
CPM 10V	-	2.45	0.50	-	0.90	5.25	9.75	-	1.30	-	Bal.

Gas atomized superalloy powder is currently being used in a number of turbine engine components. These include turbine disks, cooling plates and shafts. Engines currently using these components include General Electric's T-700, F-404, CFM-56, F101 and F110 and Pratt & Whitney Aircraft's F100. Table II gives the compositions of several superalloys currently being produced from gas atomized powder.

Table II. Composition of Superalloys Produced by Gas Atomization

Alloy	IN-100	Rene 95	MERL 76	Astroloy
Carbon	0.07	0.06	0.02	0.03
Chromium	12.5	13.0	12.4	15.0
Cobalt	18.5	8.0	18.5	17.0
Molybdenum	3.2	3.5	3.2	5.0
Tungsten	----	3.5	----	----
Hafnium	----	----	0.4	----
Niobium	----	3.5	1.4	----
Aluminum	5.0	3.5	5.0	4.0
Titanium	4.3	2.5	4.3	3.5
Vanadium	0.75	----	----	----
Boron	0.02	0.01	0.02	0.025
Zirconium	0.04	0.05	0.06	0.04
Nickel	Rem	Rem	Rem	Rem

Figure 7 shows production HIP gas turbine engine disks made from argon atomized Rene 95 powder (5).

Figure 7. P/M Rene 95 turbine disks. (Top left - as-HIP shape, Top right - machined sonic shape, Bottom left - final machined 1st stage disk, Bottom right - final machined 2nd stage disk.)

Superalloys are highly alloyed and, like the tool steels, are also subject to extensive segregation during solidification. With conventional ingot metallurgy, this may result in poor workability and non-uniform mechanical properties in the final product.

Rapid solidification through gas atomization minimizes macrosegregation resulting in improved hot-workability and property uniformity. The powder is also readily compacted to fully dense shapes which can result in reduced machining requirements.

Gas Atomized Titanium

The use of gas atomization to produce rapidly solidified titanium powder is presently in the development stage. Nevertheless, the process has been used to produce powder of a wide variety of alloys (Table III) ranging from commercially pure titanium to advanced materials such as titanium aluminide with erbium additions.

Table III. Titanium Alloys

Produced by Gas Atomization

CP Titanium
Ti-6Al-4V
Ti-10V-2Fe-3Al
Ti-10Mo-8V-2.5Al
Ti_3Al + Nb
Ti_3Al + Nb + 2Er
TiAl + Nb

The gas atomization unit (Figure 4) consists of two basic components which are a nonconsumable vacuum arc furnace, and a stainless steel atomization chamber. A tungsten electrode is used for melting. The titanium charge is melted in a water-cooled copper or uncooled graphite crucible such that a solid titanium skull is always maintained between the melt and the crucible wall. A non-reactive nozzle is used to produce a controlled stream of molten titanium which is atomized with an annular gas ring. The chamber is equipped with an oxygen analyzer for continuous monitoring of the oxygen content of the atmosphere inside the chamber.

Operation of the titanium gas atomization unit involves first placing a prealloyed charge of titanium in the furnace. All inside surfaces of the chamber are thoroughly cleaned before closing the unit. The furnace and chamber are then evacuated and backfilled with either argon or helium. An arc is struck between the electrode and the charge to initiate melting. Arc conditions are maintained such that a maximum pool size is obtained without melting through the bottom of the titanium skull. When the melt is ready for atomization, the electrode is brought closer to the surface of the melt. This drives the molten pool deeper and it melts through the bottom of the skull. The molten titanium then flows through the nozzle into the gas ring where it is atomized with either argon or helium. In its current configuration, the gas atomization unit is capable of melting and atomizing up to 1.4kg (3 lb) of titanium.

A SEM micrograph of gas atomized Ti-6Al-4V powder is shown in Figure 8. As with other gas atomized powders, titanium alloy powder is spherical and ideally suited for hot isostatic pressing. Particle size distribution of the gas atomized Ti-6Al-4V alloy is shown in Figure 9. Also plotted in this figure are particle size distributions for PREP Ti-6Al-4V powder (6) and argon atomized Rene 95, a nickel-base superalloy, powder (7). In comparison with the PREP powder, the gas atomized powder characteristically has a much wider size distribution.

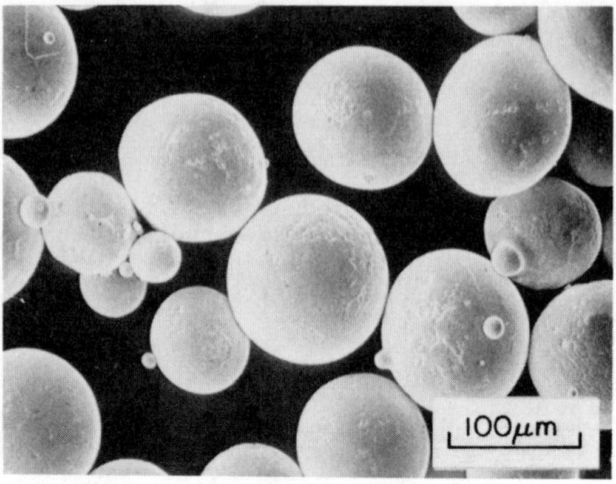

Figure 8. Gas atomized Ti-6Al-4V powder.

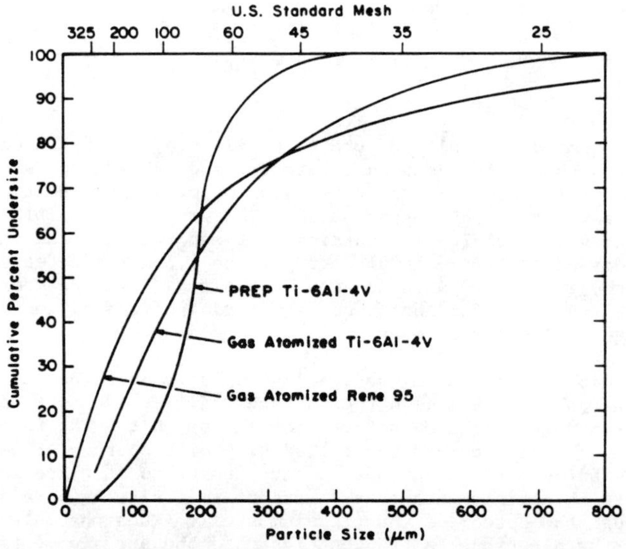

Figure 9. Particle size distribution of gas atomized Ti-6Al-4V powder compared with PREP and Rene 95 powder.

The gas atomized titanium powder is somewhat coarser than the Rene 95 powder which is primarily a result of differences in atomization conditions and equipment used for atomization of the two materials. Average particle size of the gas atomized Ti-6Al-4V powder presently being made is approximately 190μm (0.0075 in.).

The microstructure of as-atomized Ti-6Al-4V is shown in Figure 10. The powder has a martensitic microstructure which is typical for this alloy when rapidly cooled in the solid state.

Figure 10. Microstructure of gas atomized Ti-6Al-4V powder.

This martensitic transformation masks the underlying dendritic structure typical of other rapidly solidified powders.

Titanium's great affinity for oxygen makes this element a primary concern in any powder making operation. Table IV gives the oxygen content of the starting material and the powder for a typical atomization run. The results indicate a slight oxygen increase of 0.02% in the powder. Although it would be desirable for the oxygen content to remain constant, the increase in oxygen is only slightly greater than the typical 0.01% increase in conventional ingot melting.

Table IV. Chemical Analysis of Gas Atomized Ti-6Al-4V Powder

	Weight Percent					
	Al	V	Fe	C	N	O
Starting Stock	6.1	3.8	.13	.014	.01	.12
Powder	-	-	-	.016	-	.14

Another element of interest in the gas atomized powder is carbon because the molten titanium directly contacts graphite briefly when the melt is atomized. The results in Table IV show that there is very little carbon pick-up in the powder.

Hot-isostatic-pressing of the gas atomized Ti-6Al-4V powder in the alpha plus beta field develops a structure consisting of alpha platelets with intergranular beta as shown in Figure 11. Room temperature tensile properties of three typical lots of Ti-6Al-4V gas atomized powder in the hot isostatically pressed condition are given in Table V. The tensile properties are consistent with the chemistry of the powder and the properties are comparable to conventional wrought material.

It is believed that gas atomization is particularly well suited for producing advanced titanium alloys, such as those containing rare earth additions, which require rapid solidification to develop a uniform dispersion of oxide for strengthening. For example, Figure 12 shows the erbium distribution in a niobium modified alpha two titanium aluminide powder made by gas atomization. The powder contains relatively fine erbium phases which can be retained in the HIP consolidated powder product.

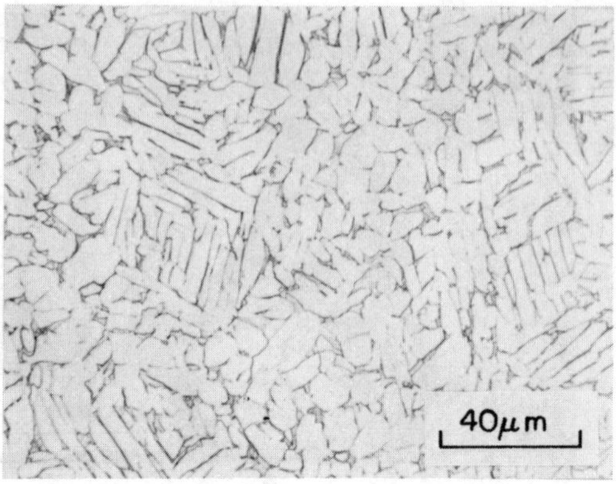

Figure 11. Microstructure of gas atomized Ti-6Al-4V powder hot-isostatically-pressed at 1750°F.

Table V. Room Temperature Tensile Properties of Gas Atomized Ti-6Al-4V*

Lot No.	Tensile Strength MPa (ksi)	Yield Strength MPa (ksi)	Elongation (%)	Reduction of Area (%)
14	909 (132)	827 (120)	17	46
	916 (133)	834 (121)	16	46
17	916 (133)	834 (121)	16	46
	909 (132)	834 (121)	17	44
25	923 (134)	841 (122)	16	40
	903 (131)	827 (120)	16	46
MIL-T-9047 (Wrought)	896 (130) Min	827 (120) Min	10 Min	25 Min

*HIP 1750°F and anneal 1300°F.

The data given in Table VI show the effectiveness of the dispersoid in increasing the tensile strength of P/M titanium aluminide. It is expected that gas atomization will be a significant factor in the development of advanced titanium alloys for extended high temperature service.

Table VI. Tensile Properties of Gas Atomized Titanium Aluminide at 1300°F

Alloy	Heat Treatment*	TS MPa (ksi)	EL (%)	RA (%)
Ti_3Al + Niobium (P/M)	A	517 (75)	6	12
Ti_3Al + Niobium + 2% Erbium (P/M)	A	648 (94)	2	4

*A = HIP 2065°F + 2200°F/1 hr/AC + 1400°F/1 hr.

As Produced Powder

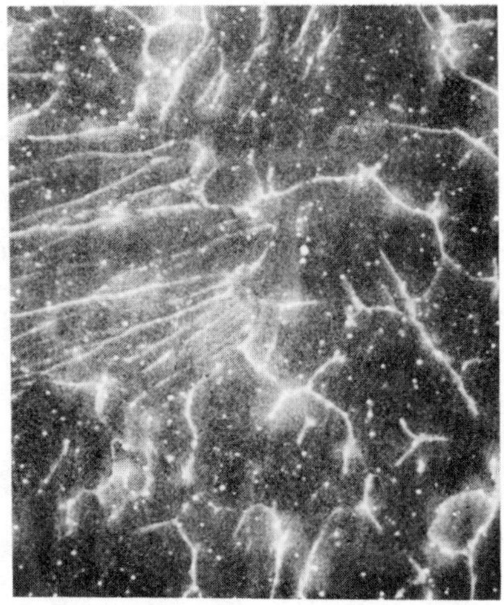

As-HIP 2065°F

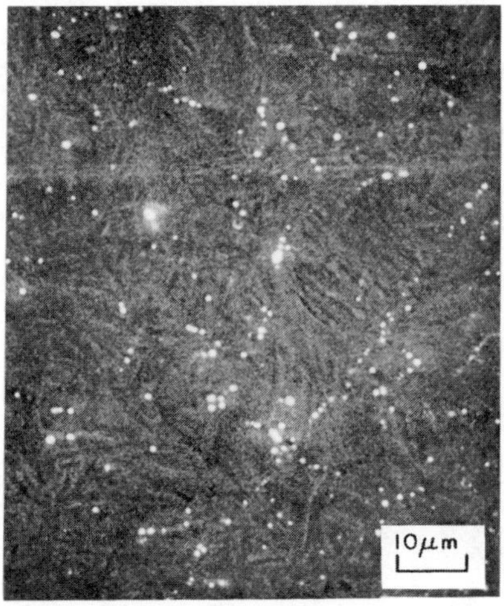

HIP + 2200°F/1 hr

Figure 12. SEM micrographs of gas atomized Ti-3Al+Nb+Er powder showing erbium-rich dispersoid (white phase).

Summary

Gas atomization produces rapidly solidified powders in steels, superalloys and titanium alloys. The fineness of the structure and the absence of macroscopic segregation result in materials which exhibit properties which are improved over those materials made by ingot metallurgy. The spherical shape and tap density of gas atomized powder make it ideally suited for consolidation to full density by hot-isostatic-pressing. Tool steels and superalloys are being routinely produced from gas atomized powder.

Gas atomization of titanium alloys is in the development stage but studies to date indicate excellent potential for the process in producing advanced titanium alloys.

Acknowledgements

This work was jointly supported by Crucible Research Center and General Electric Company, Aircraft Engine Business Group. Particular thanks are extended to Mr. G. Wasielewski, of General Electric, for his advice and technical support. The authors would also like to acknowledge the valuable technical assistance of Mr. C. L. Ruffner and C. E. Rader.

References

1. J. H. Moll, "Fully Dense P/M Shapes for Defense Applications," Powder Metallurgy in Defense Technology, C. L. Freeby and W. J. Ullrich, ed. N. J. MPIF, 1985.

2. J. H. Moll and F. J. Rizzo, "Production, Processing and Application of Products Made From Rapidly Solidified Powders," Proceedings of 1984 Vacuum Metallurgy Conference on Specialty Metal Powders, C. K. Bhat and L. W. Lherbier, ed. PA, AIME, 1985.

3. C. F. Yolton and J. H. Moll, "Gas Atomized Titanium Powder," Rapidly Solidified Materials, P. W. Lee and R. S. Carbonara, ed. ASM, 1985.

4. A. Kasak, G. Steven and T. A. Neumeyer, "High Speed Tool Steels by Particle Metallurgy," Automotive Engineering Congress, Detroit, Michigan, 1972 (SAE Paper No. 720182).

5. F. J. Rizzo, J. Lane and J. H. Moll, "Production of P/M Near-Net Shape Superalloy Hardware," Presented at the SAE Aerospace Congress and Exposition, Anaheim, CA, 1982 (SAE Paper No. 821515).

6. P. R. Roberts and P. Lowenstein, Powder Metallurgy of Titanium Alloys, F. H. Froes and J. E. Smugeresky eds., TMS AIME, Warrendale, Pa., (1980), p. 21.

7. R. H. Van Stone, F. J. Rizzo and F. J. Radavich, Proc. Second Int. Conf. on Rapid Solidification Processing: Principles and Technologies, R. Mehrabian, B. H. Kear and M. Cohen eds., Claitor's Publ. Div., Baton Rouge, LA, (1980), p. 260.

NEW RAPIDLY SOLIDIFIED

TITANIUM ALLOYS PRODUCED

BY MELT-SPINNING

Ranjan Ray and Peter Clemm

Marko Materials, Inc.
144 Rangeway Road
North Billerica, Massachusetts 01862

The principles of rapid solidification processing were used to fabricate microcrystalline titanium alloys containing finely dispersed boride phases. Chill block casting was used to prepare filaments of three titanium alloys, Ti-6Al-4V, Ti-8Al-1Mo-1V, and Ti_3Al each alloyed additionally with 1 weight percent boron. The filaments were pulverized into 60 mesh powder by a rotating hammer mill. The powders were subsequently consolidated into bar stock by hot extrusion. The alloys were tested for tensile properties at room and elevated temperatures. Microstructural investigation of the alloys were carried out using TEM and STEM techniques. Results of this study are presented in this report.

Introduction

Rapid solidification technology (RST) via powder metallurgy has been recognized as a route for the creation of cost-effective engineering materials with improved and in some cases novel, physical properties (1-2). Using the principles of rapid solidification processing, a number of new and unique alloys based on iron, nickel, aluminum, and titanium with superior mechanical properties are currently being developed in various laboratories for high performance structural applications, especially in airframe and aeropropulsion systems.

Titanium alloys are important materials for aircraft equipment; however, it is only recently that work has been done to explore and develop new titanium alloys via RST, primarily because of the well-known engineering problems associated with the melting of titanium alloys and suitable fabrication of rapidly solidified powders from prealloyed melts free of contamination. Thus, the so far untapped potential of rapid solidification technology lies in new titanium alloys distinctly different from the present commercial alloys, which have yet to be synthesized, and which will possess novel microstructures having desirable engineering properties for aerospace structural applications.

Only recently, investigations (3-4) have been undertaken to develop novel processing techniques to fabricate rapidly solidified powders of new titanium alloys. Preliminary studies indicate that novel powder metallurgy titanium alloys can be synthesized with unique combinations of microstructure and properties via RST (5-9).

Titanium alloys containing neodymium, erbium, and dysprosium can be rapidly solidified into metastable supersaturated solid solutions. Upon heat treating during hot consolidation, the metastable solid solutions decompose producing fine, incoherent dispersoids which contribute increased room temperature strength and high temperature creep resistance.

Alloys based on titanium containing small amounts of boron and carbon (1-2wt%) have the potential to be rapidly solidified into microcrystalline alloys dispersion strengthened by fine titanium boride and titanium carbide.

Alloys based on the compounds TiAl and Ti_3Al have attractive and unique high-temperature properties and are thus potential candidates for advanced gas turbine engine applications. However, conventionally processed TiAl and Ti_3Al are brittle because of a high degree of slip planarity. A potential exists for increasing the ductility of these alloys by disordering, grain refinement, and incorporation of fine incoherent dispersoids by rapid solidification processing. Rare-earth elements, boron, carbon, and silicon are some of the potential dispersoid formers.

The present paper reports the results of initial investigation of three titanium alloys prepared via a rapid solidification powder metallurgy technique.

Alloys And Powder Processing

The three titanium-base alloys studied in the present investigation are listed in Table I.

Table I. Candidate Alloys

Alloy	Composition
1	Ti-6Al-4V containing 1 wt% Boron
2	Ti-8Al-1Mo-1V containing 1 wt% Boron
3	Ti_3Al containing 1 wt% Boron

Alloys were prepared from raw materials of commercial purity as ingots in a standard non-consumable arc melting furnace. The ingots were subsequently rapidly solidified as fine filaments by Marko's proprietary chill casting technique developed for reactive titanium alloys. The key element in the processing of these alloys was a step involving rapid solidification of melt at 10^6°C/second. The liquid metal of titanium alloy is brought in contact with a rotating chill copper roll whereby a continuous narrow filament (40-60 μ thick) (Figure 1) is formed undergoing rapid solidification. The filaments were subsequently pulverized into particulates with average particle size below 40 mesh.

Hot Consolidation of Powders Via Extrusion

About two pounds of powder particulates of each of the three alloys were filled in a 2 inch diameter mild steel can and hot evacuated to a vacuum level of 1 micron. After holding for 4 hours under vacuum at 750°F to a vacuum level of 1 micron, the cans were sealed off.

The sealed cans containing RST powders of the Ti-6Al-4V-1B and Ti-8Al-1Mo-1V-1B alloys were heated at 1065°C (1950°F) for 3 hours and extruded into 0.625 inch dia bars using a 300 ton press at a reduction ratio of 10:1.

The sealed can containing RST Ti_3Al+1%B powders was heated at 1148°C (2100°F) for 3 hours and extruded into 0.625 inch diameter bar using a 300 ton press at a reduction ratio of 10:1.

Properties and Microstructures

The consolidated bars were tested for room and elevated temperature up to 1000°F tensile properties without any further heat treatment. The main objective of this research was to evaluate the potential of Marko's rapid solidification processing technique to develop improved titanium alloys.

(a)

(b)

Figure 1. (a) Titanium Filament
(b) RST Titanium Powder Made From Filament

Table II lists the tensile properties of the three candidate alloys.

Boron modified Ti-6Al-4V and Ti-8Al-1Mo-1V alloys showed attractive room and elevated temperature tensile properties. However, the RST alloys showed low ductility at room temperature. Figures 2-3 show the mechanical properties (i.e. ultimate tensile strength, yield strength, and % elongation) of RST Ti-6Al-4V+1%B alloy as compared to the properties of conventional Ti-6Al-4V alloy. Similarly, Figures 4-5 show the mechanical properties (i.e. ultimate tensile strength, yield strength, and % elongation) of RST Ti-8Al-1Mo-1V+1%B as compared to the properties of conventional Ti-8Al-1Mo-1V. Although, the RST alloys showed significant improvement in tensile and yield strength values compared to their

Table II. Room and Elevated Temperature Tensile Properties of RST Titanium Alloys

Alloy	Test Temp	0.2% Offset Yield Strength (KSI)	Ultimate Tensile Strength	% Elong	% RA
Ti-6Al-4V +1 wt% B	Room	158	161	2.5	2.2
	371°C (700°F)	132	152	3.1	4.9
	537°C (1000°F)	90	125	8.8	9.7
Ti-8Al-1Mo-1V + 1 wt% B	Room	165	170	1.8	1.4
	371°C (700°F)	124	145	8.2	10.1
	537°C (1000°F)	101	119	8.2	10.4
Ti_3Al +1 wt% B	Room	*	63	1.0	1.4
	371°C (700°F)	*	69	1.0	1.2
	537°C (1000°F)	--------- Thread Failure ---------			

*Specimen failed before reaching yielding.

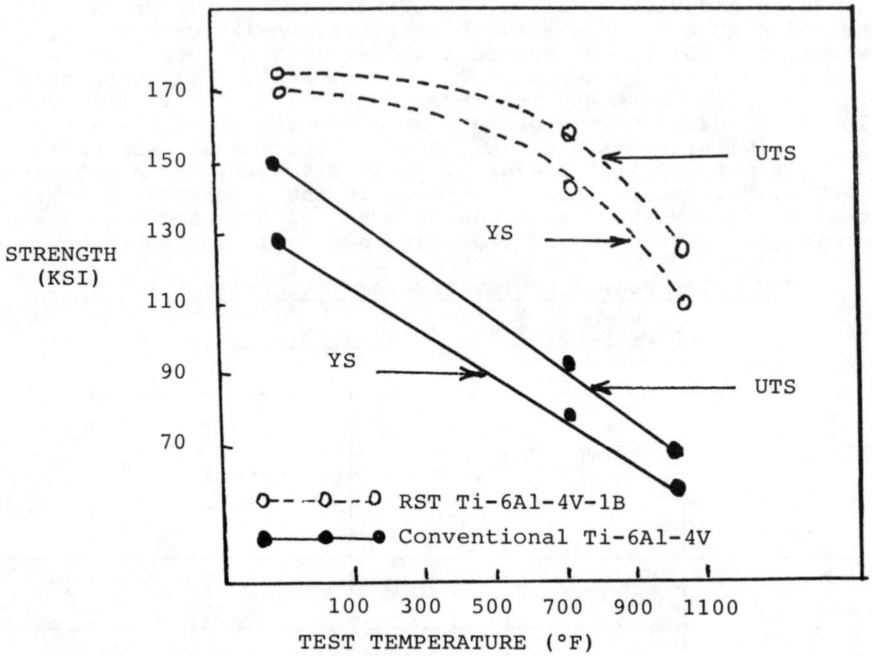

Figure 2. Yield and ultimate tensile strength of RST Ti-6Al-4V-1B and conventional Ti-6Al-4V as a function of temperature.

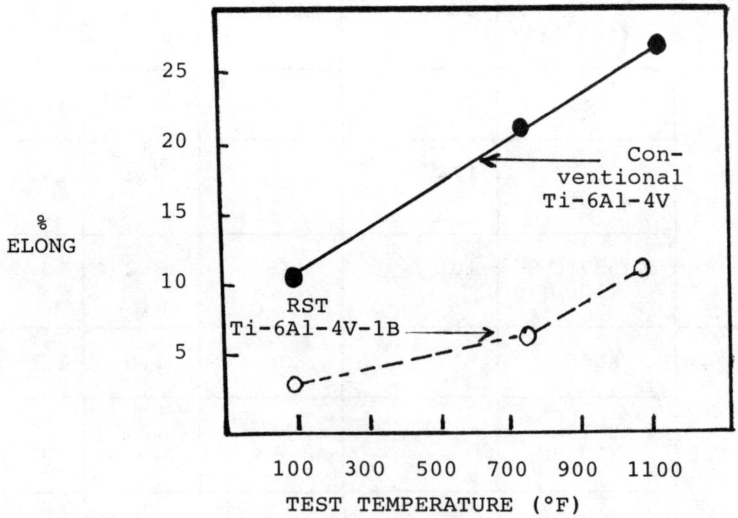

Figure 3. Tensile elongation of RST Ti-6Al-4V-1B and Ti-6Al-4V as a function of temperature.

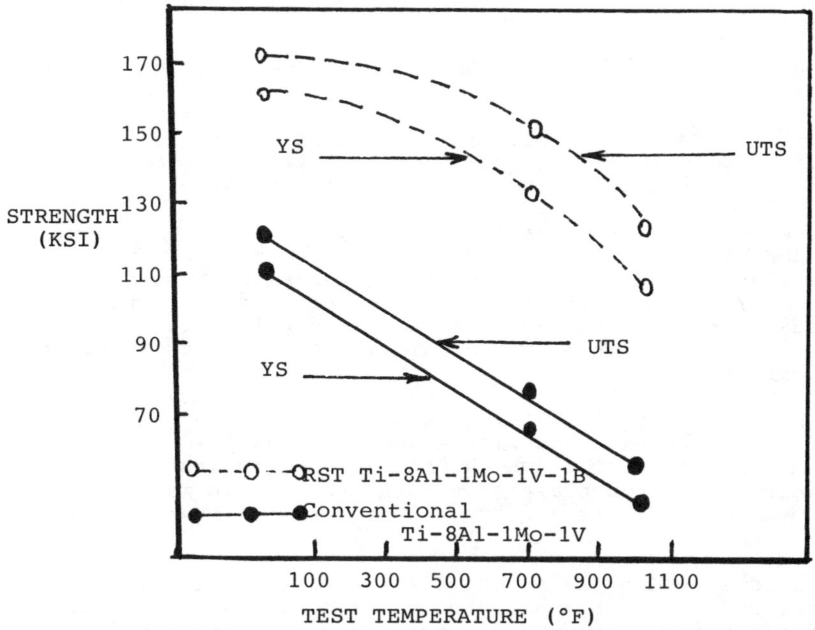

Figure 4. Yield and ultimate tensile strength of RST Ti-8Al-1Mo-1V-1B and conventional Ti-8Al-1Mo-1V as a function of test temperature.

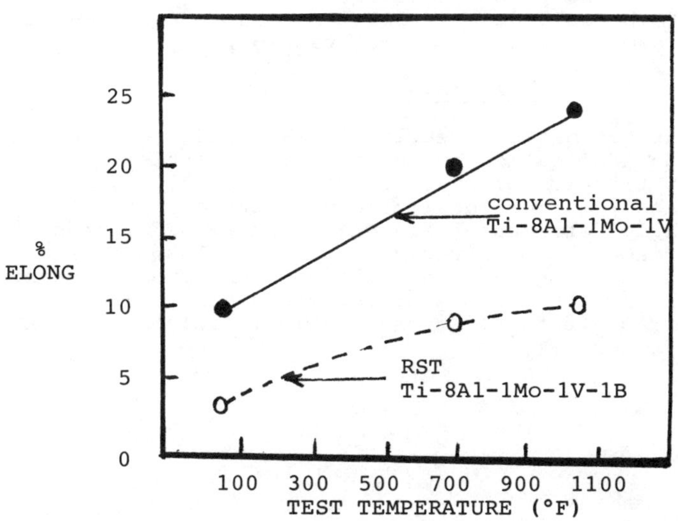

Figure 5. Tensile elongation of RST Ti-8Al-1Mo-1V-1B and Ti-8Al-1Mo-1V as a function of temperature.

conventional counterparts, the mechanism responsible for the low ductility of the RST alloys has to be clearly understood so that the principles of rapid solidification processing can be beneficially applied to such alloy systems.

Boron modified Ti_3Al was found to be quite brittle at room and elevated temperature.

Ti-6Al-4V is an alpha-beta alloy capable of attaining a variety of properties as a function of heat treatment conditions which alter the amounts of the alpha and beta phases.

Figures 6-7 show TEM images of the microstructure present in hot extruded bar of RST Ti-6Al-4V+1%B alloy. The lower magnification micrograph (Figure 6) shows the grain size to be 1-2 μ. Furthermore, it shows that the grains exhibit a substructure and second phase particles 0.05-0.2 μ in size populate the microstructure, and an equiaxed interdendritic phase exists (labelled "A" in Figure 6).

The substructure of the alpha grains consists of aligned arrays of dislocations along with acicular alpha (Figure 7). The second phase particles (which usually show growth fault that exhibit striped contrast) was found to contain mainly titanium by STEM analysis and are believed to be titanium-boron compounds.

TEM images of RST Ti-8Al-1Mo-1V+1%B are shown in Figures 8-9. The microstructure of this alloy consists of alpha grains with interstices between grains which are likely to be beta phase. Second phase particles can also be seen. The TEM images of RST Ti_3Al+1%B (Figures 10 and 11) show grains of Ti_3Al to be approximately 1 μ. Some of the Ti_3Al grains show a "speckled" structure (Figure 11). Second phase particles are present but these particles appear to be considerably smaller than in the previous two alloys and at a much smaller volume fraction.

Conclusions

The present investigation clearly demonstrated the beneficial effects of rapid solidification on mechanical properties and microstructures of titanium alloys. Conventional alloys can be produced with refined and more desirable microstructures. The addition of generally detrimental elements such as metalloids can be greatly refined and made useful via RST. Although, the preliminary alloys investigated under this present program cannot be considered optimum, the results of the study are sufficiently encouraging to warrant further intensive alloy development work using the principles of RST.

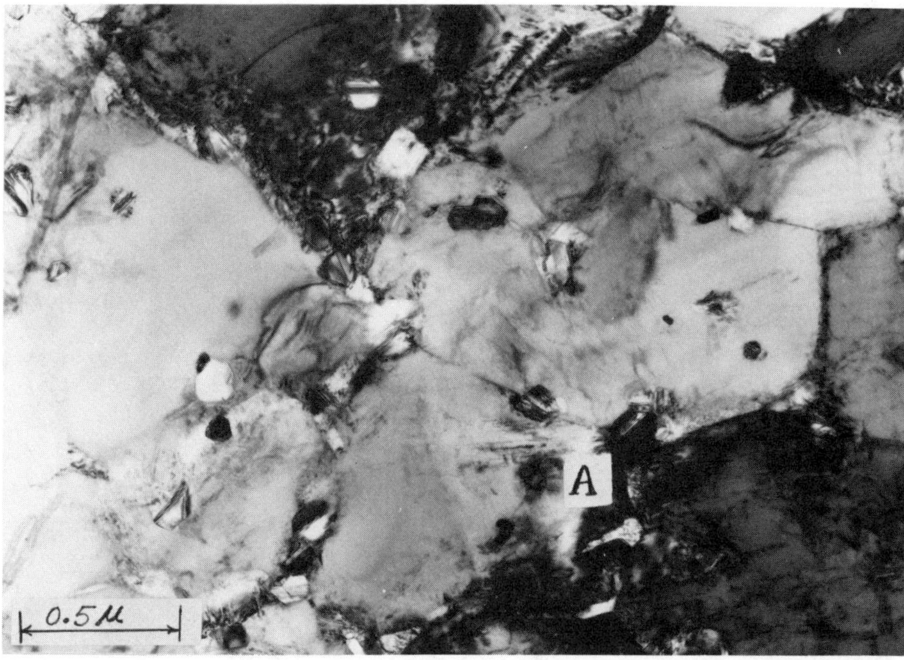

Figure 6. TEM Photograph of RST Ti-6Al-4V-1B Bar

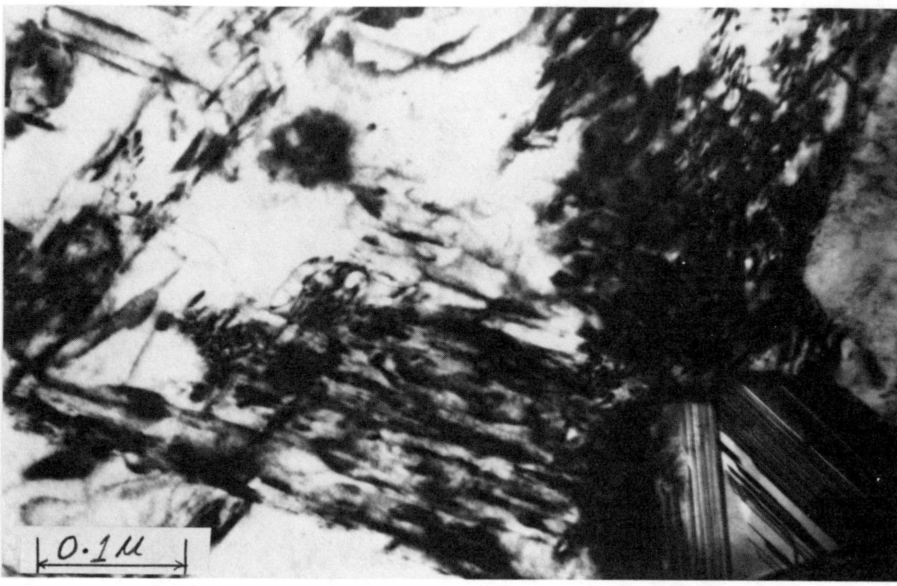

Figure 7. TEM Photograph of RST Ti-6Al-4V-1B Bar

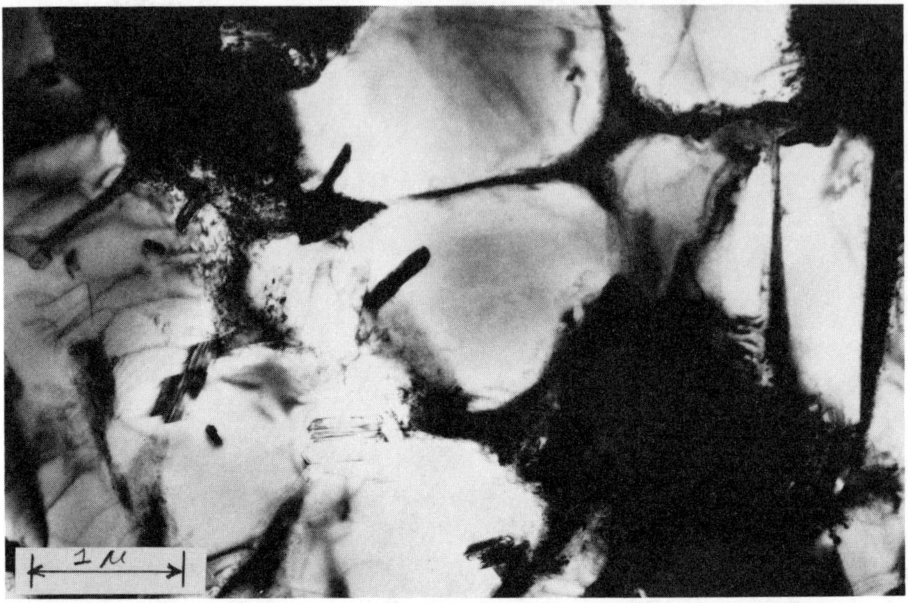

Figure 8. TEM Photograph of RST Ti-8Al-1Mo-1B Bat

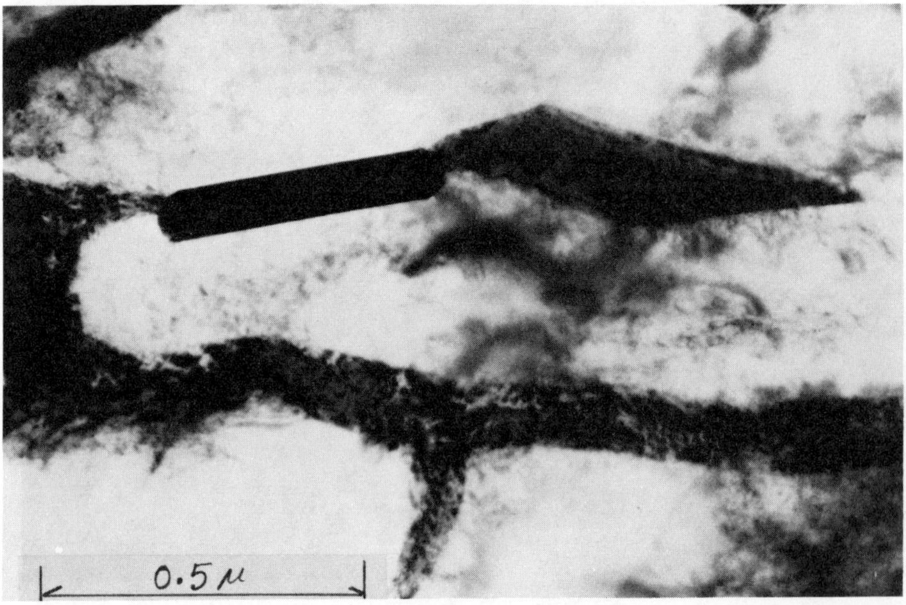

Figure 9. TEM Photograph of RST Ti-8Al-1Mo-1V-1B Bar

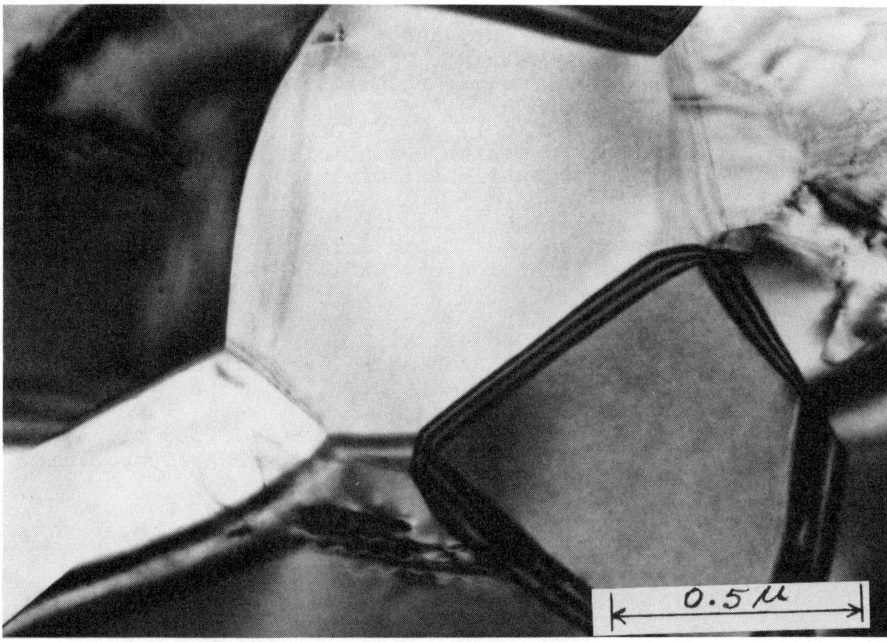

Figure 10. TEM Photograph of RST Ti_3Al+1w%B Bar

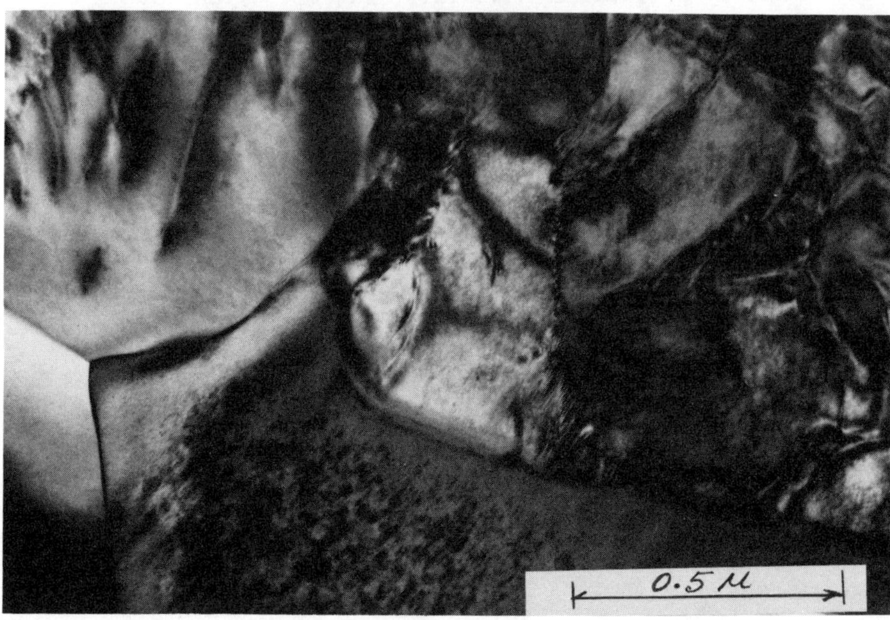

Figure 11. TEM Photograph of RST Ti_3Al+1w%B Bar

REFERENCES

1. R. Mehrabian, B.H. Kear, and M. Cohen (eds) "Rapid Solidification Processing, Principles and Technologies," Vol. 1, Claitor's Publishing Division, Baton Rouge, LA(1978).

2. R. Mehrabian, B.H. Kear, and M. Cohen (eds) "Rapid Solidification Processing, Principles and Technologies," Vol. 2, Claitor's Publishing Division, Baton Rouge, LA(1980).

3. S.M.L. Sastry, T.C. Peng, J.E. O'Neal, and L.P. Bockerman in R. Mehrabian edited "Rapid Solidification Processing, Principles and Technologies, National Bureau of Standards," p. 579 (1982).

4. A.G. Jackson, T.F. Broderick, F.H. Froes, and J. Moteff, in "Rapid Solidification Processing, Principles and Technologies," R. Mehrabian (ed), p. 585 (1983).

5. F.H. Froes and R.G. Rowe, "Rapidly Solidified Titanium" to be published in the Proceedings of the MRS, Symposium J, Boston, MA held in December 1985.

6. T.F. Broderick, A.G. Jackson, H. Jones, and F.H. Froes, Met. Trans. A, 16A 1951 (1985).

7. D.G. Konitzer, B.C.Muddle, and H.L. Fraser, Met. Trans. A 14A, 1979 (1983).

8. R.G. Rowe, T.F. Broderick, E.F. Koch, and F.H. Froes, to be presented at the First International ASM Conference on Rapidly Solidified Materials, San Diego, CA 1986 (and to be published).

9. F.H. Froes, Seminar on the Development and Use of Powder Metallurgy in Engineering Industries Minsk, (Byelorussian SSR) 25-29 March 1985.

CHARACTERISTICS OF POWDER PRODUCED

BY THE ALTi-OXY PROCESS

Dr. Joseph A. Megy

Albany Titanium, Inc.
P.O. Box 887
Albany, OR 97321

Abstract

Titanium powder is produced by sublimation of zinc from a zinc-titanium alloy particle in the ALTi-OXY process. The particle is produced from a fully molten zinc-titanium alloy solution via one of several of the known particle forming techniques or by casting and comminuting the brittle zinc-titanium alloy. The process allows prealloying with other elements and rapid solidification of the zinc-titanium solution. Several examples of novel titanium alloys produced via this process will be presented.

Introduction

ALTi has produced a number of salt-free prealloyed titanium powders using the ALTi-OXY Powder Process. These alloys have unique properties such as controlled shape and porosity which are of potential use in titanium powder metallurgy.

The "ALTi-OXY Extraction Process" is a method of producing titanium sponge from ilmenite, fluosilicic acid, salt, and aluminum raw materials. In the process, titanium metal is recovered in a zinc solvent which is recycled. Zinc is sublimed leaving a titanium sponge product which is very low in halides. The early process development was carried out at Teledyne Wah Chang beginning in 1973 (1-3), and Occidental Research Corporation beginning in 1978.(4-6)

Titanium powder metallurgy requires a titanium powder with low halide impurities to produce high quality parts. Existing titanium production processes can produce a powder by-product (e.g. "sponge fines"), which contain on the order of 1,000 ppm chloride contamination. Titanium powder with on the order of 20 ppm halide contamination or less is required to give high quality titanium PM parts which have fatigue properties comparable to parts produced from ingot titanium. Sponge fines are presently used in titanium PM parts where parts with inferior properties are acceptable.

Over the last few years NMI has produced a high quality low halide titanium powder (PREP powder) from which high quality titanium parts have been produced, thus demonstrating the viability of titanium PM technology if acceptable titanium powder is available.

In 1982, Albany Titanium acquired the rights to the patented Occidental technology and demonstrated its utility in the production of low halide, high quality titanium powder by a procedure called the "ALTi-OXY Powder Process". Although the major program at ALTi involves the commercialization of the "ALTi-OXY Extraction Process"for titanium sponge, a small program has been maintained to evaluate the type and characteristics of titanium powder which can be produced.

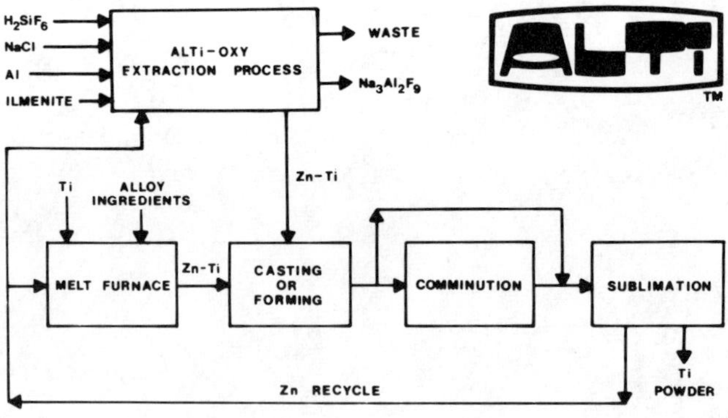

Figure 1: The ALTi-OXY Powder Process

The "ALTi-OXY Powder Process" begins with a molten mixture of titanium and zinc, which is obtained from the ALTi-OXY Extraction Process or by simply dissolving titanium in molten zinc as shown in figure 1. The molten mixture can be formed into solid zinc-titanium particles by a variety of means which give different morphologies. Since the alloy is brittle it can be cast in a mold and then sized to any desired particle size distribution by various comminuting devices. This approach leads to an angular titanium powder. By various cold surface or atomization techniques other shapes can be produced such as spheres, splats, or filaments as seen in figure 2. The zinc-titanium alloy can be solidified over a wide range of cooling rates by the various methods. Rapidly solidified zinc-titanium alloys can be formed from a solution at 900°C which can be contained more easily than pure molten titanium at 1700°C.

Figure 2: Titanium powder particles produced from water cooled rotating disc showing broken plates, spheres, and filaments (7).

Zinc is sublimed from the alloy either into a vacuum or an inert carrier gas by heating it in a furnace. The shape of the zinc-titanium alloy particles is retained during the sublimation step.

The purity of sponge and powder products produced in pilot plant facilities at Albany Titanium is shown in table 1. Note particularly low levels of nitrogen, alkali, and halide impurities.

TABLE 1.

Comparison of ALTi
Sponge Analysis with ASTM B299 Specifications

(Percent)

Impurity	ALTi	Type A	Type B	Type C	Type D
Nitrogen	0.002	0.015	0.015	0.010	0.008
Aluminum	0.010	-	-	-	-
Carbon	0.025	0.020	0.025	0.020	0.020
Sodium	-	-	-	0.190	0.010
Magnesium	0.002	0.080	0.500	-	0.080
Chloride/Fluoride	0.002	0.120	0.200	0.200	0.100
Lithium	-	-	-	-	0.090
Iron	0.040	0.120	0.100	0.050	0.040
Silicon	0.040	0.040	0.040	0.040	0.040
Hydrogen	0.005	0.005	0.030	0.050	0.020
Oxygen	0.070	0.100	0.100	0.100	0.070
Other impurities	0.050	0.050	0.050	0.050	0.050
Brinell hardness (max.)	120	120	120	120	-
Titanium, (min.)	99.6	99.3	99.1	99.3	99.6

Type A: Magnesium reduced (Kroll) and finished with vacuum distillation (ASTM).

Type B: Magnesium reduced (Kroll) and finished with acid leaching or inert gas sweep (ASTM).

Type C: Sodium reduced (Hunter) and finished with acid leaching (ASTM).

Type D: Electrolytic (National Defense Stockpile Specifications).

A novel feature of the titanium powder produced with this process is a controlled internal porosity. The zinc can be quantitatively removed from the zinc-titanium alloy at temperatures as low as 750°C. By sintering the resulting powder at various times and temperatures the internal porosity can be varied over a wide range as shown in figures 3-6. Some sintering is required if the powder is to be "passified" without contaminating it with oxygen to levels outside the specification for PM powder.

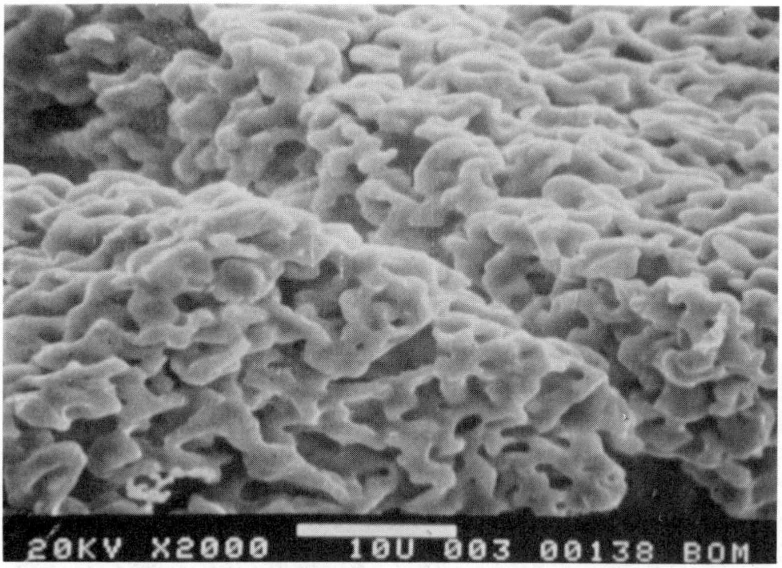

Figure 3: Titanium powder with zinc removed and particles sintered at 800°C for four hours (7).

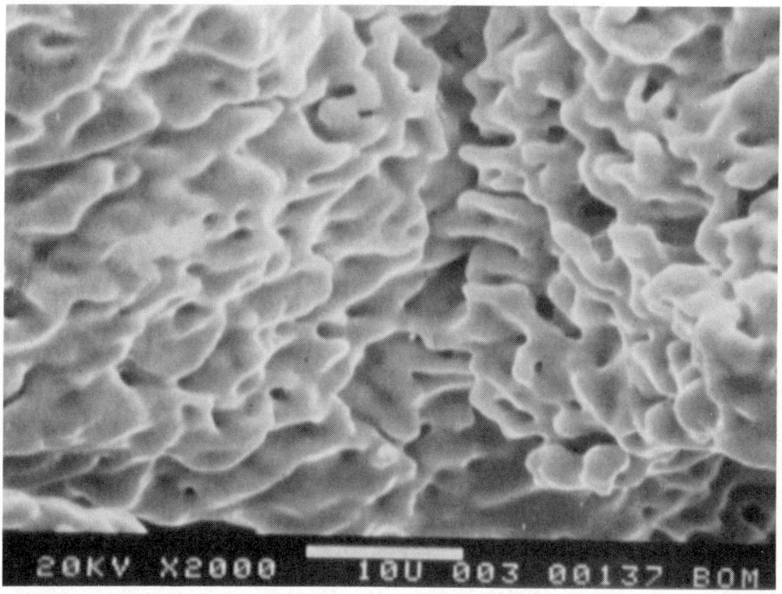

Figure 4: Titanium powder with zinc removed and particles sintered at 800°C for eight hours (7).

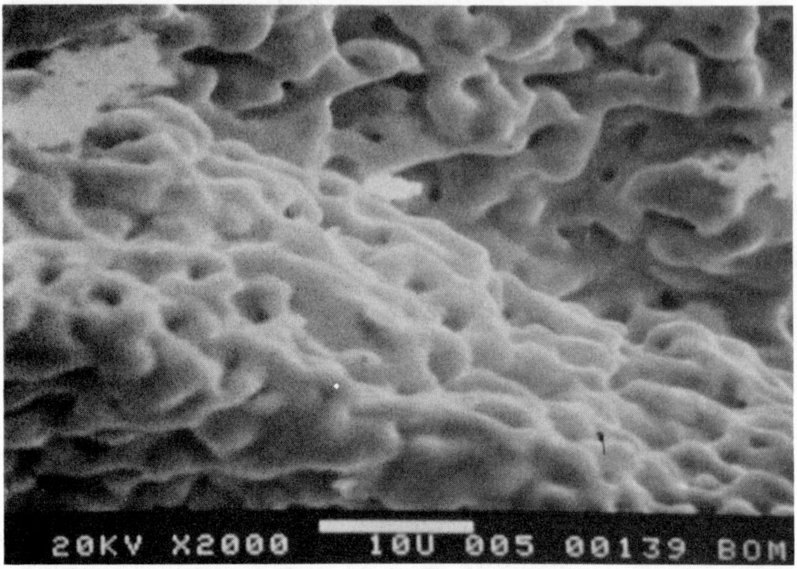

Figure 5: Titanium powder with zinc removed and particles sintered at 850°C for four hours (7).

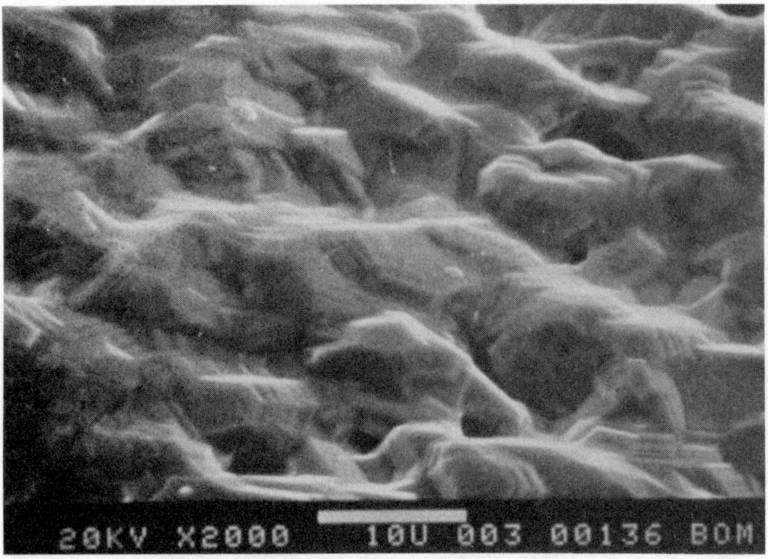

Figure 6: Titanium powder with zinc removed and particles sintered at 950°C for eight hours (7).

Prealloying

Many alloying ingredients can be added to the molten zinc-titanium solution without forming precipitates. The solution, when cast, forms a homogeneous mixture in the titanium powder product. For example, ALTi has produced a nickel-titanium alloy powder which was uniformly within the narrow Ti-Ni solid solution range with no terminal phase areas observed and exhibiting shape memory alloy properties in the powder form. Weighed amounts of titanium and alloy components can be added to the zinc solvent, so that close compositional specifications can be met.

Since the alloy ingredients are dissolved into the molten zinc at 900°C, some titanium alloys with volatile components can be produced that would be impossible with ingot metallurgy. For example, ALTi has produced titanium-lead sponge and powder.

Titanium-beryllium alloys can also be produced by this approach at sufficiently low temperatures to avoid health hazzards associated with high beryllium vapor pressure. Titanium sponge with eight weight percent beryllium as produced in which the beryllium phase was approximately five microns in diameter. This product was obtained by slow cooling during the casting step. Finer beryllium alloy phases can be obtained with rapid solidification.

Aluminum-titanium alloys are of potential interest because of their high temperature strength and oxidation resistance. The ALTi-OXY process permits the production of titanium-aluminum powder with a wide range of third phase components. Titanium and aluminum alloy phases precipitate from zinc solutions as titanium-aluminum-zinc compounds with simple ratios such as $Ti\text{-}Al\text{-}Zn_5$ and $Ti_2Al_3Zn_5$. The size of Ti-Al-Zn crystals is a function of the rate of solidification during cooling. When the zinc is sublimed away the crystal sizes are slightly smaller due to the loss of zinc from the alloy phase. Titanium aluminide powders with Ti_3Al, TiAl, and various other titanium aluminum ratios have been produced.

Third phase components can be added such as niobium or rare earths. For example Ti-2Er and Ti-15Al-2Er sponge have been obtained in the Albany Titanium laboratory. This product was obtained from a bulk zinc casting which was broken to small sponge pieces prior to sublimation of the zinc.

The ALTi-OXY Powder Process has been successful in producing quite a large variety of products, all of them with undetectable concentrations of halide contamination. The porosity, shape, and purity of the product suggest that they may be useful in titanium powder metallurgy.

REFERENCES

1. Bennedict, M., Pigford, T.H., and Levi, H.W., Nuclear Chemical Engineering (Second Edition, McGraw-Hill, 1981), 338-342.

2. Megy, J.A., "Method of Reducing Zirconium", USP 4,127,409, November 28, 1978.

3. Megy, J.A., "Method of Separating Hafnium from Zirconium", USP 4,072,506, February 7, 1978.

4. Hard, R.A., and Megy, J.A., "Process for Making Titanium, Zirconium, and Hafnium-Based Metal Particles for Powder Metallurgy", USP 4,470,847, September 11, 1984.

5. Hard, R.A., and Prieto, M.A. "Process for Making Titanium Metal from Ilmenite Ore", USP 4,390,365, June 28, 1983.

6. Megy, J.A., "Process for Making Titanium Metal from Titanium Ore", USP 4,453,243, August 28, 1984.

7. Courtesy of the US Bureau of Mines, Analytical Section, Albany Research Center.

RAPIDLY SOLIDIFIED Ti ALLOY POWDERS

J.P.A. Lofvander, S.A. Court, R. Wheeler, J.W. Sears,
D.A. Watson and H.L. Fraser

Department of Materials Science
University of Illinois
Urbana, IL 61801

Abstract

Rapidly solidified powders have been produced by centrifugal atomization (CA). Two alloys have been processed using this technique, namely Ti-0.7Er and Ti-0.9La (at.%), and the microstructures of the as-solidified powders have been fully characterized. In the case of the Er-containing alloy, a supersaturated solid solution based on α'-Ti was produced. Subsequent heat treatment at 700°C for 10 h results in the precipitation of a refined dispersion of oxide (Er_2O_3) particles. This implies that the powders have reasonably low oxygen concentrations. In the case of powders of the La-containing alloy, the microstructure in the as-solidified condition consists of a dispersion of moderately refined La particles. These become La_2O_3 upon annealing at 700°C by a process of internal oxidation. Powders of both alloys have been consolidated using hot isostactic pressing (HIP) and the microstructures of these compacts have been studied. In the case of the Er-containing samples the Er_2O_3 particles have coarsened somewhat, consistent with a low oxygen concentration. In the La-containing alloy, the HIP'd samples revealed only modest coarsening of the oxide particles. In all HIP'd samples, the microstructure contained carbide grains, which have presumably been included by erosion of the graphite crucible.

Introduction

There is considerable current interest (1) in the production of rapidly solidified Ti-alloy particulate. Of the approaches available for the generation of such particulate, atomization techniques are the most favorable in terms of powder yield, shape, size distribution and resultant microstructure.

The high reactivity of Ti normally precludes atomization by splat quenching techniques involving the use of a crucible, and consequently non-crucible methods such as the rotating electrode process have been employed. In the present study, however, powders have been produced using a centrifugal atomization technique following induction melting in a graphite crucible. It was hoped that reaction between the molten Ti and the graphite crucible could be minimized by rapid melting and also by the formation of a skull. In centrifugal atomization, molten metal is ejected as droplets from a rapidly spinning disk. Particles subsequently cool in the collection chamber or can be gas quenched in flight from the disk. High cooling rates (10^4-10^6 K/s) are achieved in the small powder particles (< 100 μm), leading to a high degree of microstructural refinement and chemical homogeneity.

Experimental Procedure

The equipment used for the production of the Ti-0.7at%Er and Ti-0.9at%La powders is shown in Fig. 1. Approximately 150-200 g of alloy was induction

Centrifugal Atomizer - Induction Melt

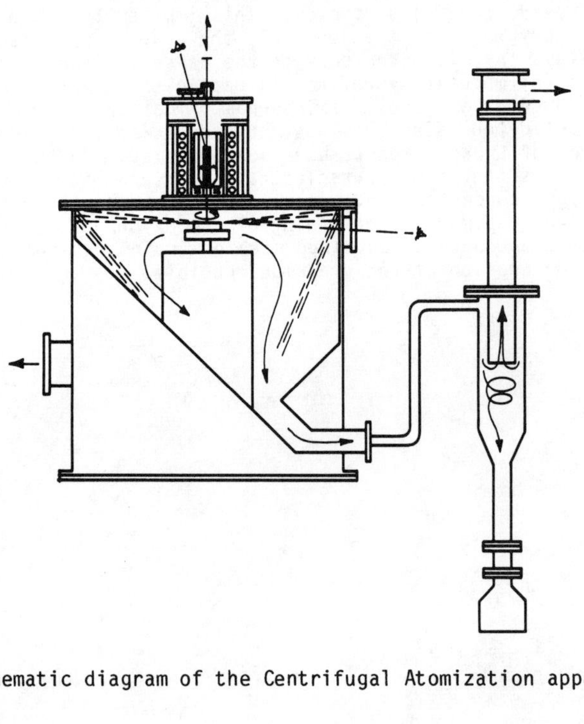

Figure 1 - Schematic diagram of the Centrifugal Atomization apparatus.

melted in the graphite crucible; the time that the alloy was molten was kept to ≤ 20 s. The graphite atomizing disk was spun at ≈ 25,000-35,000 rpm and helium gas was used to effect quenching. The yield of powder with diameters < 125 μm was generally 40-60 g (i.e. an "efficiency" of ~ 30%). Small batches of powder were heat treated in sealed quartz capsules under an argon atmosphere. Consolidation of the rapidly solidified powder was accomplished by cold compaction and hot isostatic pressing (HIP) in a stainless steel can at a temperature of 800°C and a pressure of 27 ksi (186 MPa) for 2 h, following a "pre-treatment" of 700°C for 2 h.

Thin foil samples of the as-atomized and heat treated powders were prepared using a similar technique to that originally developed by Field and Fraser (2). Optical and thin foil samples of the as-consolidated material were prepared for examination using standard techniques.

Results and Discussion

Powders

Ti-Er

As shown in Fig. 2, in the as-atomized condition the Ti-Er alloy exhibits

Figure 2 - Bright field electron micrograph of the as-atomized Ti-0.7Er (at%) alloy.

a single phase α'-Ti microstructure, with little evidence of Er precipitation, indicating that the Er is present in solid solution. This result agrees well with those of laser surface melting, where similar supersaturated solid solutions were produced (3). Evidently, the forced convective cooling provided by He gas in the atomizer permits a sufficiently high degree of undercooling prior to nucleation to be achieved so that partitionless solidification occurs. When heat treated at 700°C for 10 h, a refined dispersion of Er_2O_3 particles is formed, Fig. 3. This is again in concert

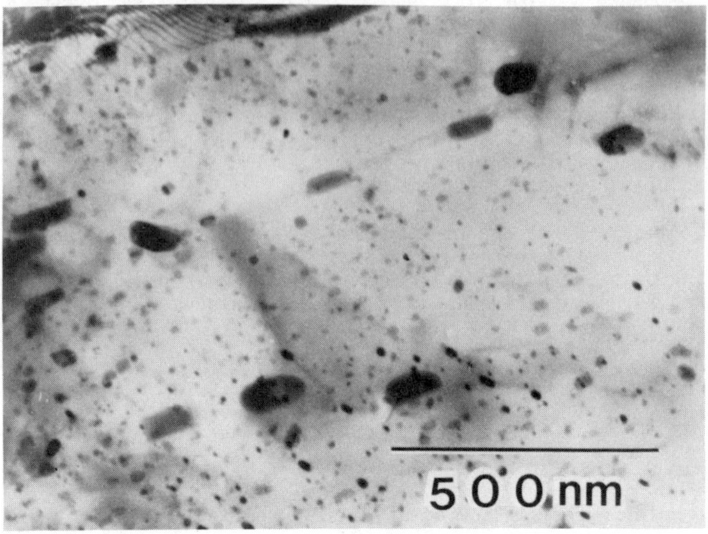

Figure 3 - Bright field electron micrograph of as-atomized T-0.7Er (at%) powder following heat treatment at 700°C for 10 h.

with the studies reported in (3), and is further evidence supporting the supposition that in the as-atomized powders the Ti matrix is supersaturated with Er. It is of interest to assess whether during processing the concentration of oxygen in the liquid Ti alloy droplets increased significantly. This may be most conveniently done by annealing powders at elevated temperatures. Thus, it has been shown (4) that the rare earth oxide particles exhibit reasonable stability providing annealing is performed in the α-Ti phase field, the stability increasing with oxygen concentration, whereas a small number of large oxide particles results when annealing is performed in the β-Ti phase field. For low oxygen concentrations in the Ti matrix, the β-transus will be close to ≈ 880°C. Therefore, powders were annealed at 700°C for 10 h (to develop a refined dispersion of oxide particles) and then subsequently at 900°C, again for 10 h, the results of these treatments being shown in Fig. 4. As can be seen, very large oxide particles are produced, implying that the powders have been annealed above the β-transus, and so it may be concluded that the oxygen concentration in the Ti matrix of the powders is very low.

Ti-La

In the case of the La-containing alloy, the as-atomized powders show very few second phase particles, Fig. 5, indicating the possible development of supersaturated solid solutions. Heat treatment at 700°C for 10 h causes a relatively low density of oxide particles to form as shown in Fig. 6. It is evident from these observations that a small supersaturation of La has been produced during rapid solidification of the liquid droplets; however, the volume fraction of the particles formed on heat treating the powders is much less than that expected for the composition of this alloy (0.9 at.% La) (e.g. (5)). Some loss of La appears to have occurred during the production of the powders. This may be explained as follows. It has been shown previously (5,6) that this alloy system exhibits a liquid phase miscibility gap; therefore, it is most probable that while the alloy was being melted prior to atomization that the La-rich liquid sunk to the bottom of the graphite

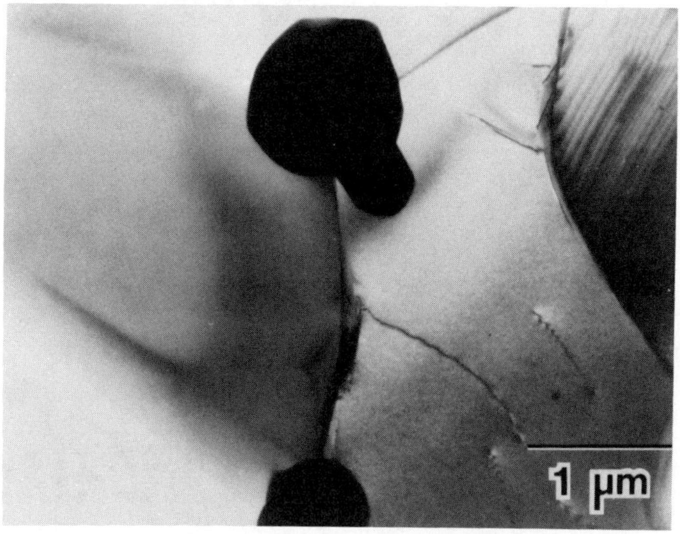

Figure 4 - Bright field electron micrograph of as-atomized Ti-0.7Er (at%) powder after heat treatment at 900°C for 10 h following heat treatment at 700°C for 10 h.

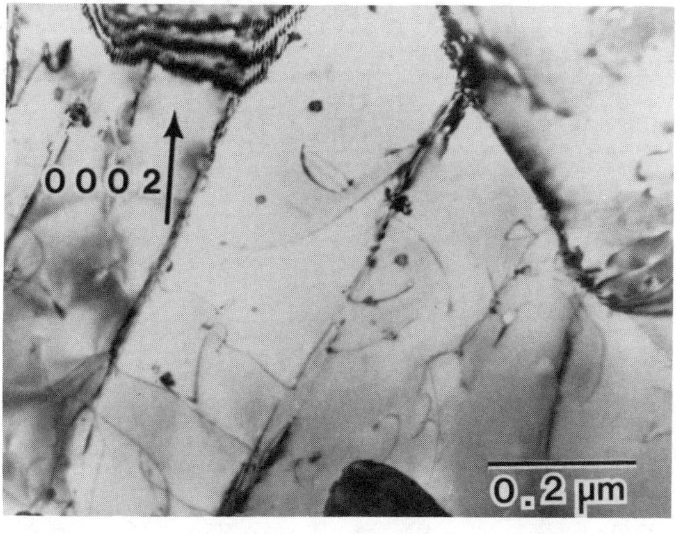

Figure 5 - Bright field electron micrograph of the as-atomized Ti-0.9La (at%) alloy.

crucible. In this way, the concentration of La in the remaining liquid would be reduced significantly.

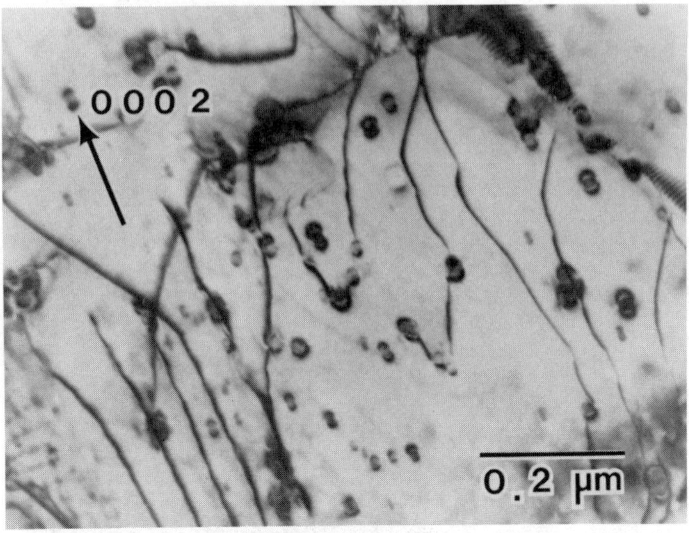

Figure 6 - Bright field electron micrograph of as-atomized Ti-0.9La (at%) powder following heat treatment at 700°C for 10 h.

Consolidated Samples

Powders of the two alloys were hot isostatically pressed at 800°C and 186 MPa (27 ksi) for 2 h. Optical micrographs of the consolidated Ti-Er and Ti-La alloys are shown in Figs. 7a and 8a, respectively. As can be seen in both cases, essentially full density has been achieved. The microstructures of these compacted samples have been characterized using analytical transmission electron microscopy, and typical micrographs are shown in Figs. 7b and 8b, for the Er- and La-containing alloys, respectively. For the case of the Er-containing alloy, Fig. 7b, the finer particles have coarsened quite considerably (average diameters ≈ 100-150 nm). This is consistent with the coarsening behavior expected for an alloy with a relatively large excess concentration of Er over that required to form Er_2O_3 (4). Of course, the excess concentration in the particular sample studied here is not that large, and so the coarsening exhibited by the particles is most probably due to the combined effects of stress and temperature. These effects of stress are being studied currently (7).

In contrast to these observations, the particles in the HIP'd Ti-La alloy have not coarsened significantly, Fig. 8b. This should not be taken to mean that the La_2O_3 particles are more stable than those of Er_2O_3 but rather that the La-containing alloy contains significantly more oxygen relative to La. This relative increase in oxygen over La is a direct consequence of the liquid phase separation that causes the amount of La in the remaining Ti-rich material to be markedly reduced.

In addition to the features described above, in consolidated samples of both alloys microstructures were observed such as those shown in Fig. 9a and b for the Er-containing alloy. As can be seen, grains of one phase are present together with lenticular precipitates of a second phase. Selected area diffraction analysis (Fig. 10) may be used to show that the grain is Ti_xC_{1-x}, with the second phase being α-Ti. These relatively large carbide particles are presumably formed as a result of the interaction between the graphite crucible and the liquid Ti alloy.

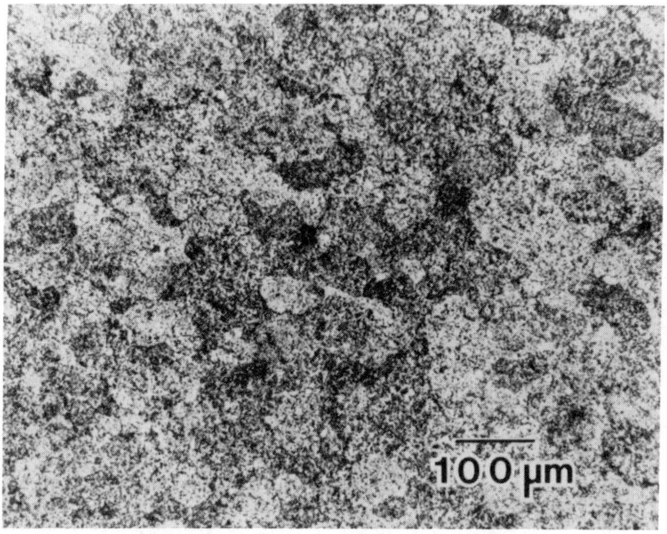

(a)

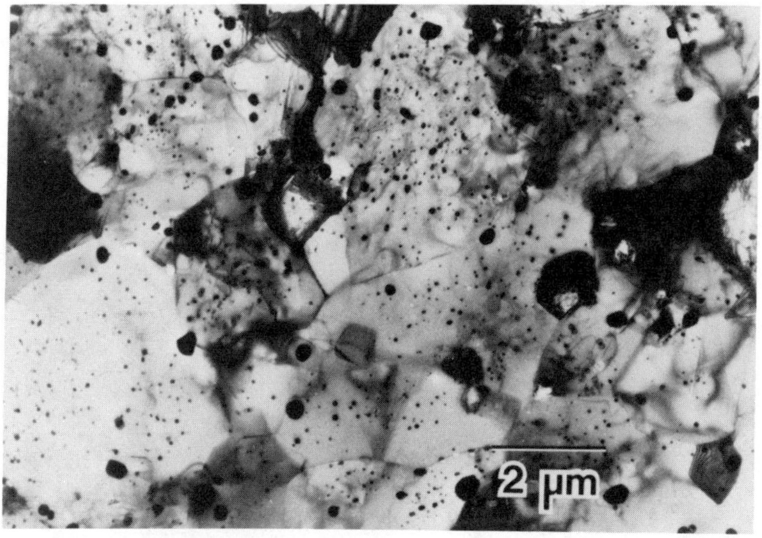

(b)

Figure 7 - (a) Optical micrograph, and (b) bright field electron micrograph of the Ti-0.7Er (at%) alloy in the as-HIP'd condition.

Summary

Centrifugal atomization has been used to produce rapidly solidified powders of Ti-Er and Ti-La alloys. It is found that in the case of Ti-Er,

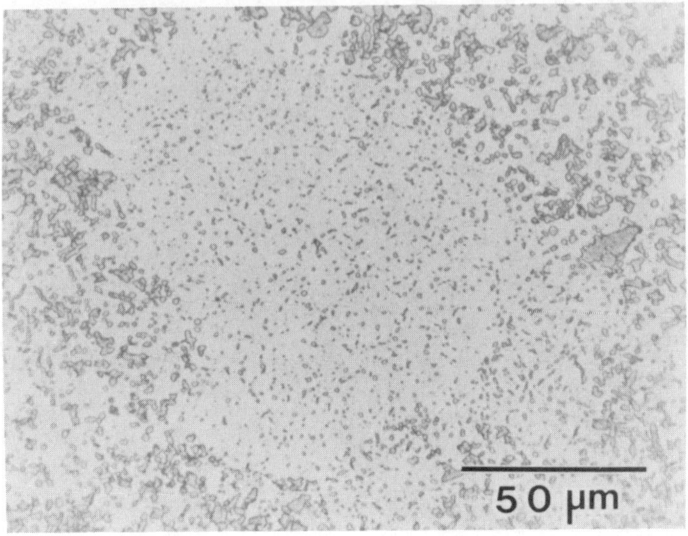

(a)

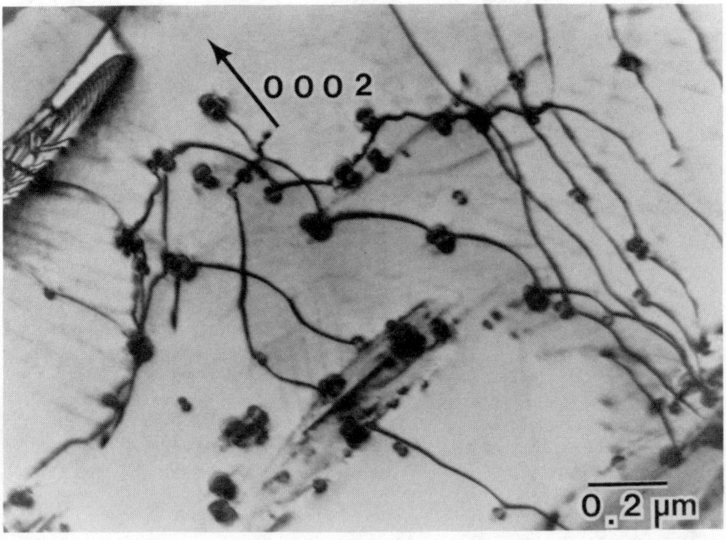

(b)

Figure 8 - (a) Optical micrograph, and (b) bright field electron micrograph of the Ti-0.9La (at%) alloy in the as-HIP'd condition.

supersaturated solid solutions of Er in Ti may be produced, and that subsequent heat treatment at 700°C results in the formation of a refined dispersion of Er_2O_3. In contrast, for the Ti-La alloy there is only a slight supersaturation of La produced; this is thought to be due to the liquid phase separation that occurs in this alloy system, causing the overall concentration

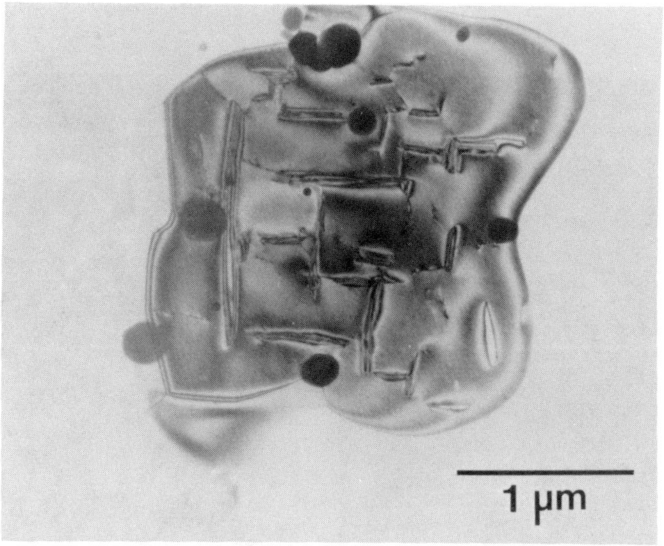

(a)

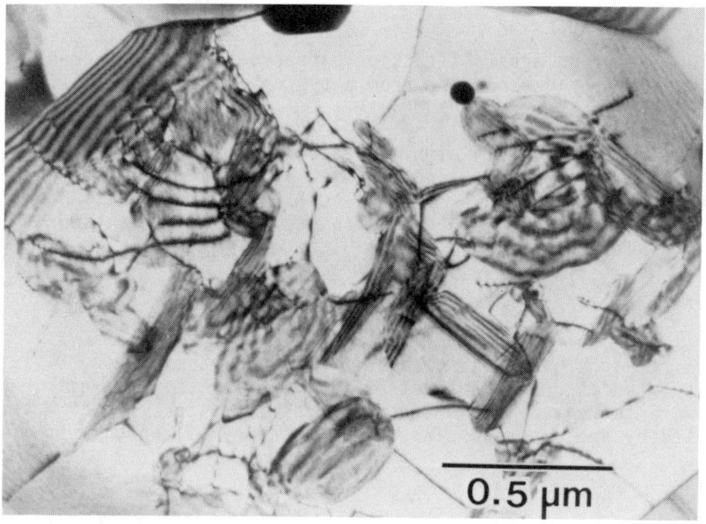

(b)

Figure 9 - Bright field electron micrographs taken from a sample of the as-HIP'd Ti-0.7Er (at%) alloy, showing carbide grains containing lenticular α-Ti precipitates.

of La to be reduced. The microstructures of the two alloys following HIP'ing consisted of dispersions of rare earth oxides; that in the Er-containing alloy has coarsened, somewhat more than that expected from theory. In contrast, the dispersion in the La-containing alloy has not coarsened significantly, this

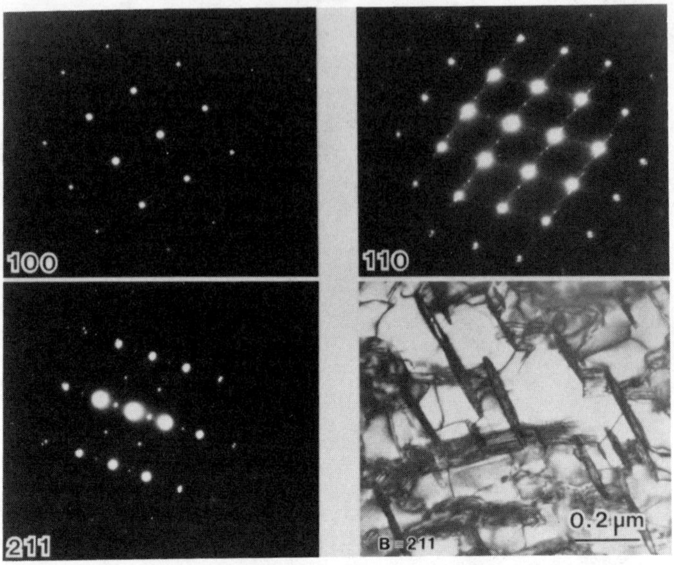

Figure 10 - Selected area diffraction patterns and a corresponding bright field electron micrograph taken from a typical carbide grain within the as-HIP'd Ti-0.7Er (at%) alloy.

being due to the relatively high oxygen concentration in these samples resulting from the reduction in La caused by the liquid phase separation described above. In HIP'd samples of both alloys relatively large grains of carbide are seen, which has been attributed to the interaction of liquid Ti with the graphite crucible.

Acknowedgements

This research has been supported in part by the US DoE, Division of Materials Sciences under contract DEAC02-76ER01198. The electron microscopy studies were performed in the Center for the Microanalysis of Materials, which is operated by the Materials Research Laboratory of the University of Illinois.

References

1. F.H. Froes and D. Eylon, in "Titanium, Science and Technology", G. Lukjering, U. Zwicker and W. Bunk, eds., DGM, Oberursel, W. Germany, 1 (1985) 267.

2. R.D. Field and H.L. Fraser, Metall. Trans., 9A, 1 (1978) 131-134.

3. D.G. Konitzer, B.C. Muddle and H.L. Fraser, Scripta Metall. 17 (1983) 963-966.

4. D.G. Konitzer, J.P.A. Lofvander, S.A. Court, R. Kirchheim and H.L. Fraser, submitted to Acta Metall.

5. S.A. Court, D.G. Konitzer, M.H. Loretto and H.L. Fraser, submitted to Acta Metall.

6. "The Handbook of Binary Phase Diagrams", General Electric Company, Schenectady, NY (1976).

7. D.J. Miller, unpublished research, University of Illinois.

RAPIDLY SOLIDIFIED TITANIUM ALLOYS PRODUCED BY THE MELT EXTRACTION AND MELT OVERFLOW PROCESSES

T. Gaspar*, Y. Sahai** and L. E. Hackman*

* Ribbon Technology Corporation, Gahanna, Ohio

** Department of Metallurgical Engineering
Ohio State University
Columbus, Ohio

Abstract

A new single roller rapid solidification process called melt overflow for direct casting titanium alloy strip, fiber or particulate is described. A mathematical model is developed to predict as-cast product thickness from process parameters. A mass balance is used to calculate the increase in the melt pool meniscus length from viscous drag against the chill block. A heat transfer model is developed to calculate the strip thickness based on the length of time the melt pool is in contact with the substrate. The model assumes a thermal contact resistance exists at the melt pool/substrate interface. Experiments were performed to cast Ti-6Al-4V alloy strip by pendant drop melt extraction and melt overflow. A technique for induction melting of titanium alloys is described. Melt overflow produced 450 g (1 lb) quantities of continuous strip while melt extraction produced short lengths of ribbon. The as-cast microstructures of the Ti-6Al-4V alloy strips consist of acicular alpha, primary alpha and beta. The thickness of direct cast Ti-6Al-4V strip agreed with the thickness predicted by the mathematical model.

Introduction

A number of rapid solidification processes have been developed over the last three decades to produce rapidly solidified ribbon and foil. The generic term "melt spinning" refers to techniques in which molten metal is extruded through an orifice and either solidifies in flight to form filaments (free flight melt spinning) or the extruded jet of liquid metal is quenched against a moving chill surface to cast ribbons or foils (chill block melt spinning).

Ribbon Technology Corporation (RIBTEC) was incorporated in 1972. The company originally licensed melt spinning technology to cast rapidly solidified metal ribbons for reinforcing plastics. In 1973, RIBTEC licensed the crucible melt extraction process from Battelle Development Corporation and began producing rapidly solidified carbon steel and stainless steel fibers for reinforcing concrete and castable refractories, respectively. RIBTEC currently produces over 1000 tons of stainless steel fibers annually by melt extraction.

In 1983, RIBTEC developed a rapid solidification process called melt overflow (Figure 1). Molten metal overflows a reservoir onto the surface of a rotating chill block. The melt stream is not extruded through an orifice to contact the chill block like melt spinning techniques, rather, the melt pool overflows a reservoir and is channeled by a runner or tundish to contact the moving chill surface. Rapidly solidified metallic sheet, strip, foil, filaments, fibers, flakes and particulate are cast depending on the surface geometry of the chill block. The adaptability of melt overflow in direct casting such a diversity of product sizes, shapes, and alloys is, perhaps, the greatest advantage of the process. To date, alloys based on aluminum, magnesium, titanium, nickel, copper, and iron have been cast by melt overflow.

This paper describes a simple mathematical model of Ribtec's melt overflow process. Experiments to direct cast Ti-6Al-4V alloy strip by both the melt extraction process and melt overflow process are compared and discussed. Finally, the mathematical model and experimental strip thickness will be compared.

Mathematical model

A simple mathematical model was developed for NASA to predict the thickness of solidified strip as a function of melt overflow process parameters (1). The solidified thickness is predicted from the length of time the molten metal is in contact with the chill roll using equations for heat transfer. The extent of increase of metal depth (and solidification time) due to viscous drag is first calculated using a force balance.

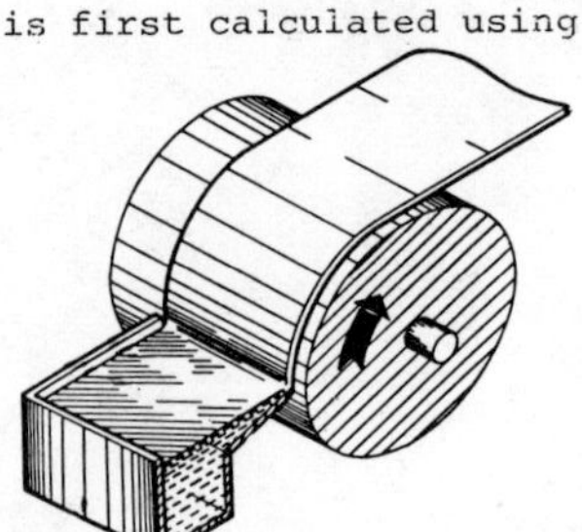

Figure 1. Schematic of RIBTEC's Melt Overflow Process

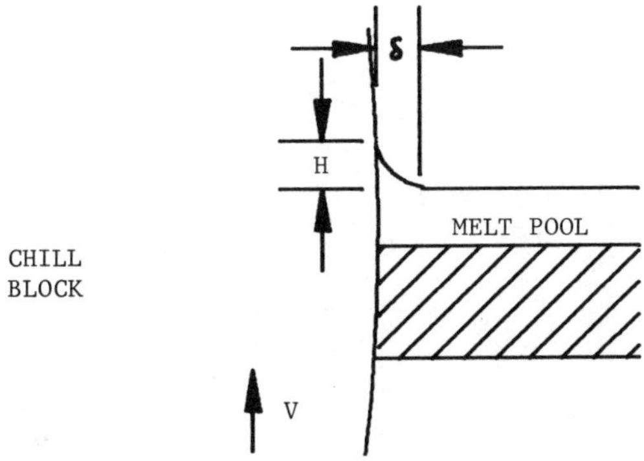

Figure 2. Model geometry at melt pool/substrate interface

Consider a chill block surface moving at velocity, V (Figure 2). The liquid metal is dragged up to a height, H, and the affected distance from the roll is δ . The force acting on this liquid in an upward direction, F_μ is:

$$F_\mu = H\ W\ \mu\ \frac{dv}{dx} \tag{1}$$

or,

$$F_\mu = H\ W\ \mu\ \frac{(V-0)}{\delta} \tag{2}$$

where μ is the viscosity of the liquid metal and W is the width of the strip.

The force will be balanced by the weight of the metal:

$$F_g = \frac{H\,\delta\,W}{2}\ \rho\, g \tag{3}$$

where ρ is the density of the metal and g is acceleration due to gravity.

Equating equations (2) and (3) yields:

$$\delta^2 \quad \frac{2\,\mu\, V}{\rho g} \tag{4}$$

The affected region is very small, therefore it can be assumed that . Substituting H for δ in equation 4 results in:

$$H = \sqrt{\frac{2\,\mu\, V}{\rho\, g}} \tag{5}$$

Equation (5) can be used to calculate the increase of metal depth H at the "lip" for any casting speed V. Values for increased melt depth H calculated from equation (5) are added to the depth of liquid at the casting lip to estimate the contact time t of the liquid metal against the chill roll.

A heat transfer model is developed, based on the assumptions that the

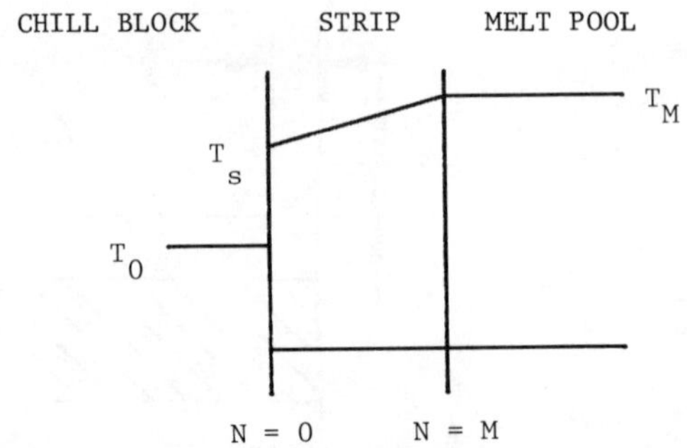

Figure 3. Model Temperature Profile of Solidifying Strip

thermal contact resistance exists at the metal-substrate interface and predominates over the resistance offered by the metal; the temperature at the substrate surface is constant, To; the metal freezes as a plane at N=M where the temperature is at the melting point of the metal, Tm; and the thermal conductivity of the solidifying metal contributes to the overall resistance to heat flow and results in a thermal gradient within the solidifying strip that is a linear function.

Figure 3 presents a schematic diagram of the model temperature profile in the solidifying strip. The solidified metal temperature at the interface, Ts, is is higher than the roll temperature, To due to thermal contact resistance.

Heat transfer due to conduction in the solidified layers is:

$$q = k\,\frac{Tm - Ts}{M} \tag{6}$$

Where k is the thermal conductivity of the metal and M is the thickness of the solidified layers. Heat transfer at the solidifying metal-chill block interface is given by:

$$q = h\,(Ts - T_0) \tag{7}$$

Where h is the heat transfer coefficient. Combining equations (6) and (7) yields:

$$q = \frac{Tm - T_0}{\frac{1}{h} + \frac{M}{k}} \tag{8}$$

The total heat evolved due to latent and superheat is given by:

$$q = \frac{\rho H'}{\rho}\, f\, \frac{dM}{dt} \tag{9}$$

Where H'_f is effective latent heat: $H'_f = H_f + Cp_1\ \Delta Ts$
and, H'_f is the latent heat of fusion, Cp_1 is the heat capacity of the molten

metal and T is the superheat of the melt. The densities of the liquid and solid metal at the melting temperature are assumed to be equal.

Equating equations (8) and (9), we get:

$$\rho H'_f \frac{dM}{dt} = \frac{Tm - T_0}{1/h + M/k} \quad (10)$$

Upon integrating with M=0 at time t=0 and M=M at time t=t;

$$M + \frac{M^2 h}{2k} = \frac{h\,(Tm - To)\,t}{\rho\, H'_f} \quad (11)$$

Equation 11 can be used to determine the thickness of solidified layer, M for any time t that the molten metal is in contact with the substrate.

A computer program was written to calculate the thickness of a solidified layer M according to equation 11. Values for the physical properties of titanium used in the model were: k = 0.2 W/cm K (11.6 Btu/ft hr F); = 4.51 g/cm (281.56 lb/ft); and H = 428 J/g (184.5 Btu/lb). The heat transfer coefficient, h, used for all calculations is 10 W/cm K (17610 Btu/ft hr F) and the length of contact 2.8 cm (1.1 in).

Experimental Procedure

A series of experiments were performed at Battelle, under contract to RIBTEC, to evaluate rapid solidification processing of titanium alloys. The pendant drop melt extraction (PDME), melt spinning (MS) and melt overflow (MO) rapid solidification processes were evaluated and compared (2). This discussion will focus on the pendant drop melt extraction and melt overflow techniques.

The pendant drop melt extraction machine operated inside a glass bell jar, under vacuum. A filament type electron beam source melted the tip of a titanium alloy rod. The rod was suspended vertically over a rotating chill block. The tip of the rod was melted until the liquid droplet fell from it onto the chill block to produce short lengths of titanium alloy ribbon.

The chill block measured 20 cm (8 in) diameter and 1.3 cm (0.5 in) thick and was made from copper. The surface speed of the chill block was 10 m/s (33 ft/s). The droplet fell approximately 0.6 cm (0.25 in) to contact the chill block surface. A counter-rotating wire brush abraded the chill block surface.

The melt overflow experiments were performed in a vacuum induction melting furnace. A 0.45 kg (1 lb) charge of titanium was melted in a graphite crucible coated with yttria mold wash. A 8 mm (5/16 in) hole in the bottom of the crucible allowed liquid metal to flow out the bottom of the crucible. The hole was covered with titanium foil to delay the release of the molten metal. The induction melted titanium alloy melted the titanium foil in the bottom of the crucible and streamed into a tundish coated with yttria to overflow onto the chill block. Casting times were 3 seconds to 4 seconds.

The chill block used to cast titanium alloy foil consisted of a solid copper cylinder that measured 25 cm (10 in) diameter by 3.8 cm (1.5 in) wide. Ti-6Al-4V alloy strips up to 2.5 cm (1 in) wide were cast at speeds up to 10 m/s (33 ft/s). A wire brush wiper was used to remove any particles of titanium that might stick to the wheel during the strip casting process.

Results

The pendant drop melt extraction (PDME) experiments were performed at casting speeds of approximately 10 m/s (33 ft/s). The ribbons cast at this speed were approximately 200 micrometers (0.008 in) thick and 1.3 cm (0.5 in) wide in random lengths. A wire brush was needed to cast titanium alloy ribbon by PDME: polished chill block surfaces using 600 grit abrasive paper and 8 micrometer diamond paste yielded little or no strip.

The substrate cast surface of the PDME strip is shown in Figure 4. The surface appears to be thermally etched and fine-grained. The free cast surface is dimpled and has longitudinal flow lines that are parallel to the casting direction (Figure 5). The dimples appear to be gas bubbles at higher magnification (Figure 6).

The Ti-6Al-4V alloy strip was mounted, polished and etched in a solution of 95 H O + 3.5 HNO + 1.5 HF. The as-cast microstructure consists of acicular alpha surrounded by primary alpha and beta (Figure 7). The absence of an alpha case implies that there is little surface oxidation.

Figure 4. PDME Substrate Cast Surface

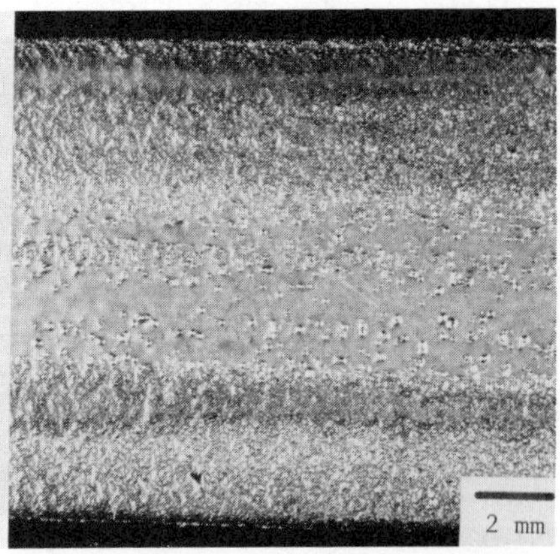

Figure 5. PDME Free Cast surface

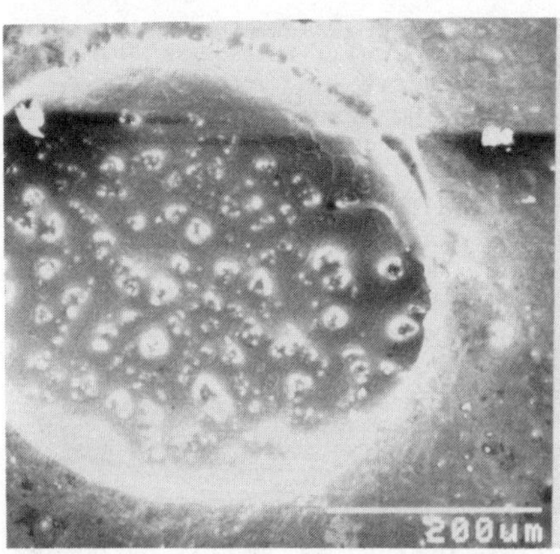

Figure 6. Gas Bubble in PDME Free Cast Surface

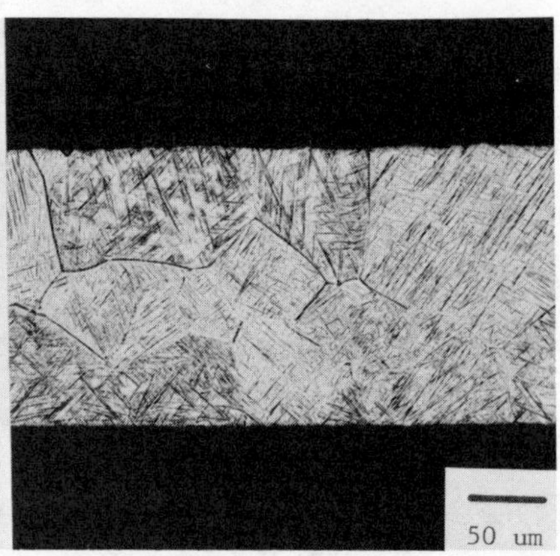

Figure 7. Microstructure of PDME Strip

A typical melt overflow Ti-6Al-4V alloy continuous strip measured over 10 m (35 ft) long, 0.18 mm (0.007 in) thick and 18 mm (0.7 in) wide. The surface that contacted the substrate appears smooth (Figure 8) while the free cast surface has longitudinal flow lines (Figure 9). There is no evidence of the dimples or gas bubbles on the surface of the melt overflow strip that were found on the PDME strip.

Samples of the melt overflow Ti-6Al-4V alloy strip were mounted, polished and etched. The as-cast microstructure (Figure 10) is similar to the PDME ribbons (Figure 7) consisting of acicular alpha, primary alpha, and beta. The primary alpha and acicular alpha is finer in melt overflow ribbon than PDME ribbon indicating that the melt overflow strip may experienced higher secondary cooling.

Measurement of strip thickness is imprecise because the free cast surface is not always flat. Thickness measurements were made over a 79 cm (31 in) length of strip with both a pointed anvil micrometer and a flat anvil micrometer. The 79 cm (31 in) strip length equals the periphery of the chill block and therefore represents casting during one revolution of the chill roller. Figure 11 shows the mean thickness of Ti-6Al-4V alloy strip cast by melt overflow during ten different experiments. The solid line in Figure 11

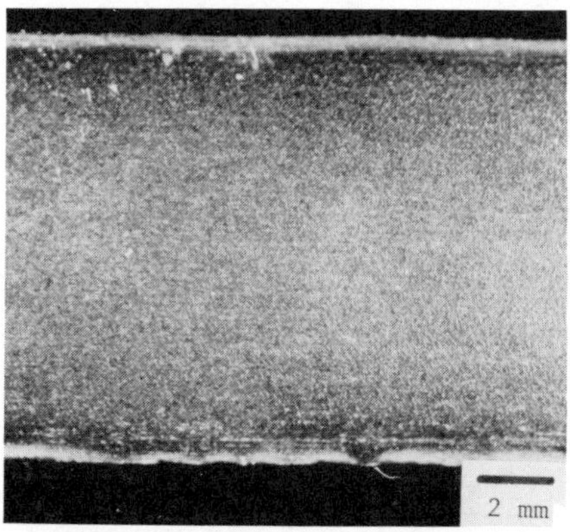

Figure 8. Melt Overflow Strip Substrate Cast Surface

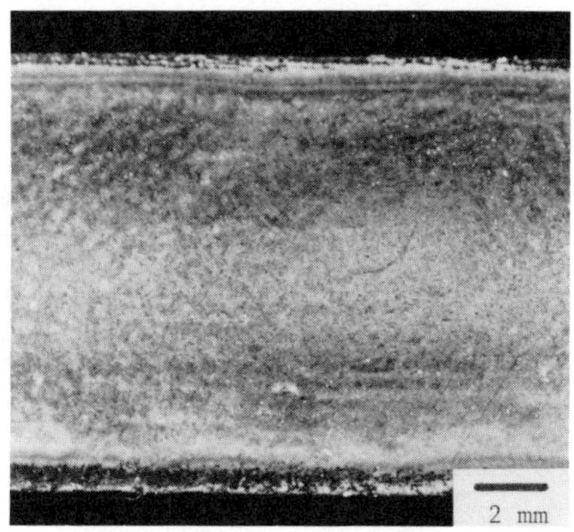

Figure 9. Melt Overflow Strip Free Cast Surface

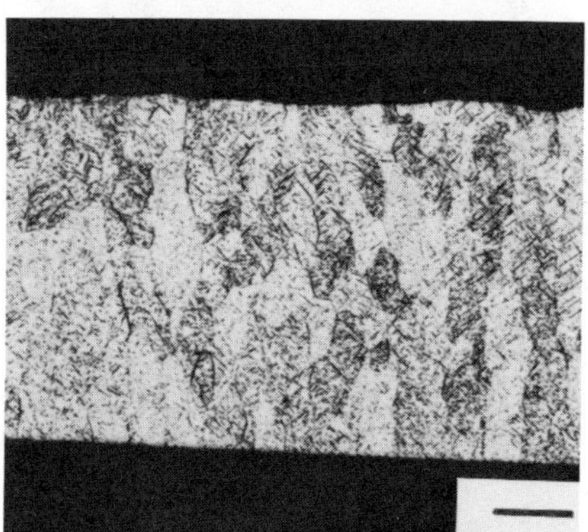

Figure 10. Melt Overflow Strip Microstructure

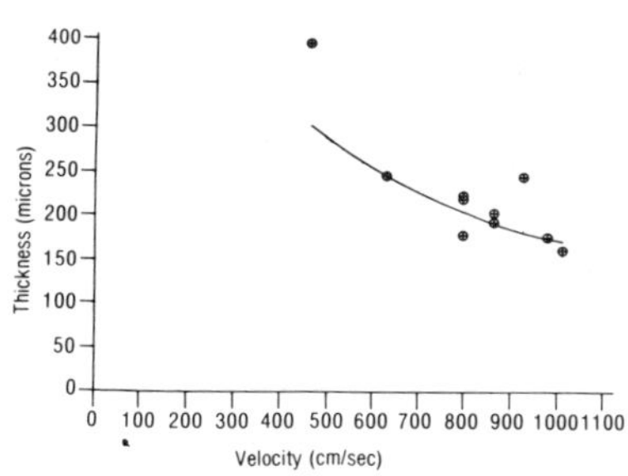

Figure 11. Strip Thickness as a Function of Casting Speed

represents the thickness of pure titanium strip calculated from Equation 11. The thickness predicted by the heat transfer model agrees reasonably well with the thickness measured from melt overflow Ti-6Al-4V strip.

Conclusions

RIBTEC's melt overflow process is capable of producing rapidly solidified titanium alloy strip of continuous length. The melt overflow process directly cast rapidly solidified titanium alloy strip in 450 gram (one pound) quantities. The as-cast surfaces appear suitable for secondary processing. Melt extraction is not well suited to producing continuous strip, but is capable of producing titanium alloy ribbons of short, random lengths. Melt extraction appears limited to producing titanium alloy fiber and filament.

A simple heat transfer model was developed to describe RIBTEC's melt overflow process. The thickness of the melt overflow cast strip agrees with the thickness predicted by the mathematical model. The model assumes a heat transfer coefficient between the solidifying strip and chill block. Work is beginning on a more sophisticated model of the melt overflow process.

The feasibility of direct casting rapidly solidified titanium alloy strip has been successfully demonstrated. Future research will focus on the feasibility of casting titanium alloy strip, filament, fiber and particulate by melt overflow with continuing support from NASA. RIBTEC is scaling up Melt Overflow to cast wider titanium alloy strip in larger quantities. Wider strip, may be used in honeycomb structures or metal matrix composites.

REFERENCES

1. Gaspar, Thomas, Hackman, Lloyd, Sahai, Yogeshwar, Mozhi, T.A., and Clark, W.A.T., NASA Contract NAS1-17978, Final Report, June, 1985.

2. Maringer, R. E. (private communication, 18 March 1985).

SOLIDIFICATION CHARACTERISTICS OF LASER SURFACE MELTS

IN ADVANCED TITANIUM ALLOYS

W. A. Baeslack III*, S. Krishnamurthy** and F.H. Froes***

*Department of Welding Engineering
The Ohio State University
Columbus, OH 43210
**Metcut Materials Group
Wright-Patterson AFB, OH 45433
***Materials Laboratory
Air Force Wright Aeronautical Laboratories
Wright-Patterson AFB, OH 45433

ABSTRACT

Rapidly-solidified (RS) microstructures were generated and characterized for several dispersoid-strengthened and eutectoid-former titanium alloys using the laser surface melting (LSM) technique. A comparison of the LSM microstructures with those of RS products produced using alternate methods identified microstructural similarities and differences which should be considered when utilizing LSM for the screening of potential new RS titanium alloy compositions. Solidification-related characteristics specifically examined and discussed in this paper include beta grain and subgrain morphology, solidification substructure transitions and segregation effects.

Introduction

The application of rapid solidification technology has resulted in the development of a variety of new and novel titanium alloys (1,2). Compositions include eutectoid-former and rare earth dispersion-strengthened alloys which cannot be produced via ingot metallurgy due to alloying element segregation and the formation of coarse, equilibrium constituent particles which degrade mechanical properties. Through RS processing, fine-grained, homogeneous microstructures can be obtained which exhibit a high degree of metastability. Subsequent consolidation and heat treatment promote the precipitation of extremely fine second phase particles which provide increased room-temperature strength and more importantly significant improvements in elevated-temperature strength and creep resistance.

Laser surface melting has been widely used as a rapid screening technique for the evaluation of potential new RS alloy compositions. The process involves rapid scanning (5 to 50 mm/s) of the material with a continuous laser focused to a power density of approximately 10^5 W/mm^2. Impingement of the laser beam on the surface melts an extremely thin layer which subsequently solidifies under rapid cooling rate conditions. Through systematic variations in laser beam power and traversing rate, LSM cooling rates of between 10^3 and 10^6 °C/s can be reproducibly obtained. Previous investigators (1-5) have utilized this technique to generate RS microstructures in advanced titanium alloys for microstructure examination, cooling rate/dendrite arm spacing correlations and hardness determination. Although cooling rates through the solidification range obtained with LSM parallel those achieved in RS processes used for bulk alloy production (eg., atomization, melt extraction, melt spinning), there exist inherent differences in the solidification characteristics of these RS products. The purpose of the present study was to identify how these solidification related differences influence the RS product microstructures. This was accomplished through the generation and characterization of LSM structures in a variety of advanced RS titanium alloys and the comparison of these structures with those of RS products (ribbon, splat and melt extracted fiber) produced for identical alloys using alternate techniques.

Experimental Procedure

Two general alloy systems were examined in the present work: 1) alloys containing significant amounts of eutectoid-former elements and 2) commercial alpha-beta and metastable-beta alloys containing the rare earth Er for dispersion strengthening. The nominal compositions of these alloys and the respective alloy types are provided in Table I. Alloys were produced by vacuum arc melting, with the initial melt stock comprised of elemental additions, master alloys and/or commercial alloys. Subsequent to melting, the alloys were homogenized via a supertransus vacuum heat treatment and swaged in a two-step operation to obtain a final swaged rod diameter of between 8 and 10 mm (total reduction of approximately 50%). Commercial alloy compositions containing Er were given an additional homogenization heat treatment at 1100°C for 24 hours.

Wafer specimens for LSM were sectioned transversely from the swaged rod in 3 mm thicknesses. Laser surface melting was performed using a 550 watt CO_2 laser in the continuous mode. A high speed, computer-controlled traversing system provided uniform melt zone widths and argon shielding prevented atmospheric contamination of the melt zone. Cooling rates between approximately 1×10^3 and 2×10^5 °C/s were obtained for laser powers between 425 and 550 watts and traversing rates between 0.4 and 80 mm/s. (Cooling rates were calculated through the solidification range using an analytical 3-D heat flow model assuming a point heat source and a heat

Table I. Compositions and Characteristics of Alloys Investigated

Alloy Composition*	Alloy Type	LSM Microstructure
Ti-6Al-4V	α/β	α′
Ti-6Al-4V-0.5Er	α/β + DS**	α′
Ti-6Al-4V-1.5Er	α/β + DS	α′ + Disp.
Ti-6Al-4V-3Er	α/β + DS	α′ + Disp.
Ti-10V-2Fe-3Al	Metastable β	β + α″
Ti-10V-2Fe-3Al-1.5Er	Metastable β + DS	β + α″
Ti-15V-3Al-3Sn-3Cr	Metastable β	β
Ti-15V-3Al-3Sn-3Cr-1.5Er	Metastable β + DS	β + Disp.
Ti-9Co	Eutectoid	β
Ti-5.5Ni	Near-Eutectoid	Martensite + Bainite
Ti-22Fe	Hyper-Eutectoid	β
Ti-40W	Hyper-Eutectoid	β
Ti-15Cr	Eutectoid	β
Ti-20V	Metastable β	β
Ti-20V-1.5Er	Metastable β + Disp.	β

*Weight Percentages
**Dispersion Strengthened

transfer efficiency of 25% (2).

Following laser melting, specimens were mounted with the melt surface exposed or were sectioned transverse to the melt longitudinal direction and mounted such that the transverse section could be examined. Metallographic preparation involved grinding, polishing down to 0.05 micron alumina and etching with standard Kroll's etch for 15 seconds to one minute. Characterization of the laser melt structures was performed using light and scanning-electron microscopy (SEM) and energy-dispersive X-ray analysis (EDX).

Results and Discussion

Figure 1 illustrates transverse LSM macrostructures for a Ti-20V-1.5Er alloy. The relatively uniform, conduction-limited melt zone geometries were typical of melts produced in all of the titanium alloys. As is evident in Figure 1b, slower traversing rates (below 4 mm/s) promoted depression of the melt pool surface due to laser beam/melt interactions and related convective fluid flow effects. Also associated with fluid flow in the melt were ripples observable on the as-solidified melt surfaces. Figure 2 illustrates the as-solidified surface of a laser melt produced on the as-solidified fusion zone surface of a gas tungsten-arc weldment in Ti-15V-3Al-3Sn-3Cr. Note the presence of ripples on both melt surfaces and the much finer spacing in the laser versus the much larger arc weld pool. Investigators of similar ripples in laser melted Udimet 700 (6) have proposed their origination from surface tension driven convective fluid flow in the melt pool. Alternate mechanisms include perturbations on the pool surface resulting from mechanical effects or fluctuations in the laser heat source.

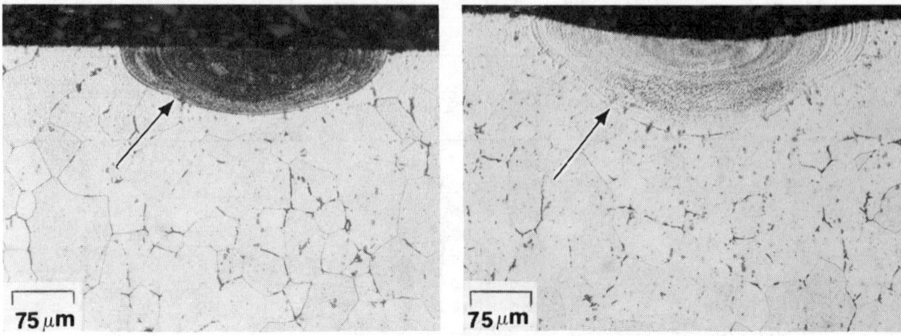

Figure 1 - Laser surface melts in Ti-20V-1.5Er produced at 450 watts and traversing rates of a) 40 mm/s and b) 4 mm/s. Arrows indicate melt/substrate interface.

Beta Grain Morphology

The solidification of RS particulate (such as ribbon, splat, or fiber) generally initiates at an ideally-inert chill surface. Whether or not solidification is completed while the particulate is in contact with the chill is both process and process-parameter dependent. LSM differs from RS particulate generation in that the metal itself serves as the chill surface and heat sink. Despite this fundamental difference, both types of RS processes promote steep, highly-directional temperature gradients during solidification and the associated growth of columnar beta grains in a crystallographic direction most parallel to the maximum temperature gradient at the solid-liquid interface. As expected, examination of the titanium alloy splat, ribbon and fibers using SEM and TEM revealed predominantly columnar beta grain growth inward from the chill surface(s). Such columnar beta grains completely traversing a Ti-6Al-4V-1.5Er ribbon and converging toward the center of a Ti-15Cr splat are shown in Figure 3. In several alloys, equiaxed beta grains were observed in a chill zone directly adjacent to the chill interface. TEM micrographs of typical equiaxed grains in Ti-9Co

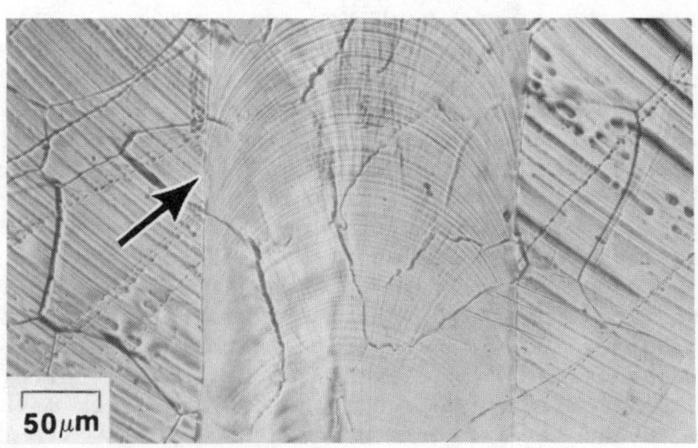

Figure 2 - Light micrograph of as-solidified laser surface melt produced on gas tungsten-arc weld fusion zone in Ti-15V-3Al-3Sn-3Cr (500 watts, 80 mm/s). Arrow indicates melt/substrate interface.

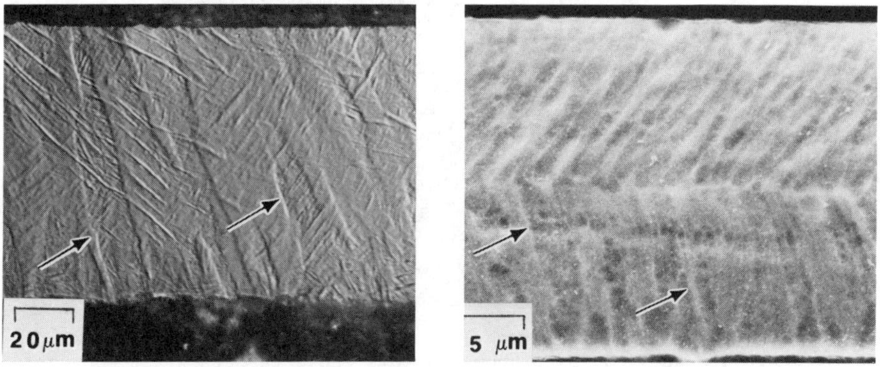

a b

Figure 3 - Light (a) and SEM (b) micrographs showing columnar beta grain structures in Ti-6Al-4V-1.5Er ribbon (a) and Ti-15Cr splat (b). Arrows indicate beta grain boundaries.

splat and ribbon are shown in Figure 4. Equiaxed grains were also observed at the center of splat cooled Ti-40W and Ti-5.5Ni. The origin of these grains were equiaxed dendrites formed in this relatively shallow temperature gradient region. Metallographic examination of the eutectoid-former alloys showed the observed beta grain morphology to be dependent on specimen orientation. For example, columnar grains in ribbon and splat commonly appeared equiaxed when examined with TEM perpendicular to the flat side of the product. Consequently, caution and thorough characterization was required in order to completely define beta grain morphology in many of the RS products.

In certain RS particulate, the room-temperature beta grain structure was noticeably changed from that expected at solidification due to grain boundary migration, grain growth or possibly recrystallization. The tendency for these phenomena was found to be greatest in products which separated from the chill

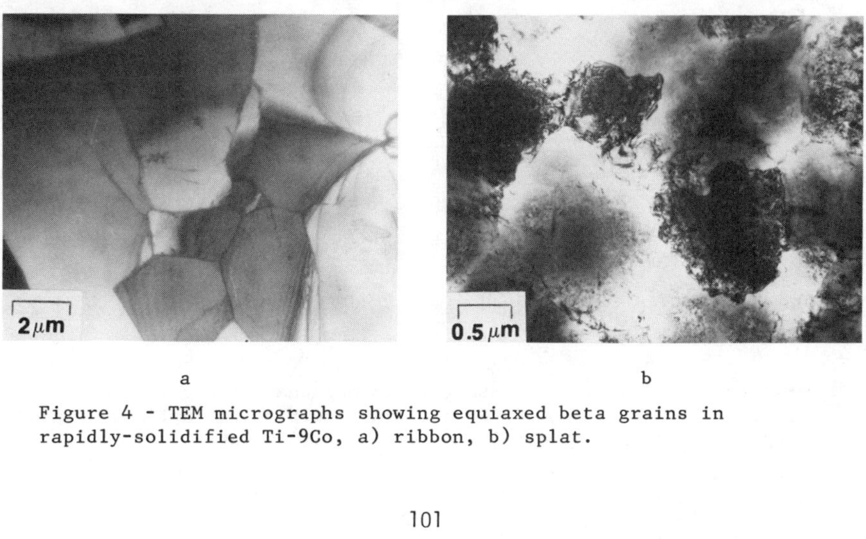

a b

Figure 4 - TEM micrographs showing equiaxed beta grains in rapidly-solidified Ti-9Co, a) ribbon, b) splat.

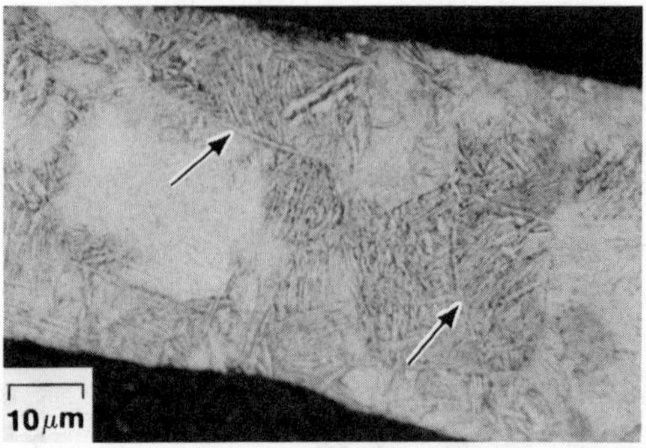

Figure 5 - Light micrograph showing coarse, equiaxed beta grain structure in Ti-5.5Ni ribbon. Arrows indicate beta grain boundaries.

prior or immediately subsequent to solidification. Evidence of such grain growth is illustrated in Figure 5 by the nearly equiaxed beta grains exhibited in the transverse section of a ribbon in Ti-5.5Ni. As indicated above, these effects were a function of RS alloy chemistry and alloy production methods.

The as-solidified Ti-15V-3Al-3Sn-3Cr laser melt surface in Figure 2 clearly shows the continuous growth of coarse beta grains from the arc weld into the laser melt zone. Evidence of such "epitaxial" beta grain growth dominated the titanium alloy melt zones. Figure 6 shows light and SEM micrographs of a transverse laser melt section in Ti-15V-3Al-3Sn-3Cr. The epitaxial and competitive growth of columnar beta grains are apparent. ("Competitive" growth involves the preferred growth of grains whose crystallographic growth directions are most closely aligned with the maximum temperature gradient, and similarly the elimination of poorly aligned grains).

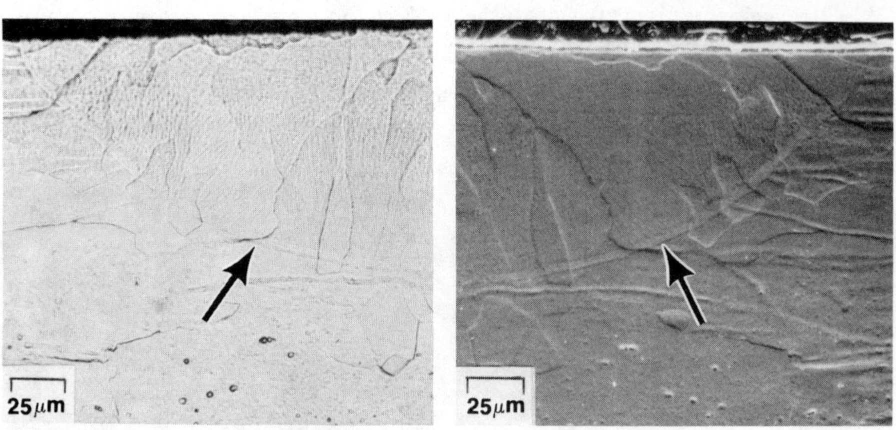

a b

Figure 6 - Light (a) and SEM (b) micrographs (mirror images of identical region) showing beta grain growth across melt/substrate interface (arrows) in Ti-15V-3Al-3Sn-3Cr (450 watts, 40 mm/s).

Comparisons of the LSM and RS particulate beta grain structures consistently showed upto an order of magnitude difference in the grain size. These differences resulted in part from the slightly higher cooling rates experienced by the splat and ribbon (10^5 - 10^6 °C/s) versus the laser surface melts (10^5 °C/s). However, the primary difference appeared to be the much coarser epitaxial nucleation sites in the laser melts (equal to the substrate beta grain size) as compared to the finer heterogenous nucleation sites provided by the chill surface in the RS particulate. It is of interest to note that despite the beta grain size differences, competitive growth of the columnar grains in the splat and ribbon was often observed but on a much finer scale (Figure 3).

Figure 7a illustrates the top surface of a laser melt produced in Ti-5.5 Ni which exhibited an extremely coarse beta grain size such that the entire laser melt zone was confined to a single beta grain. As shown, fine columnar grains or subgrains can be observed oriented predominantly along the laser traversing direction. SEM characterization of these subgrains at increased magnification showed them to consist of "packets" of aligned cells or cellular-dendrites (Figure 7b). If complete epitaxial growth is assumed, these grains should exhibit different varients of the primary growth directions of the substrate beta grain. To date, TEM analysis has been inconclusive in confirming this possibility. Previous investigators(6) of laser melt grain nucleation have attributed the nucleation of non-epitaxial grains from oxide at the trailing edge of the melt zone. The possibility of this mechanism in the titanium laser melts is less likely considering the negligible oxidation experienced during the LSM process.

Solid-state beta grain nucleation and growth was not restricted to the RS particulate, but was also observed near the surface of laser surface melts produced with low cooling rates. Figure 8 illustrates evidence of recrystallization near the surface of a laser melt in Ti-15V-3Al-3Sn-3Cr which experienced a cooling rate of about 2 x 10^3 °C/s. Similar structures were observed in Ti-5.5Ni and Ti-9Co.

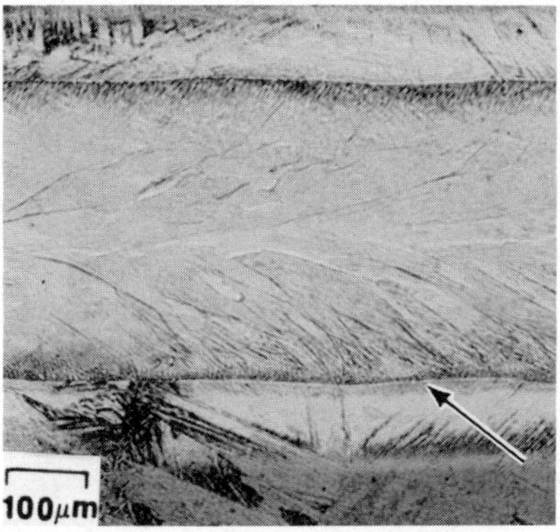

a

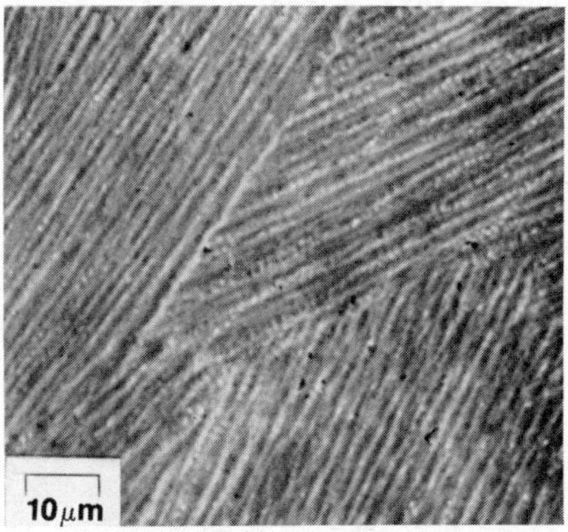

b

Figure 7 - Light (a) and SEM (b) micrographs showing beta subgrain structure in Ti-5.5Ni laser surface melt (550 watts, 40 mm/s).

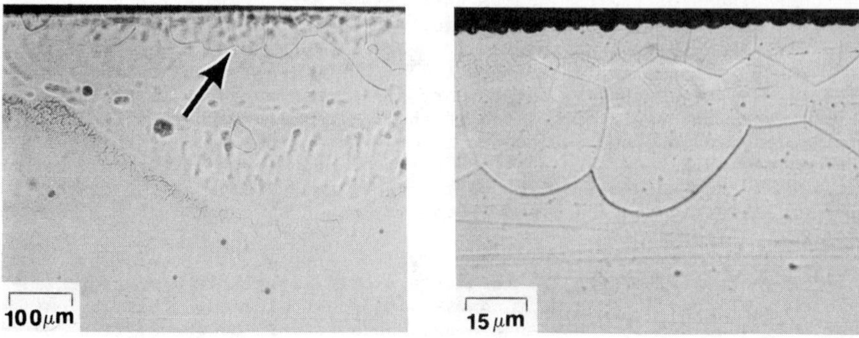

a b

Figure 8 - Light micrographs showing beta grain structure in Ti-15V-3Al-3Sn-3Cr laser surface melt (600 watts, 0.4 mm/s).

Solidification Substructure Characteristics

The laser melt solidification substructures were readily observable on the as-solidified laser melt surfaces and polished and etched specimens of all but the Ti-6Al-4V alloys. Solidification substructure morphologies generally included cellular and cellular-dendritic for the Ti-Ni, Ti-Co and Ti-Fe eutectoid-former alloys and the metastable-beta dispersoid-containing alloys and predominantly columnar-dendritic in the highly-alloyed Ti-40W alloy. As shown in Figure 9 for Ti-9Co, solidification substructure coarseness (eg., cell or dendrite arm spacings) were dependent on cooling rate. This exponential relationship was determined to be similar for the Ti-Co, Ti-Ni and Ti-Fe eutectoid-former alloys and equal to $\lambda \simeq 50\varepsilon^{-0.35}$.

Of the alloys investigated and directly compared with the RS particulate, the Ti-40W alloy provided the clearest delineation of solidification substructure characteristics and transitions. Figure 10a illustrates the microstructure near the center of a splat cooled at approximately 10^6 - 10^7 $^oC/s$. At the chill surfaces, where the liquid undercooling was greatest,

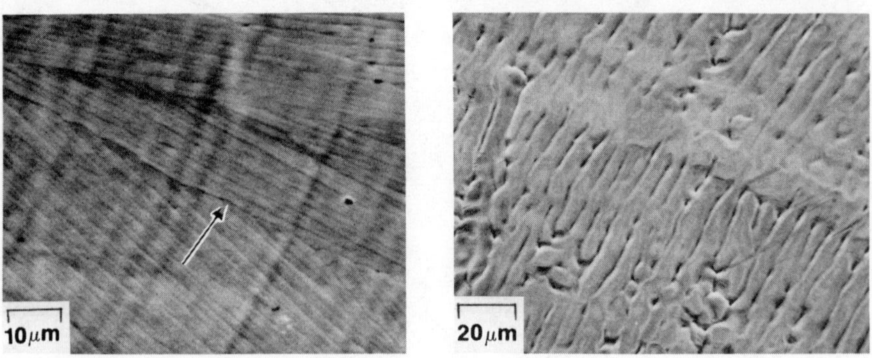

a b

Figure 9 - SEM micrographs of as-solidified laser surface melts produced in Ti-9Co at 550 watts and traversing rates of a) 40 mm/s and b) 0.4 mm/s.

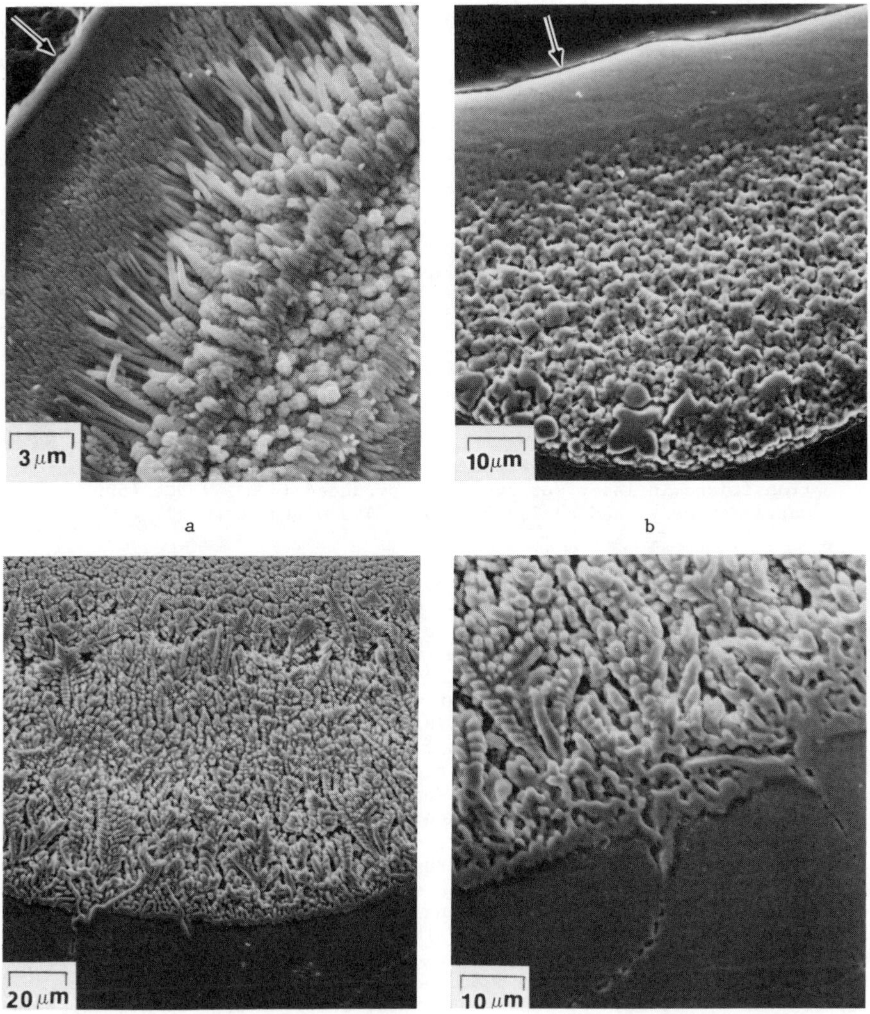

Figure 10 - SEM micrographs comparing solidification substructures in Ti-40W produced by a) splat quenching, b) melt extraction and c,d) laser surface melting (500 watts, 4 mm/s). Arrows indicate melt/substrate interfaces.

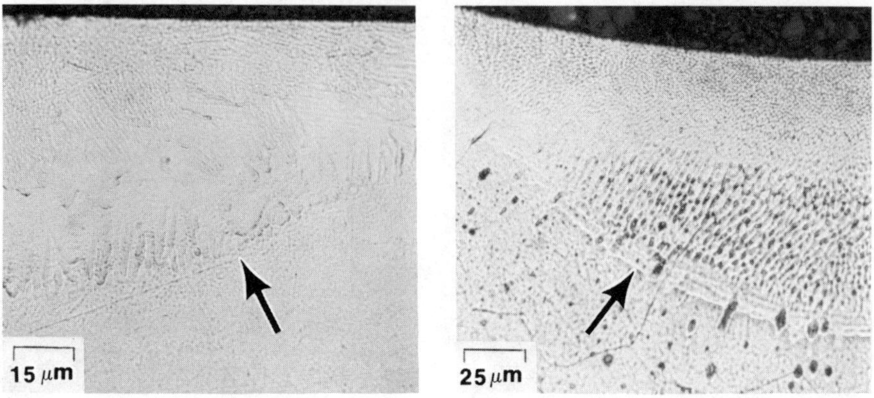

a b

Figure 11 - Light micrographs showing solidification substructure transitions in laser surface melts produced in a) Ti-9Co (500 watts, 40 mm/s) and b) Ti-20V-1.5Er (450 watts, 4 mm/s).

solidification proceeded by a planar growth mode which was evidenced by a featureless microstructure. Throughout most of the splat cross section, a breakdown of the solidification front was observed first to a cellular growth mode and finally to columnar-dendritic and equiaxed-dendritic modes. It is well known that these solidification structure transitions are associated with constitutional undercooling phenomena. A decreasing ratio in the temperature gradient in the liquid at the solid-liquid interface to the growth rate of this interface as solidification proceeded inward from the chill surfaces promoted constitutional undercooling and the observed structure transitions. Figure 10b illustrates the cross section of a fiber produced by melt extraction at cooling rates of approximately 10^5 °C/s. At a distance of 5-10 microns from the chill surface, the solidification structure rapidly transitioned from planar to cellular, cellular-dendritic, columnar-dendritic and finally equiaxed, with the substructure coarseness increasing toward the outer fiber surface. The observation of coarser equiaxed dendrites at the outer surface suggested slower cooling rates and shallower temperature gradients in this region and the possibility that solidification was incomplete prior to separation from the disk. Solidification substructure characteristics of laser melts produced between 10^4 and 10^5 °C/s are represented by Figures 10c and d. As shown, complete wetting and epitaxial growth of the melt from the base metal substrate promoted highly directional nearly exclusively columnar-dendritic growth, with single dendrites often traversing the entire width of the melt. Despite variations in the temperature gradient from the edge to the center of the melt zone, substructure transitions comparable to those in the splat and fiber were not evident. Note that the fine equiaxed dendrite morphology at the laser melt surface did not result from equiaxed growth, but rather from a change in the direction of grain growth to a direction nearly parallel to the traversing direction.

Figure 11 illustrates light micrographs of transverse LSM sections in Ti-9Co and Ti-20V-1.5Er. Both of the melts exhibited similar, predominantly cellular and cellular-dendritic substructures common to most of the alloys investigated. In contrast to the Ti-40W laser melts, these alloys exhibited a distinct transition from coarser, more cellular-dendritic appearing structures near the fusion boundary to a finer, cellular structure near the upper surface. The structure transitions are normally related to changes in the temperature gradient and growth rate along the solid-liquid interface which influence constitutional undercooling and cooling rate. However, observed structures were the reverse of what is commonly observed in fusion welds, which exhibit transitions from planar or cellular growth near the interface to more dendritic structures in the central melt regions (due to increasing constitutional supercooling). Indeed, such transitions indicate the complex nature of heat and fluid flow in the laser surface melt.

Segregation Effects

An important advantage of RS processing is the generation of chemically homogeneous alloys which promotes the uniform precipitation of second phase particles during consolidation and heat treatment. The laser melts examined in the present study were generally found to be homogeneous, particularly for those produced at the most rapid cooling rates. Certain alloy melts did, however, exhibit macrosegregation. Figure 12 illustrates evidence of Er macrosegregation during laser melt solidification of a Ti-15V-3Al-3Sn-3Cr-1.5Er alloy in the form of transverse solute banding. This type of macrosegregation has been previously reported for fusion welds in titanium alloys (7) and appears as curvilinear contours on the polished and etched melt surface. It has been shown to result from periodic changes in the velocity of the solid-liquid interface which influence the solute-enriched liquid ahead of the interface (assuming an equilibrium partitioning ratio of less than unity) or the solute in the mushy region adjacent to this interface. In Figure 12, the solute bands are revealed by Er-rich dispersoids which formed either during the final stages of solidification or more likely in the solid state on cooling. This form of macrosegregation was observed primarily in the high energy input/low cooling rate laser surface melts, and was minimized for the high cooling rate structures.

Microsegregation effects (cell and dendrite scale) were observed via etching response and EDX analysis in all laser melts except for the Ti-6Al-4V

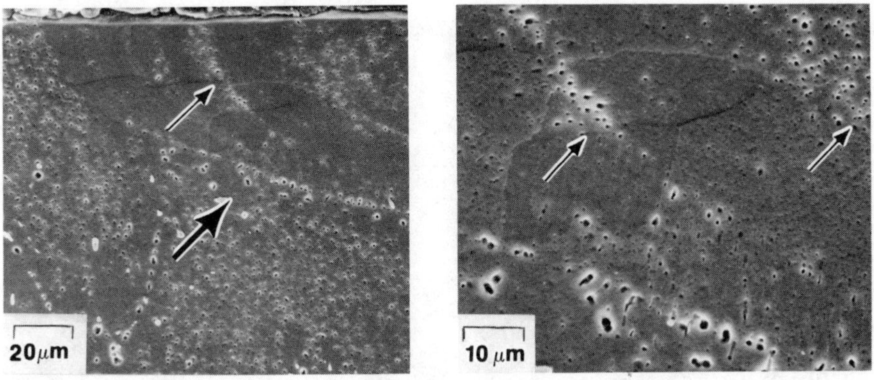

a b

Figure 12 - SEM micrographs showing solute banding in Ti-15V-3Al-Sn-3Cr-1.5Er (450 watts, 40 mm/s). Arrows indicate Er-rich dispersoids.

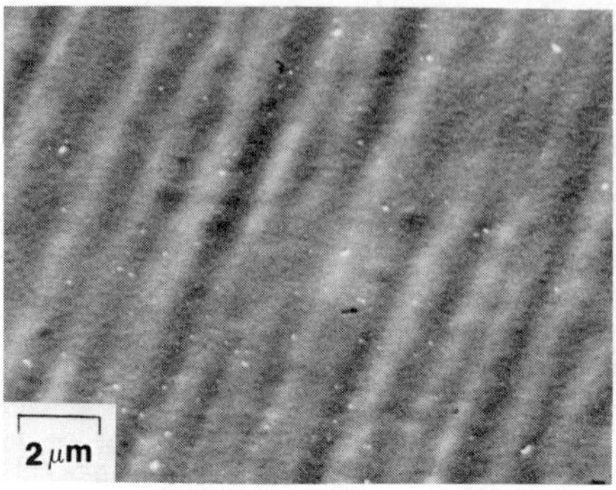

a

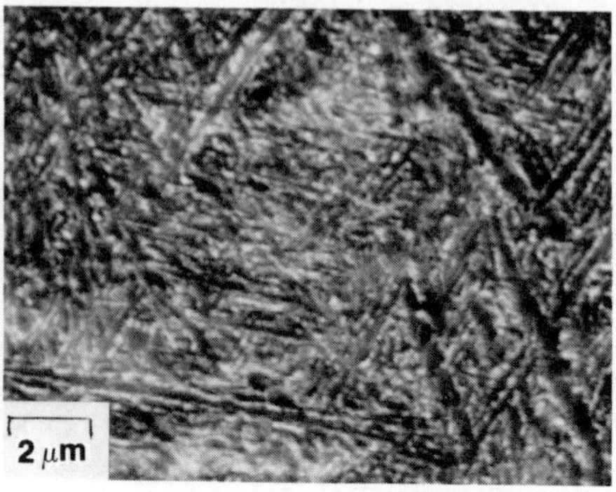

b

Figure 13 - SEM micrographs of laser surface melts in Ti-5.5Ni produced at 450 watts and traversing rates of a) 40 mm/s and b) 4 mm/s.

alloys. The levels of segregation residual to room temperature were a function of alloy type and alloying level, solidification conditions and cooling rate following solidification. These parameters influenced both the extent of initial segregation at the completion of solidification and the degree of homogenization which occurred on cooling. As expected, the most rapidly cooled laser surface melts and the splat-cooled particulate exhibited the greatest residual microsegregation due to their rapid solid-state cooling, while the slow solid-state cooling rates in the ribbon and fibers promoted greater homogenization. The presence of residual microsegregation in the slower cooled products was also difficult to identify in the Ti-6Al-4V and the Ti-5.5Ni alloy due to solid-state phase transformation to martensite or bainite. A comparison of laser melts produced in the Ti-5.5Ni alloy at 40 and 4 mm/s exemplifies structure differences due to such

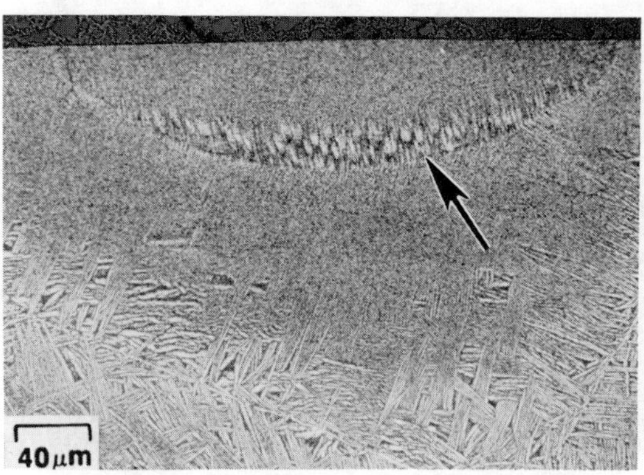

a

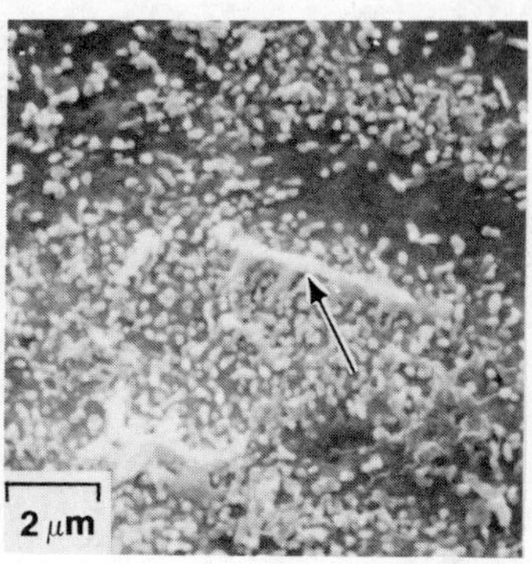

b

Figure 14 - Light (a) and SEM (b) micrographs in laser melt in Ti-5.5Ni aged at 625 °C/10 minutes. Arrow in (a) indicates melt boundary, arrow in (b) indicates Ti_2Ni at inter-dendritic region near melt boundary.

transformation effects. Effects of microsegregation on second phase constituent particle (eg., intermetallics in eutectoid-former alloys, and complex Er-oxide phases in dispersoid-former alloys) were not clearly observed using SEM analysis. However, in eutectoid-former alloys the segregation of alloying elements to dendrite boundaries near the fusion boundary was found to influence precipitation during subsequent heat treatment. This is exemplified in Figure 14 by interdendritic Ti_2Ni present in a Ti-5.5Ni laser melt aged at 625°C for 10 minutes. In the fine cellular structure which characterized the bulk of this melt zone (similar to the Ti-9Co alloy in Figure 11), intermetallic formation was extremely fine and uniform.

Summary

Solidification-related structure characteristics of laser melts exhibit both similarities and differences as compared to RS particulate. From the representative structures discussed above, several general conclusions can be made:

1) Columnar beta grain structures are common to both the LSM and RS particulate structures, although significant differences in grain size are observed,

2) beta subgrain structures originating during the solidification process are commonly observed in the LSM's,

3) solidification substructures are observed in nearly all laser melts and in the more rapidly cooled or highly-alloyed RS products, substructure transitions are observed in the LSM and RS particulate but differ with respect to the chill location,

4) macrosegregation in the form of transverse solute banding can be observed in LSM structures, particularly in the slowly cooled melts.

Continuing work on these alloys is concentrating on the effects of consolidation and heat treatment thermal cycles on the LSM structure and hardness. In addition, TEM and STEM examination of the as-solidified and heat-treated laser surface melts is being conducted to provide a more thorough understanding of these microstructures.

References

1. S.M.L. Sastry, T.C. Peng, P.J. Meschter and J.E. O'Neal, "Rapid Solidification of Titanium Alloys," Journal of Metals, 35(9) (1983), 21-28.

2. T.C. Peng, S.M.L. Sastry and J.E. O'Neal, "Exploratory Studies of Laser Processing of Titanium Alloys," (MDRL 81-06, McDonnell Douglas Corp, 1981).

3. D.G. Konitzer, B.C. Muddle and H.L. Fraser, "Formation and Thermal Stability of an Oxide Dispersion in a Rapidly Solidified Ti-Er Alloy," Scripta Metallurgica, 17, (1983) 963-966.

4. W.A. Baeslack III, S. Krishnamurthy and F.H. Froes, "Rapid Solidification and Aging of a Hypoeutectoid Ti-Ni Alloy," Strength of Metals and Alloys, vol. 3, (New York, NY, Pergamon Press, 1985)

5. W.A. Baeslack III, "A Comparison of Rapidly-Solidified Microstructures in a Hypereutectoid Ti-40W Alloy," Journal of Materials Science Letters, 5 (1986) 315.

References (cont.)

6. S.M. Copley, D.G. Beck, O. Esquivel and M. Bass, "Solidification of Continuous Laser-Melted Trails," Lasers in Metallurgy, ed., H. Mukherjee and J. Mazumder, (Warrendale, PA: The Metallurgical Society, 1981) 11-20.

7. W.A. Baeslack III, "Effect of Solute Banding on Transformations in Dissimilar Titanium Alloy Weldments," Journal of Materials Science Letters," 1 (1982) 229.

SECONDARY COOLING EFFECTS IN RAPIDLY SOLIDIFIED TITANIUM ALLOYS

S. Krishnamurthy[a] and F. H. Froes[b]

[a]Metcut-Materials Research Group
P.O. Box 33511
Wright-Patterson AFB, OH 45433

[b]Air Force Wright Aeronautical Laboratories
Materials Laboratory, AFWAL/MLLS
Wright-Patterson AFB, OH 45433

Abstract

Both pendant drop melt extraction (PDME) and electron beam splat quenching (EBSQ) techniques for rapid solidification of titanium utilize an electron beam for melting the alloys and, consequently, require a vacuum for their operation. In these processes, rapid solidification of the molten metal occurs by conduction of heat into a rotating metallic substrate (wheel or disk). The rapidly solidified material resides on the heat extracting substrate for a very short period and then separates from it. Further cooling of the solidified material after this breakaway from the substrate occurs by a radiative heat loss. The radiative or secondary cooling rates can be very slow and may lead to grain coarsening as well as diffusional phase transformations in the rapidly solidified material. The calculation of these post solidification cooling rates is presented and examples of the effects of secondary cooling on microstructures are given from our studies on titanium-eutectoid former alloys.

Introduction

Pendant drop melt extraction (PDME) is one of the most common laboratory techniques used for rapid solidification of titanium alloys [1]. In this process, a small melt puddle is created at the end of a vertically held alloy rod by means of an electron beam. Filament or ribbon is extracted from this melt puddle by a rotating metal wheel. Another process employed for titanium rapid solidification is the electron beam splat quenching (EBSQ) which produces flakes rather than ribbons [2]. Like PDME, this process also utilizes an electron beam to produce droplets at the end of a titanium alloy rod. These droplets are allowed to fall on a rapidly rotating horizontal metal plate and are solidified into horseshoe-shaped flakes. The thickness of EBSQ flakes is comparable to that of ribbons produced by the PDME process. Because of the need for an electron beam heating source, both PDME and EBSQ methods require a vacuum for their operation.

There are two different stages in the cooling of rapidly solidified ribbons or flakes formed by these techniques. First, there is a period during which the material stays in good contact with the chilling metal substrate and heat extraction from the alloy occurs predominantly by conduction cooling. The second stage involves a breakaway of the rapidly solidified material from the quenching substrate and further cooling of the alloy occurs in vacuum by radiation. In the past, much attention has been paid to the primary cooling stage in which heat loss occurs by conduction. The objective of this paper is to illustrate the effects of secondary cooling in which heat loss occurs mainly by radiation. The examples used in this paper are taken from our recent work on titanium alloys containing beta-eutectoid forming additions.

Calculation of Radiative Cooling

Radiative heat loss rate:

$$\frac{dQ}{dt} = -e\,\sigma\,A\,(T^4 - T_o^{\,4})$$

where e = emissivity of the material
σ = Stefan-Boltzmann constant
A = surface area of the material
T = temperature of the material
T_o = temperature of the surroundings

Since the alloy temperature T is greater than the ambient temperature T_o, $T^4 >> T_o^{\,4}$ may be assumed. Further, substituting $dQ = MC_p dT$, we can write an equation for the cooling rate:

$$\frac{dT}{dt} \cong \frac{-e\,\sigma\,A\,T^4}{MC_p}$$

where M = mass
C_p = specific heat of the material

The previous equation may be integrated to obtain the time "t" required for the material to cool by radiation from an initial temperature T_i to any temperature T:

$$t \cong \frac{MC_p}{3\ e\ \sigma\ A}\left(\frac{1}{T^3} - \frac{1}{T_i^{\,3}}\right)$$

In these equations, we can substitute

$\sigma = 5.67 \times 10^{-8}\ J/(s \cdot m^2 \cdot K^4)$
$e \cong 0.4$ (assumed)
$C_p \cong 525\ J/(kg \cdot K)$

Further, for a titanium ribbon with a thickness of 50μm and a surface area (A) of $10^{-4}\ m^2$, the calculated value of its mass (M) is 1.125×10^{-5} kg. Substitution of these values results in the following equations:

$$t \cong 8.658 \times 10^{8}\left(\frac{1}{T^3} - \frac{1}{T_i^{\,3}}\right) \text{ seconds} \qquad (1)$$

$$\frac{dT}{dt} \cong -3.85 \times 10^{-10} \cdot T^4 \text{ K/s} \qquad (2)$$

Results and Discussion

As Equation 1 indicates, the time (t) elapsed during radiative cooling depends on the temperature T_i of the ribbon when it separates from the wheel (or disk) surface. This temperature is generally unknown, but a range of values may be assumed for the present calculations. Figure 1 illustrates a plot of Equation 1 for two different values of T_i, 1400 K and 900 K, respectively. As expected, different values of T_i lead to different thermal excursions for the ribbons. However, for T_i values in the range 900-1400 K, differences in thermal excursion are noticeable only in the first 5 or 6 seconds (Figure 1). At longer times, the variation of ribbon temperature with time is nearly independent of the initial temperature T_i. It should also be noted from this plot that the time required for the ribbon to cool to ambient temperature is more than 30 seconds.

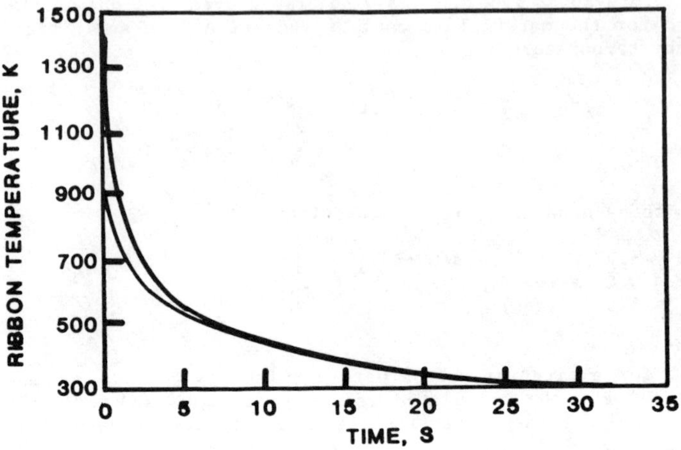

Figure 1 - The variation of ribbon temperature with time after the breakaway of ribbon from wheel surface. The two curves correspond to initial temperatures (T_i) of 1400 K (top curve) and 900 K (bottom curve), respectively.

The cooling rate of titanium ribbon after its breakaway from the quenching substrate is obtained by plotting Equation 2 as shown in Figure 2. This cooling rate decreases from about 10^3 K/s to very low values ($<10^2$ K/s) as the ribbon cools.

The ramifications of the results (Figure 2) may be illustrated by using examples from our recent work on beta-eutectoid forming titanium alloys. Based on the rate of eutectoid decomposition of the beta phase, the alloying elements are generally classified as either active (e.g., Cu, Ni, etc.) or sluggish (e.g., Fe, Cr, etc.) eutectoid formers. The present discussion of the role of secondary cooling will first address the former class of alloys and then point out many general effects which will be applicable to all titanium alloys.

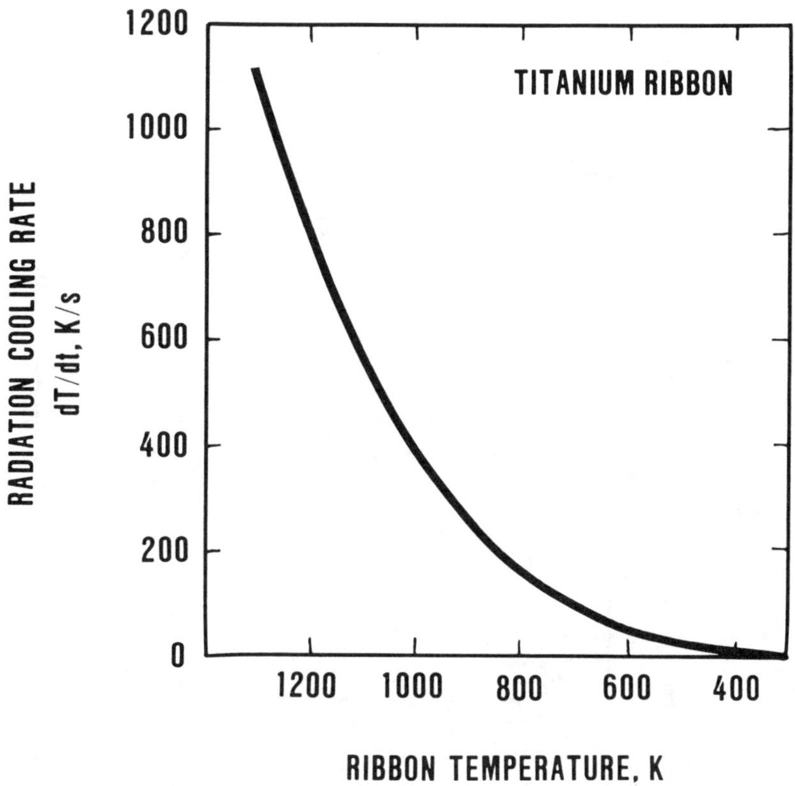

Figure 2 - Calculated variation of radiative cooling rate as the ribbon cools to ambient temperature.

The Ti-Cu alloy system is known to exhibit a very active eutectoid reaction. The alloy Ti-7Cu represents a near-eutectoid composition (Figure 3). Rapid cooling of this alloy from the melt using the hammer-and-anvil splat quenching process produced a martensitic microstructure (Figure 4). This result implies that the alloy was quenched rapidly to a temperature below M_s. In contrast, the microstructure of Ti-7Cu alloy ribbons prepared by the pendant drop melt extraction process is shown in Figure 5. Both lamellar and nonlamellar alpha+Ti_2Cu eutectoid structure are clearly observed in this material. The occurrence of this diffusional transformation product instead of a martensitic structure is a direct result of the slower secondary cooling conditions in the PDME ribbons. Previous work [5] on wrought Ti-Cu alloys showed that the nose of the TTT curve for the reaction $\beta \rightarrow \alpha+Ti_2Cu$ occurs at about 650°C (923 K). This reaction was reported to occur within 10 seconds at the nose temperature, but the data for t <10 seconds were not available due to the rapid kinetics [5]. In the present case of the PDME ribbons, however, the time for completion of the eutectoid decomposition may be estimated by using Equation 1. Substituting T_i = 1400 K and T = 925 K, the duration (t) of secondary cooling is obtained to be ~0.8 seconds. These results therefore suggest that the eutectoid reaction in Ti-7Cu alloy PDME ribbons occurred within 0.8 seconds.

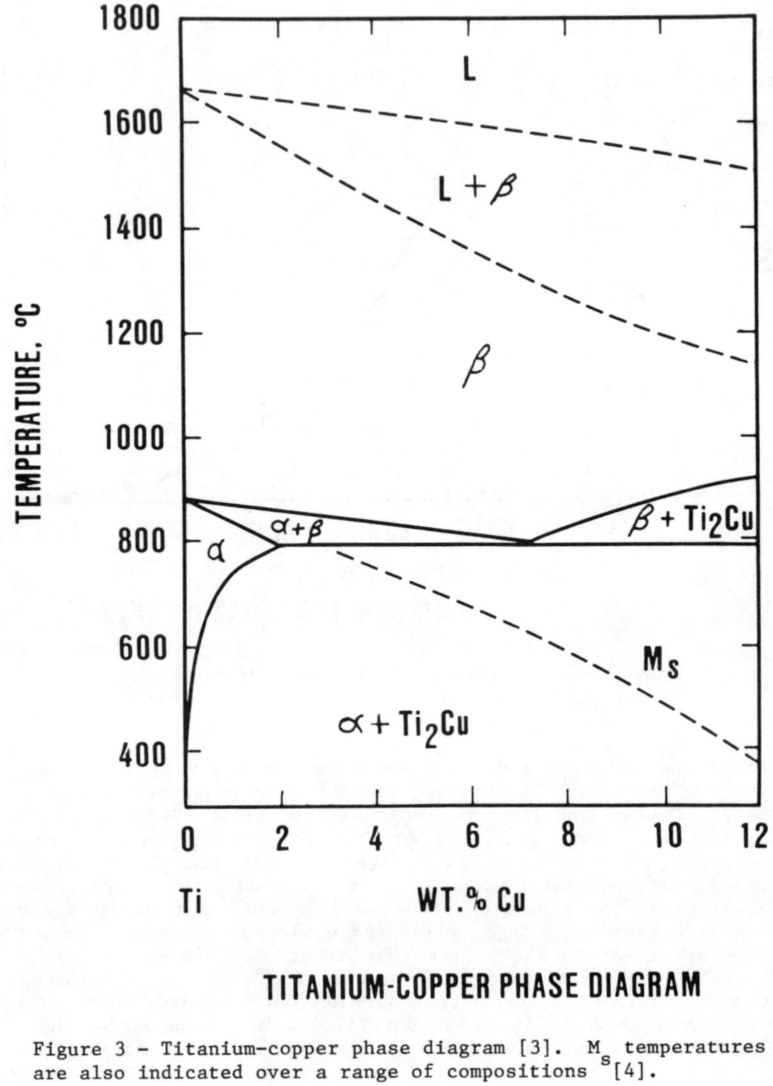

Figure 3 - Titanium-copper phase diagram [3]. M_s temperatures are also indicated over a range of compositions [4].

Figure 4 - Transmission electron micrograph of a Ti-7Cu splat showing martensitic structure.

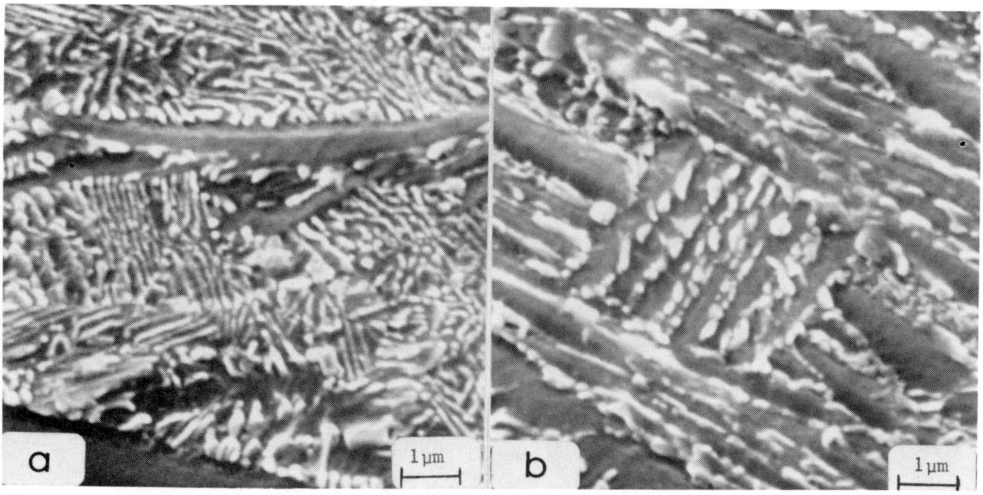

Figure 5 - Scanning electron micrographs of a Ti-7Cu PDME ribbon showing (a) lamellar and (b) nonlamellar alpha+Ti_2Cu eutectoid.

Figure 6 shows the microstructure of a bar of Ti-7Cu alloy which was solution treated in the beta phase field at 925°C (1700°F) for 30 minutes and then quenched in a brine solution. The scanning electron micrograph (Figure 6a) indicates a martensitic structure. TEM characterization (Figure 6b) shows Ti_2Cu particles at plate boundaries which appear to have formed due to autotempering. Figures 5 and 6 indicate that the solid state cooling rate is considerably slower in the PDME ribbons as compared to the cooling rate of a beta solutionized and brine quenched bulk sample.

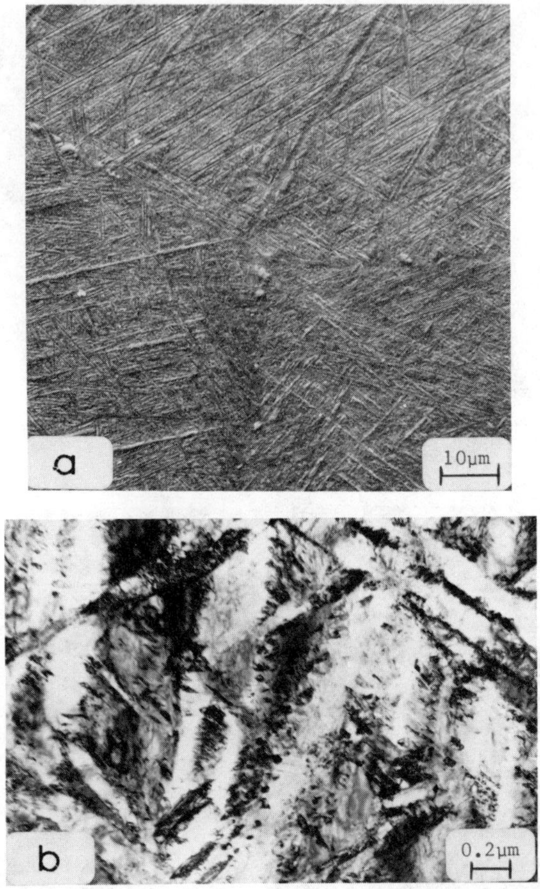

Figure 6 - Microstructure of beta solution treated (925°C/30 min/brine quench) Ti-7Cu bar (a) SEM and (b) TEM.

Similar to the above results, the eutectoid reaction $\beta \rightarrow \alpha + Ti_2Ni$ has been previously observed in as-produced PDME ribbons as well as EBSQ flakes of near-eutectoid titanium-nickel alloys [6]. Again, calculations based on Equation 1 suggest that the eutectoid decomposition occurred within $\sim$1.4 seconds in these rapidly solidified alloys. The eutectoid product was seen even in titanium-nickel alloy ribbons prepared by chill block melt spinning under argon atmosphere [6]. Evidently the convective cooling by argon was inadequate for suppressing the active eutectoid decomposition in this alloy.

Although the above discussion has considered examples of active eutectoid forming titanium alloys, many of the effects of secondary cooling are common to all alloys produced by the PDME or EBSQ techniques. For example, the slower cooling rate can lead to grain coarsening. In certain titanium alloys, the precipitation of omega phase may be promoted during secondary cooling. Other effects such as diffusion bonding of the rapidly solidified ribbons or fibers may also occur (Figure 7).

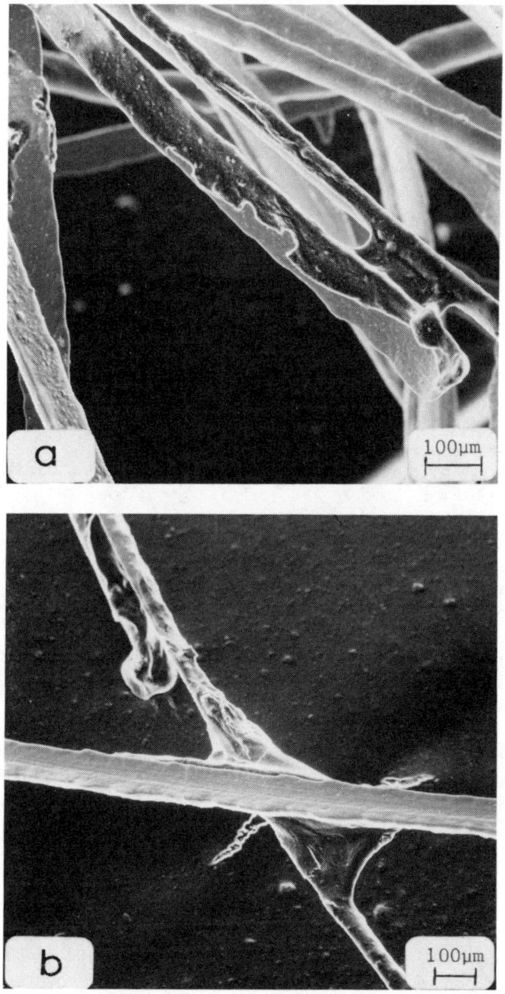

Figure 7 - Scanning electron micrographs showing bonding of PDME fibers (a) Ti-3W and (b) Ti-2Si.

Conclusions

The PDME and EBSQ techniques provide the advantage of melting and rapid solidification of titanium alloys without utilizing a special crucible for containment of the molten metal. However, as the present results have shown, rapid solidification by these techniques is generally followed by a slow secondary cooling involving radiative heat loss in vacuum. The slower radiative cooling rates can lead to undesirable effects such as microstructural coarsening, solid state phase transformations, and diffusion bonding of the rapidly solidified material. Therefore, attempts should be made to obtain maximum cooling during the primary or conduction-dominated cooling regime, and minimize the extent of the radiative cooling.

Acknowledgments

One of the authors (S.K.) gratefully acknowledges the financial support provided under U.S. Air Force Materials Laboratory Contract No. F33615-85-C-5072.

References

1. R. E. Maringer, C. E. Mobley, and E. W. Collings, "Preparation and Properties of Compacts of Melt Extracted Staple Fibers of Ti-6Al-4V Alloy," AIChE Symposium Series No. 180, 74 (1978), 111-116.

2. S. M. L. Sastry, T. C. Peng, P. J. Meschter, and J. E. O'Neal, "Rapid Solidification Processing of Titanium Alloys," Journal of Metals, 35(9)(1983), 21-28.

3. J. L. Murray, private communication with author, National Bureau of Standards, 1985.

4. Y. C. Huang, S. Suzuki, H. Kaneko, and T. Sato, " Thermodynamics of the M_s Points in Titanium Alloys," in The Science, Technology and Application of Titanium, ed. R. I. Jaffee and N. E. Promisel (Oxford: Pergamon Press, 1970) 691-693.

5. G. W. Franti, J. C. Williams, and H. I. Aaronson, "A Survey of Eutectoid Decomposition in Ten Ti-X Systems," Met. Trans. A, 9A(11) (1978), 1641-1649.

6. S. Krishnamurthy, A. G. Jackson, H. Jones, and F. H. Froes, "Beta-Eutectoid Decomposition in Rapidly Solidified Titanium-Nickel Alloys," submitted to Metallurgical Transactions A, 1984.

CHARACTERISTICS OF OXIDE DISPERSIONS IN

RAPIDLY SOLIDIFIED TITANIUM ALLOYS

R. E. Anderson and J. L. Larson

Pratt & Whitney - Engineering Division South
P. O. Box 2691
West Palm Beach, Florida 33402

Abstract

Many studies have shown that rapidly solidified titanium alloys with rare earth elemental additions yield fine dispersions of rare earth oxides. These dispersions have potential for property enhancement in conventional as well as novel alloys. Rapidly solidified powder was produced and characterized of conventional alloys with erbium additions. The particulate was screened to several size fractions and vacuum hot compacted. The resulting dispersions in each size fraction were evaluated according to average diameter, spacing, volume percent, and thermal stability.

Summary

Pratt & Whitney (P&W) has undertaken the development of a high temperature titanium alloy using a dispersion strengthening approach to enhance the properties of otherwise attractive compositions. A unique device has been produced at P&W to facilitate the production of a dispersion strengthened matrice. As described herein, powder produced in P&W's new device using prealloyed Ti-8Al-1Mo-1V (Ti-8-1-1) barstock and erbium metal was found to exhibit several desirable characteristics of a rapidly solidified microstructure. Work is continuing to fully characterize the as solidified powder microstructure as well as the properties of fully consolidated and heat treated barstock.

Introduction

The most commonly used measure of the performance of a gas turbine engine is the ratio of engine thrust to engine weight. Thrust is defined by such factors as air flow rate, combustor exit temperature, and engine rotational speed. Weight is obviously controlled by such factors as engine size, material strength, and material density. The drive in the high performance gas turbine industry is to maximize the thrust-to-weight ratio. To the metallurgist this means a constant effort must be maintained to maximize the temperature capability of the materials of construction while, simultaneously, searching for lower density materials to replace those already in service. Titanium, with its density of approximately 4.5 g/cc (0.16 lb/in3) and melting point in excess of 1650°C (3300°F), appears to be an excellent candidate to replace today's widely used nickel-base superalloys leading to a substantial increase in engine thrust-to-weight ratio. However deficiencies in such secondary properties as oxidation behavior, creep strength and hot tensile strength have restricted the use of titanium to applications experiencing temperatures well below about 650°C (1200°F).

P&W, with elements of the Air Force and other members of the propulsion and aerospace community, has initiated a major program to increase the temperature capability of titanium and its alloys. Fundamental to this effort is the application of the principles of rapid solidification in conjunction with appropriate sound matrix alloying principles and surface protection schemes. P&W has constructed and is operating a unique device for converting prealloyed titanium barstock to clean, rapidly solidified metal powder of a modified composition. The work discussed below is funded by the Air Force and is aimed at characterizing the product of this device in terms of particle morphology, size distribution, solidification factors, chemical and microstructural homogeneity, and cleanliness. Ultimately, this program will produce full density mill products for mechanical property characterization and potential component testing.

The work is in its early phases and has centered on powder made from prealloyed Ti-8-1-1 barstock, both directly converted and modified with elemental erbium additions to produce a fine dispersion of erbium oxide. Titanium alloys typically have a high solubility for oxygen and low solubility for rare earth elements. By using rapid solidification techniques, the rare earth elements can be held in solution in the titanium alloy matrix upon quenching from the liquid state (1,2). With subsequent thermal exposure the rare earth elements getter oxygen from the matrix to form an oxide dispersion (2,3). Uniformly dispersed, fine precipitates have the potential of improving tensile strength by impeding dislocation motion (3-6). Presented below is a description of the powder produced to date. Data is presented on compositional retention, contamination (both interstitial and bulk), microstructure and powder morphology. Work is ongoing to develop an appropriate consolidation and heat treatment sequence.

Powder Production

The P&W device, shown in Figure 1, was sized for laboratory purposes to melt approximately one kg (2.2 lbs) of titanium barstock per minute. The entire concept of the device is linked to the maintenance of a metal skull, which isolates the molten pool from the structural materials, preventing contamination. As is usual for P&W powder metallurgy operations, an attempt is made to maintain an all inert environment throughout the processing sequence to prevent contamination of either the molten metal or the loose powder.

Figure 1 P&W Titanium Powder Device

The prealloyed meltstock, in the form of 2.5 to 5.0 cm (1 to 2 in) diameter barstock, is fed into a consummable arc melter where it is contained in a water cooled copper hearth. Late alloying additions can be made by secondary additions. For this program, erbium was added to the Ti-8-1-1 feed stock to approximate Ti-8-1-1-0.75 Er. Also, baseline Ti-8-1-1 was atomized for comparison.

Once a molten pool has been generated by the arc melter, the molten alloy is allowed to flow over the hearth and into the atomizer. For this series of experiments, atomization has been accomplished through the action of a classical gas atomizer using helium as the working medium, as shown in Figure 2. The solidified particulates are then conducted by the helium quench gas to the cyclone separators and collector cans. The cans may be valved off under helium to protect the powder during storage.

Figure 2 Gas Atomization of Titanium Alloy Powder

Powder sizing is accomplished in a helium filled dry box, used exclusively for titanium powder processing. The particle size distribution obtained during this experimentation is shown in Figure 3. The powder was screened to -50 mesh (< 300 microns) for consolidation. Individual heats were evaluated for particle morphology, inclusions, O2 content, and other common interstitials. Figure 4 shows scanning electron microscope (SEM) micrographs of typical as produced powder particles. Table 1 gives the interstitial content of the powder as compared to the starting meltstock. A sample of the rapidly solidified base alloy, Ti-8-1-1, was evaluated for phase identification by X-ray analysis. Figure 5 identifies the phases which were observed in the as produced powder, which was separated into various size fractions (-325 mesh, -140 +325, -80 +140, -50 +80), as alpha and alpha prime. A -50 mesh sample of the Ti-8-1-1-0.75Er powder was evaluated by transmission electron microscopy (TEM) using a Phillips EM 400 Scanning Transmission Electron Microscope (STEM). The powder matrix was free of any erbia "dispersion". Evaluation of the matrix composition by KEVEX revealed that the erbium was in solution. A few isolated erbia particles were observed near the edge of the foil and identified by Electron Energy Loss Spectroscopy (EELS) as erbium oxide.

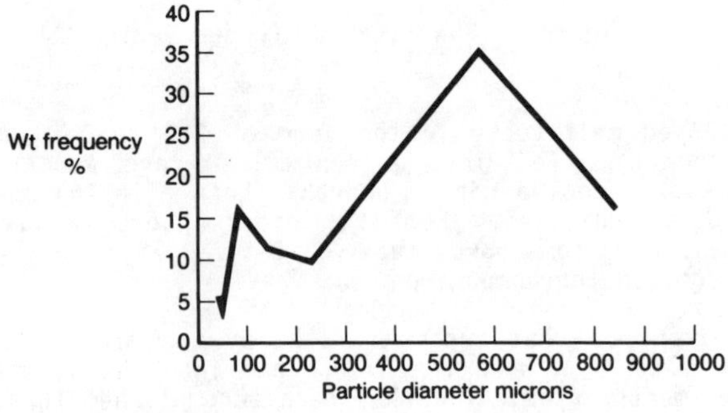

Figure 3 Particle Size Distribution of Ti-8Al-1Mo-1V

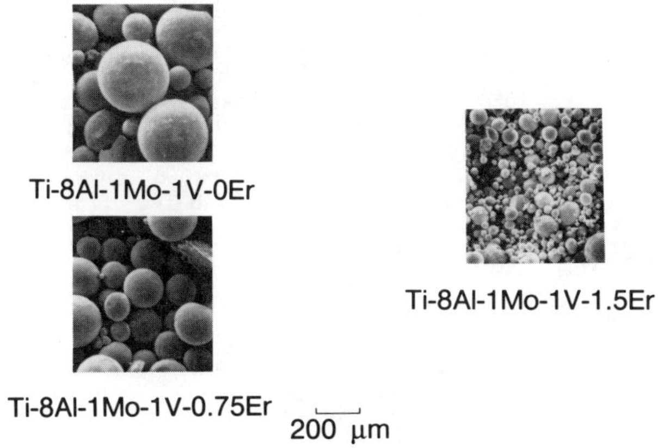

Figure 4 Particle Morphology

Barstock	O	N	H
Ti-8Al-1Mo-1V Meltstock*	0.12	0.05	0.015
Ti-8Al-1Mo-1V Extruded powder	0.16	0.01	0.01
Ti-8Al-1Mo-1V-0.75E$_r$ Extruded powder	0.20	0.01	0.01
Ti-8Al-1Mo-1V-1.5E$_r$ Extruded powder	0.19	0.01	0.01

* Maximum allowed

Table 1 Chemical Analysis of Common Interstitials

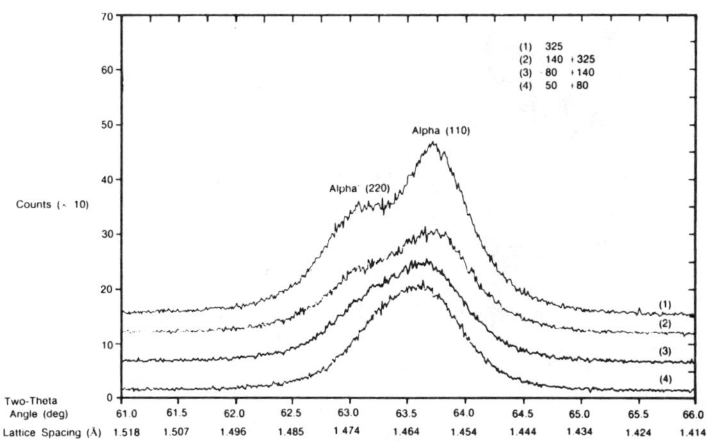

Figure 5 Diffractometer Scan Results of Ti-8Al-1Mo-1V-0.75Er

Powder Consolidation

The powder produced for this study has been consolidated by both extrusion and vacuum hot compaction (VHC). Three pound samples were canned in 3 inch diameter 1020 steel extrusion cans under inert conditions. The cans were evacuated and sealed prior to extrusion. Extrusions were accomplished using P&W's 850 ton Lombard extrusion press at 900°C (1650°F) with an extrusion ratio of 8:1. Figure 6 shows the as extruded microstructure of both the Ti-8-1-1 and Ti-8-1-1-0.75Er. TEM evaluation, Figure 7, revealed the precipitation of erbia particles in the Ti-8-1-1-0.75Er extrusion. Their identity was verified by X-ray phase analysis.

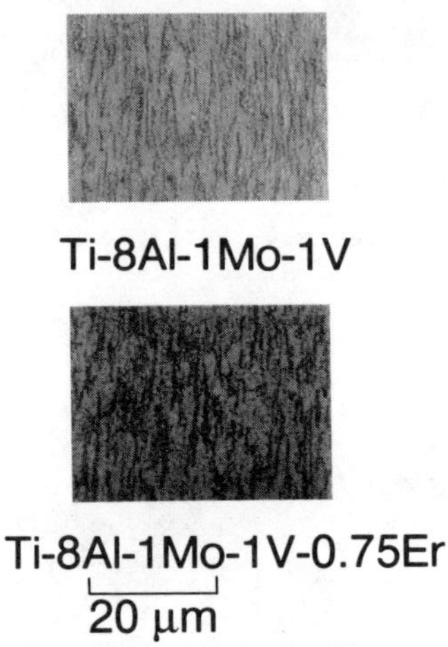

Figure 6 Microstructure of As-Extruded Barstock

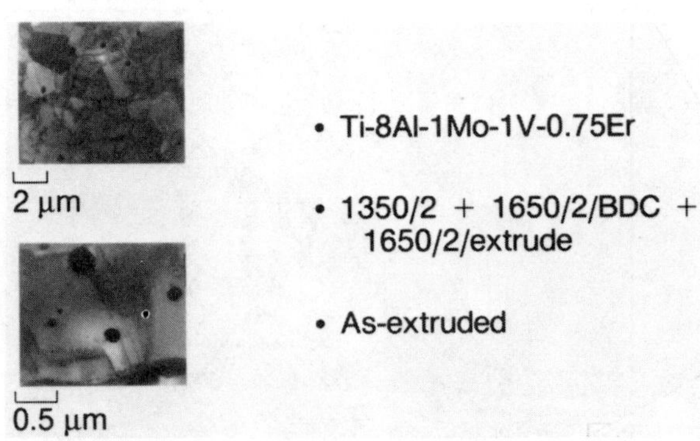

Figure 7 Transmission Microstructure of Extruded Barstock

Since the average size and spacing of the erbia dispersion in the extrusion appeared to be coarser than desired, a vacuum hot compaction study was initiated to determine the times and temperatures necessary for full consolidation, without excessive coarsening of the erbia precipitates. It was found that full consolidation was accomplished when compacting at 900°C (1650°F/30 ksi/2 hr load). Evaluation of the microstructure revealed a somewhat coarse erbia dispersion, similar to the extrusion. Various lower temperature pre-compaction heat treatments are presently being evaluated to see if it is possible to nucleate a finer dispersion at a lower temperature, that remains stable after full consolidation. While the results are inconclusive, Figure 8 shows that the 538°C (1000°F) and 621°C (1150°F) heat treatment may in fact nucleate a finer dispersion. The 704°C (1300°F) preheat treatment did not help to refine the resulting erbia dispersion.

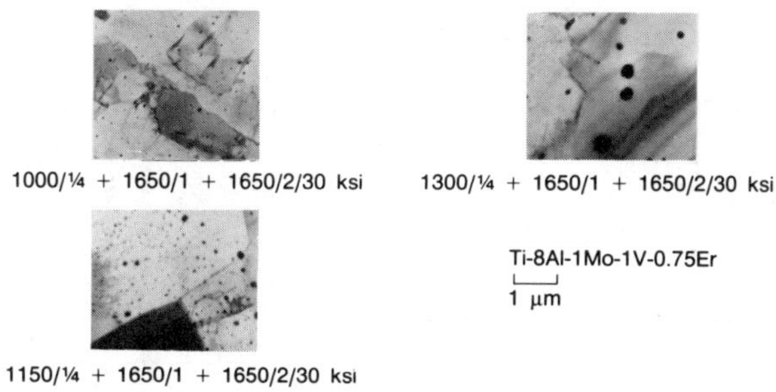

Figure 8 Transmission Microstructure of Compacted Powders

The four compactions were compared statistically using TEM micrographs and a Bausch and Lomb Omnicon automated image analysis system. Figure 9 shows the erbia particle size distribution for each compaction. Although inconclusive, these results also indicate that the 538°C (1000°F) and 621°C (1150°F) pre-heat treatment helped to nucleate a finer dispersion.

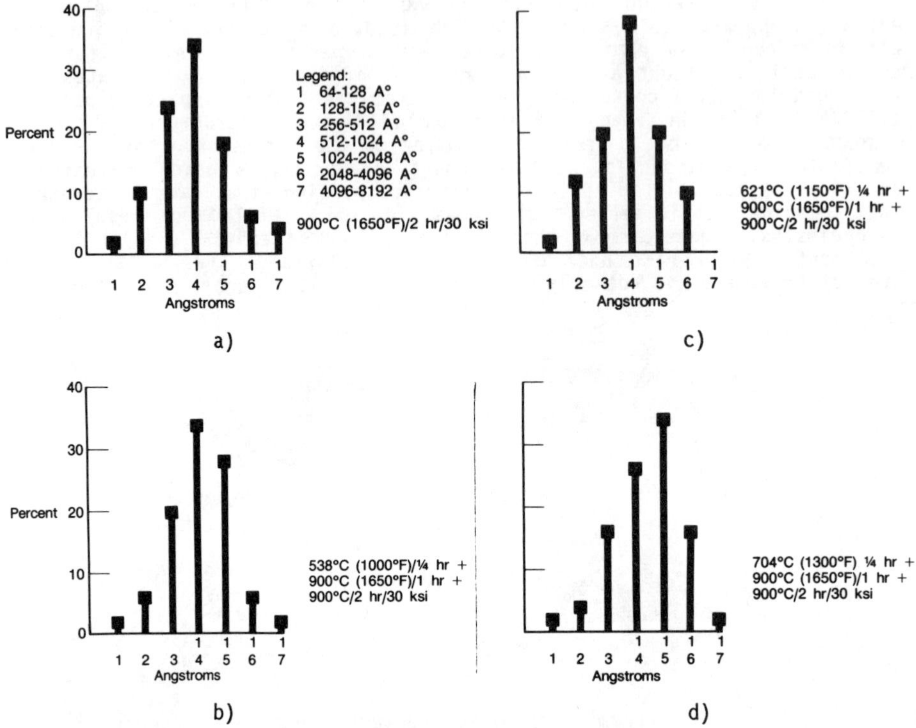

Figure 9 Erbia Particle Size Distribution in Ti-8Al-1Mo-1V-0.75 Er

Conclusion

P&W's newly developed titanium atomization device has demonstrated its capability of producing laboratory quantities of clean titanium powder. The powder morphology, chemical and microstructural homogeneity are indicative of a high quality rapidly solidified product. The ability to quench erbium into solution in the Ti-8-1-1 alloy is evidence that the apparatus is truly a rapid solidification device.

The precipitation of erbia during thermal exposure has yet to be optimized. Further experimentation is needed to find a suitable set of parameters which will control coarsening and still yield a fully consolidated billet. Mechanical testing is also necessary to identify what dispersion size and spacing yields the best combination of properties. Various levels of erbium additions need to be investigated for identification of the optimum volume percent of erbia.

Acknowledgement

This work was performed under Air Force Contract F33615-85-C-5067.

References

1. S.M.L. Sastry, P.J. Meschter, and J.E. O'Neil, "Structure and Properties of Rapidly Solidified Dispersion - Strengthened Titanium Alloys: Part 1. Characterization of Dispersoid Distribution, Structure, and Chemistry", Metallurgical Transactions A, Vol-15A (1984) p 1451-1463.

2. D.G. Konitzer, B.C. Muddle and H.L. Fraser "Formation and Thermal Stability of an Oxide Dispersion in a Rapidly Solidified Ti-Er Alloy" Scripta Met, Vol 17 (1983) p 963-966.

3. S.M.L. Sastry, T.C. Peng, and L.P. Beckerman, "Structure and Properties of Rapidly Solidified Titanium Alloys: Part II. Tensile and Creep Properties" Metallurgical Transactions A Vol 15A (1984) p 1465-1474.

4. G.A. Ansell, T.D. Cooper, and F.V. Lenel, eds.: "Oxide Dispersion Strengthening: Proceedings of the Second Bolton Landing Conference", (Gordon and Breach Science Publishers, New York, NY, 1968).

5. E. Orowan, Symposium and Internal Stress in Metals and Alloys, Institute of Metals, (London, 1948) p 45.

6. M.F. Ashby, "Physics of Strength and Plasticity", edited by A.S. Argon (M.I.T. Press, Cambridge, MA 1968) p 113.

SOLIDIFICATION STRUCTURE AND DISPERSOIDS IN RAPIDLY SOLIDIFIED Ti-Al-Sn-Zr-Er-B ALLOYS

R. G. Rowe*, T. F. Broderick**, E. F. Koch*, and F. H. Froes**

*General Electric Corporate Research and Development
PO Box 8
Schenectady, NY 12301

**Air Force Materials Laboratory, AFWAL/MLLS,
WPAFB, OH 45433

Abstract

The microstructure of melt extracted and melt spun titanium alloys containing erbium and boron revealed a duplex solidification structure of columnar grains leading to equiaxed and dendritic structures near the free surface of melt extracted and melt spun alloys. The solidification structure was revealed by apparent boride segregation to cellular, interdendritic, and grain boundaries. Precipitation of needle or lath-like TiB particles occurred adjacent to Er_2O_3 dispersoid particles in as-rapidly solidified ribbon. On aging, TiB reprecipitated independent of the Er_2O_3 dispersion. Er_2O_3 dispersoid particles exhibited severe grain boundary depletion and coarsening after aging at 925°C (1700°F).

Introduction

Rapid solidification of titanium alloys has been shown to produce alloys with much greater homogeniety than alloys solidified by conventional ingot techniques (1,2). This is particularly true for titanium alloys containing elements which exhibit low equilibrium solubility such as erbium and boron. In this study, we have studied the effect of rapid solidification on the solidification structure of titanium alloys containing erbium for dispersoid development and boron for grain refinement and delineation of the solidification structure.

Experimental Procedure

Two substrate quenching processes, melt extraction and melt spinning were used to rapidly solidify titanium alloys (3). Schematics of each system are shown in Figures 1 (a) and (b). Pendant drop melt extraction (4,5) was performed in vacuum, using an electron beam heat source for melting, and a wheel surface speed of 15 m/sec. The thickness of melt extracted filament ranged from 25 to 175 μm. The melt extraction capacity was limited to 30 to 50 gm of alloy per run.

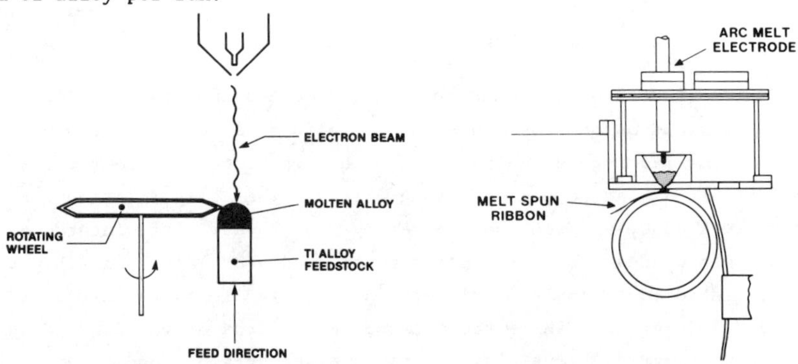

Figure 1. Schematics of (a) titanium melt extraction and (b) titanium melt spinning.

Melt spinning was by cold hearth arc melting and bottom ejection onto a wheel with a surface speed of 60 m/sec, similar to systems described by Tutzauer (6) and Whang (7). The thickness of melt spun ribbon ranged from 30 to 60 μm. Melt spun ribbon was collected in a large (1.25 m diameter) collection chamber in an argon gas atmosphere. Ribbon accumulated in the chamber following melt spinning is shown in Figure 2. The capacity for melt spinning was in excess of 1/2 kg per run.

Several base alloys were evaluated: Ti-6Al-3Sn-3Zr (Ti-633), Ti-6Al-2Sn-4Zr-2Mo (Ti-6242), Ti-6Al-2Sn-4Zr (Ti-624) and Ti-6Al-6Zr (Ti-6-6), all wt.%. Each was highly alloyed, characteristic of high temperature creep-resisting alpha titanium alloys. The base alloys were studied with additions of either erbium or boron or both.

Experimental Results

Both melt extraction and melt spinning produced particulate which was uniform in thickness within a given set of operating parameters. Under constant conditions of melt-substrate contact, this resulted in ribbon or filament which was solidified at a uniform cooling rate.

Figure 2. Melt spun ribbon in 1.25 m diameter collection chamber.

Solidification Structure: We found that there was little difference in the solidification microstructure or dispersoid stability among the three alpha titanium alloy matrices which were studied. Melt extracted alloys such as Ti-6Al-4V with no dispersoid additions showed no dendritic segregation, only cellular structure. Alloys with less than 3 wt.% Er or 0.02 wt.% B had a similar solidification structure. Figure 3 is an optical micrograph of a transverse cross section of melt extracted filament of Ti-6Al-6Zr-1Er-0.02B (wt.%). The grain structure was columnar from the contact surface (bottom) to the free surface (top). The transgranular striations within the columnar grains are from martensitic transformation of the beta phase. Their presence helped to identify the prior beta grain structure. In alloys with Er additions, the degree of segregation during solidification was often best revealed by the pattern of second phase precipitation after aging. The alloy Ti-6Al-2Sn-4Zr-1Er (wt.%) was melt extracted and annealed at 950°C (1740°F) for 1 hour, Figure 4. The uniform precipitation of Er_2O_3 particles in this grain interior region demonstrated that there was little segregation of erbium during solidification. In other regions, the Er_2O_3 particles were arrayed in strings, suggesting some cellular segregation. Some elongated, rod-shaped dispersoid particles were present in addition to spherical particles. These have been seen in several different alloy systems, and have been identified as Er_2O_3, the same as the spherical particles (8,9).

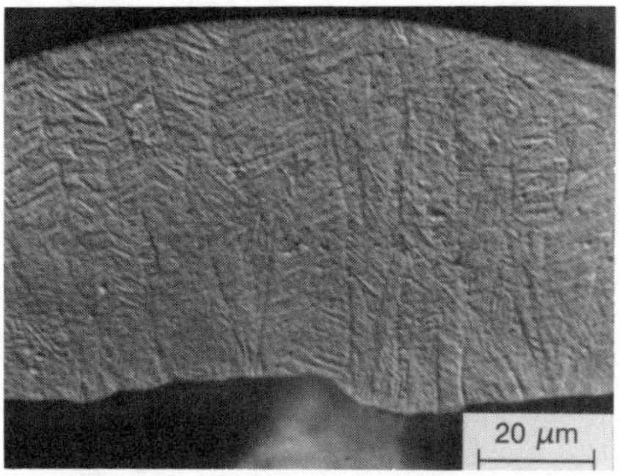

Figure 3. Transverse metallographic section of as-melt-extracted filament of Ti-6Al-6Zr-1Er-0.02B (wt.%)

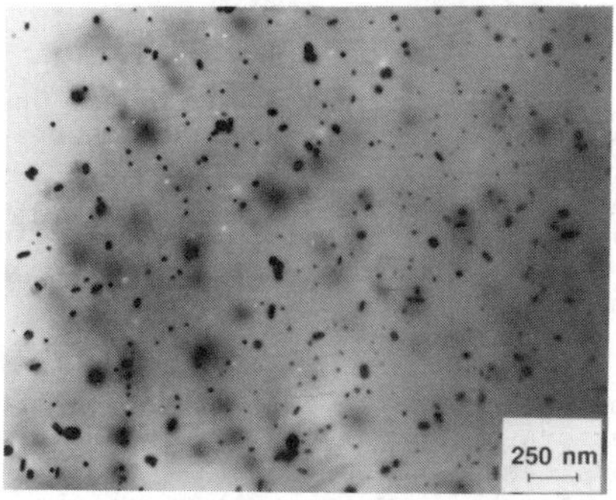

Figure 4. TEM micrograph of a grain interior region of Ti-6Al-2Sn-4Zr-1Er (wt.%) after annealing one hour at 950°C (1740°F).

In the case of boron-containing alloys, segregation to internal boundaries clearly revealed the solidification structure. The solidification structure of melt extracted alloy Ti-6Al-3Sn-3Zr-0.25B (wt.%) is shown in

Figure 5. This is a transverse cross section of a melt extracted filament with the wheel contact surface at the bottom left. Solidification was columnar, radiating from the point of contact with the extraction wheel. Midway up from the contact surface, parallel etch lines within the columnar grains suggested cellular segregation. The region of cellular solidification ended at the top quarter of the filament thickness. Above this, the microstructure had features of dendritic solidification and was generally equiaxed. This structure has been described by the term "equiaxed dendritic" in studies of rapidly solidified aluminum alloys (10-12). This terminology will be used here to describe this structure.

Characteristic dendritic structures were observed in powder particles of this alloy, but the transition from segregationless columnar solidification → cellular segregation → branched columnar dendrites → multiple side-branched dendrites that is observed in superalloys (13,14) was not seen in melt spun or melt extracted titanium alloys containing boron. Columnar grains were 3 to 4 μm wide. The cell spacing in the region of cellular solidification was 0.5 μm. The characteristic feature size in the top part of the ribbon with equiaxed dendritic structure was the same as the columnar grain width, 3 to 4 μm. This was much finer than the cast stick microstructure which had thick boride platelets in interdendritic regions spaced 20 μm apart.

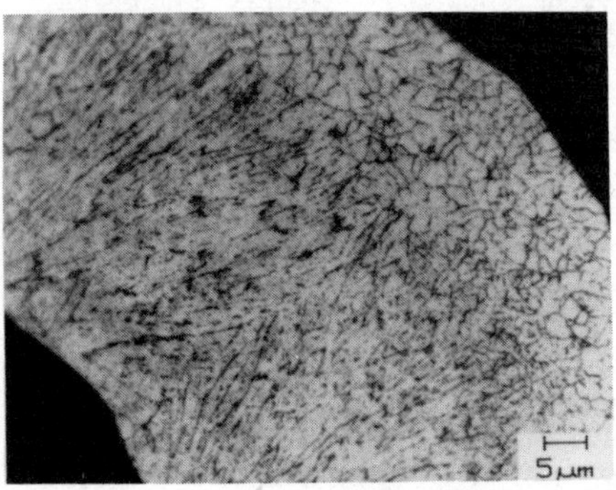

Figure 5. Transverse section of as-melt-extracted filament of Ti-6Al-3Sn-3Zr-0.25B (wt.%).

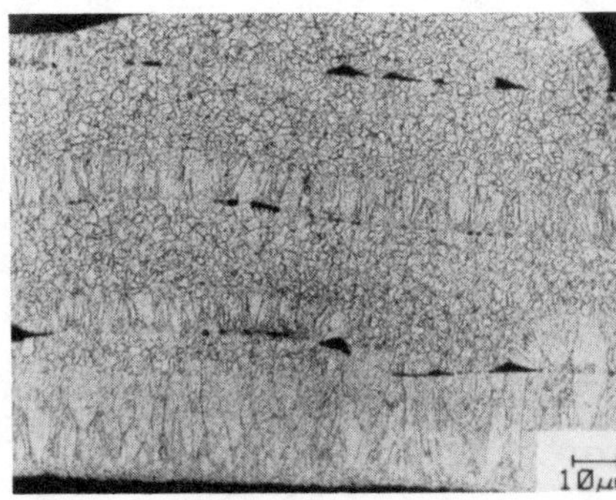

Figure 6. Transverse section of as-melt spun multilayer ribbon,

Multilayer Melt Spun Ribbon: One possibility which has been considered a source of two-zone microstructures is the departure of filament from the melt quenching wheel. Multi-layer melt spun ribbons of Ti-6Al-2Sn-4Zr-2Mo-.15Si-0.3B (wt.%) were prepared which had a two-zone structure in each layer. This is shown in Figure 6, and confirms that the two zone structure is a feature of the solidification process rather than due to loss of contact with the wheel.

Multilayer ribbon also gave some insight into the effect of substrate composition on the solidification process. Close examination of the microstructure, Figure 6, revealed that upper layer columnar grains were continuations (and hence probably epitaxial) of the equiaxed dendritic layer below it. Furthermore, the extent of the columnar zone was greater in the first layer than in succeeding layers which suggested that earlier nucleation, and lower substrate thermal conductivity, reduced the undercooling of the melt in upper layer ribbons relative to the first layer.

Unmelted Particle Inclusions: Both pendant drop melt extraction and arc melt spinning have a melt pool which is supported by a solid skull. Because of turbulence in the melt, this occasionally leads to incorporation of unmelted particles into filament or ribbon. Figure 7 shows an included particle which has the coarse dendritic structure of the starting ingot imbedded in a rapidly solidified filament of Ti-6Al-6Zr-1Er-0.08B (wt.%). The rapidly solidified alloy in the rest of the filament had a martensitic microstructure characteristic of the rapid solid state quench after solidification.

The top surface of melt extracted filament was often decorated with surface perturbations which appeared to be related to local nucleation of the alloy ahead of the melt front, Figure 8 (15,16). Inclusions carried into the melt appear to be trapped at the free surface by surface tension. Similar protrusions have been observed at the surface of laser spin atomized powder (17).

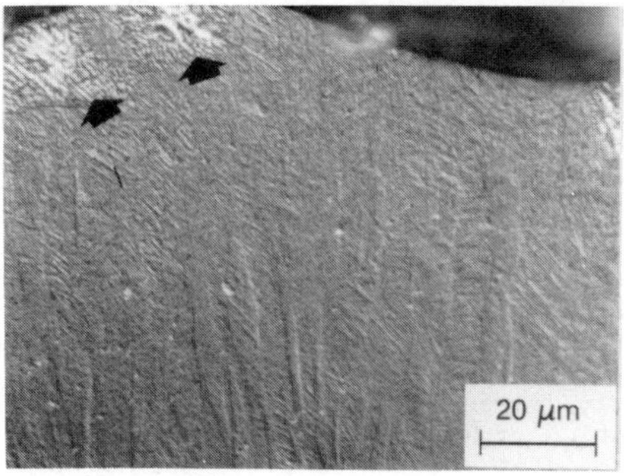

Figure 7. Transverse metallographic section of melt extracted Ti-6Al-6Zr-1Er-0.08B (wt.%) showing an unmelted particle with the ingot dendritic structure.

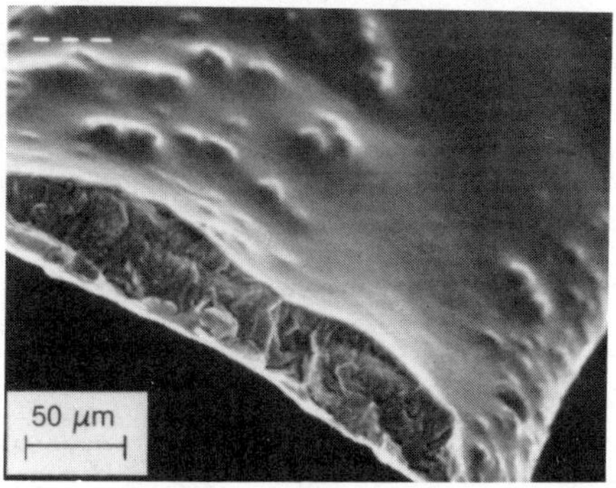

Figure 8. Surface protrusions at the free surface of melt spun ribbon caused by the inclusion of small unmelted particles.

It is believed that all rapidly solidified titanium alloys produced by cold hearth melting will have unmelted particles in the solidified product. This is because of the inherent nature of skull containment that solid particles are always present in the melt pool. Fortunately, the size of the particles was generally very small, 0.5 to 2 μm, below the size for which concern over included defects should be significant (18).

<u>Er_2O_3 and TiB Dispersoid Formation</u>: A single phase alpha-titanium alloy, Ti-6Al-6Zr-1Er-.08B (wt.%) was rapidly solidified by both melt extraction and by melt spinning. Melt extracted filament of this alloy had a uniform columnar microstructure. The melt spun ribbon microstructure was variable. Some ribbon was entirely equiaxed similar to the equiaxed dendritic structure near the upper surface of Figure 5. Other ribbon was columnar with no equiaxed structure. The microstructural variability suggested that uniform interfacial contact was not achieved in melt spinning this alloy.

Transmission electron microscopy from the center of a melt spun ribbon revealed an as-melt-spun microstructure with heavy particle precipitation along internal boundaries, Figure 9. Dispersoid particles were identified as Er_2O_3 (near spherical particles) and TiB (lath or needle-like particles) (9). In a number of cases, TiB needles appeared to be growing from Er_2O_3 particles, suggesting heterogeneous nucleation of one of the species at the interface of the other.

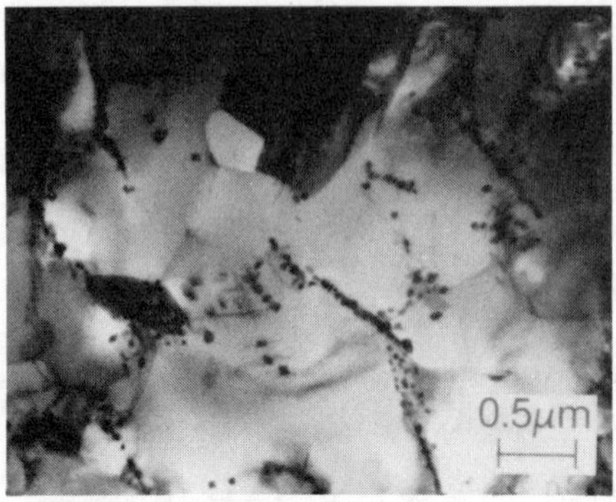

Figure 9. TEM micrograph of as-melt-spun ribbon of Ti-6Al-6Zr-1Er-0.08B (wt.%).

Heat treatment at 925°C (1700°F) for 2 hr. produced the microstructures in Figures 10(a) and (b). Laths or needles were no longer associated with the Er_2O_3 particles, but were scattered randomly throughout the matrix. The Er_2O_3 dispersoid showed some coarsening after 2 hrs at 925°C (1700°F). Figure 10(a) shows the distribution of TiB needles and Er_2O_3 particles in a region where the Er_2O_3 dispersoid was fairly coarse and localized. Figure 10(b) shows the same independent precipitation of TiB needles and Er_2O_3 particles, but from a region with a homogeneous distribution of Er_2O_3 particles in grain interiors. Enhancement of coarsening or dissolution processes at grain boundaries appears to have depleted the areas near the grain boundaries of Er_2O_3 particles (9).

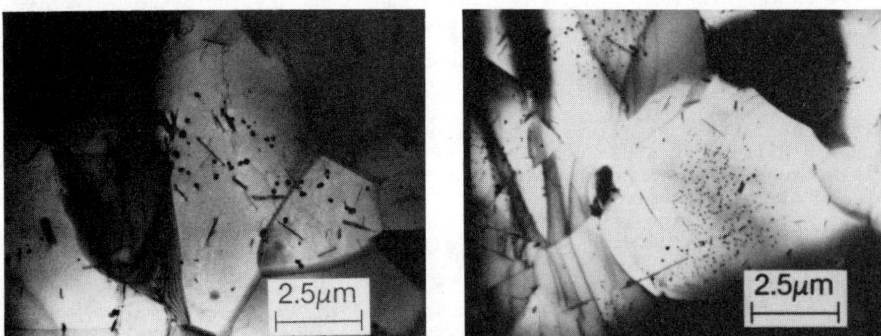

Figure 10. TEM micrographs of ribbon of Ti-6Al-6Zr-1Er-0.08B (wt.%) after annealing 1 hr at 925°C. (a) A region of localized coarse Er_2O_3 dispersoid. (b) A region of fine Er_2O_3 precipitation showing coarsening near grain boundaries.

Discussion of Results

For Ti-B alloys and Ti-15V-3Al-3Sn-3Cr, a solidification speed of 20 to 40 cm/sec has been shown to be necessary to achieve submicron dendrite arm or cell spacings (19,20). This corresponded to a maximum cooling rate of 10^6 K/sec. Comparison of these results with the cell spacings midway through the thickness of melt extracted Ti-633-.25B, Figure 5, suggested that the solidification speed was at least 20-40 cm/sec. Closer to the wheel, the speed should have been higher since cellular segregation appeared to have been supressed. The apparent primary dendrite arm spacing of the coarse equiaxed dendritic structure in Figure 5, suggested that the solidification speed in the equiaxed zone was less than 10 cm/sec.

Differences in Solidification Structure: A loss of thermal gradient due to recalescence at the top of the columnar zone has been suggested as a means of explaining similar microstructures in melt spun aluminum alloys (21,22). This could explain the evolution of the solidification structure in these titanium alloys, but additional factors must also be significant since the alloy with only an Er addition did not exhibit any microstructural evidence of recalescence, while the alloys containing boron did. In equilibrium with a titanium solid solution, the standard free energy for the reaction (Ti) + (B) → TiB has been estimated to be on the order of -10 to -15 kcal/mole at the eutectic temperature of 885°C (1625°F) (2,23,24). The reaction (Er) + (O) → Er_2O_3 has been estimated to have a free energy of reaction of -113 kcal/mole at 950°C (1740°F) (9). Both estimates were based upon estimated excess free energies of solution of the constituent elements in a binary titanium solid solution. Comparing these reactions, it can be seen that for an alloy containing erbium and oxygen, the driving force for compound formation is the highest of the two, yet this system exhibited the least segregation. Because of this, differences in the kinetics of segregation between the two alloy systems, or differences in the degree of undercooling of the substrate quenched particulate are likely to be behind this difference in recalescence behavior.

Examination of the extent of the columnar solidification zone in multilayer melt spun ribbon, Figure 6, for the bottom and succeeding layers leads to the inference that the bottom layer, with the longest columnar zone, had the highest undercooling. This may reflect differences in the time period for nucleation of solidification at the liquid-wheel contact surface and the liquid-ribbon interfaces, and this suggests nucleation differences as one possible reason for the difference in solidification behavior between alloys containing only erbium, and alloys with boron additions. If boron additions enhance nucleation of solid at the melt-substrate interface, then boron-containing alloys would achieve less undercooling before solidification begins, reducing the extent of the columnar solidification zone before recalescence, and producing a two zone microstructure.

The other main effect of boron was to strongly influence the distribution of erbium oxides in the coarse equiaxed zone. TEM examination of the melt spun ribbon suggested that Er_2O_3 precipitation occurred only along solidification boundaries, and that no additional precipitation from a metastable solid solution occurred. A similar distribution has been shown in Ti-8Al-1.5Er-0.25 (25). This suggests enhancement of erbium segregation at high temperatures by boron in the alloy, or a change in solidification conditions. The comparison of the solidification structures of alloys with and without boron reinforces the idea that the effect of boron may be to promoting earlier nucleation of the solidification front, and consequently lower undercooling at the onset of solidification.

Dispersoid Depletion at Grain Boundaries: Dispersoid particles which originated in the columnar solidification zone may have initially had a more uniform or at least spatially more refined distribution. Particles in these regions coarsened into a pattern which suggested enhancement of the coarsening process at grain boundaries, Figure 10(b). The estimated alpha transus temperature for Ti-6Al-6Zr was 980-1000°C (1795-1830°F) 50 to 75°C, above the annealing temperature which minimizes but does not rule out the possibility of coarsening due to beta phase formation at boundaries (26).

Conclusions

- Melt spun and melt extracted ribbon of titanium-aluminum base alloys with boron additions exhibited a two zone mixed columnar-equiaxed dendritic solidification structure.
- Differences in the kinetics of segregation or the nucleation of solid at the melt-substrate interface have been suggested as the basis for the differences in solidification structure between alloys with and without boron.
- A cell spacing of 0.5 μm in the columnar growth region of melt spun and melt extracted boron-bearing alloys implied a solidification velocity of greater than 20 cm/sec.
- Boron increased the degree of segregation of Er_2O_3 dispersoids in alloys with both Er and B additions.
- TiB precipitates occurred in association with Er_2O_3 particles as-melt-spun, but were randomly distributed, independent of Er_2O_3, after aging.
- Enhancement of the rate of dispersoid coarsening near grain boundaries was suggested by the observation of zones of dispersoid depletion near grain boundaries in annealed alloys.

Acknowledgments

The authors would like to acknowledge the contributions of Messrs. R.O. Auer, R.E. Ettinger, P.G. Frischmann and L.C. Perocchi for their major roles during development of the melt extraction and melt spinning processes used in this study. We would also like to acknowledge Messrs. G.K. Scarr and R.A. Amato of the General Electric Aircraft Engine Business Group for sharing optical and transmission electron micrographs of rapidly solidified titanium alloys which helped form our understanding of the problem.

References

1. F.H. Froes and R.G. Rowe, "Rapidly Solidified Titanium", in Rapidly Solidified Materials and Their Mechanical and Magnetic Properties, Edited by B.C. Giessen, D.R. Polk and A.I. Taub, Proc. Conf., Boston 1985, Mat. Res. Soc., To be published

2. R.G. Rowe and F.H. Froes, "Titanium Alloy Rapid Solidification", in Rapid Solidification of Crystalline Alloys, Edited by J.L. Wood, The Institute of Metals (London), London, To be published.

3. S.J. Savage and F.H. Froes, "Production of Rapidly Solidified Metals and Alloys", Journal of Metals, 36, April, 1984, pp. 20-33.

4. R.E. Maringer and C.E. Mobley, Jr., "Centrifugal Method of Forming Filaments from an Unconfined Source of Molten Material", U.S. Patent #3,896,203, July 22, 1975.

5. R.E. Maringer and C.E. Mobley, "Casting of Metallic Filament and Fiber", J. Vac. Sci. Tech. 11(6), pp. 1067-1071, 1974.

6. H. Tutzauer, P. Esquinazi, M.E. De La Cruz, and F. De La Cruz, "Production of Ribbon Quenched Samples Using an Arc Furance", Rev. Sci. Instr. 51, pp. 546-547, April 1980.

7. S.H. Whang, "Rapidly Solidified Titanium Alloys Containing Novel Additives", J. of Metals, 36, April, 1984, p. 34.

8. D.G. Konitzer, J.T. Stanley, M.H. Loretto, and H.L. Fraser, "The Nature of Dispersed Phases in Ti-0.7 at.% Er Prepared by Rapid Solidification Processing", To be published, Acta Metallurgica.

9. R.G. Rowe, T.F. Broderick, E.F. Koch, and F.H. Froes, "Microstructural Study of Rapidly Solidified Titanium Containing Erbium and Boron", in Rapidly Solidified Materials, Edited by P. W. Lee and R. S. Carbonara, Proc. 1986 Int. Conf., San Diego, CA, Feb., 1986, Amer. Soc. for Metals, Metals Park, Ohio, 1986, pp. 107-114.

10. L.J. Masur, J.T. Burke, T.X. Katamis, and M.C. Flemings, "Microsegregation of an Aluminum and Magnesium Alloy at High Solidification Rates", in Rapidly Solidified Amorphous and Crystalline Alloys, Ed. B. H. Kear, B.C. Giessen and M. Cohen, Proc. of the MRS Annual Meeting, 1981, Boston, MA, North Holland, NY, 1982, pp. 185-189.

11. C.Y. Barlow and B. Ralph, "Microstructures and Grain Growth in Rapidly Solidified Aluminum", Metallography, 8, pp. 171-181, (1985).

12. P. Van Mourik, M. Van Rooijen, N.M. Van Der Pers, T.H. De Keijser, and E.J. Mittemeijer, "Rapidly Solidified (Melt Spun) Aluminum Alloys: Morphology, Texture and Excess Vacancies", Acta Cryst. Vol. A40, p. C192, (1984).

13. H.A. Davies, "Metallographic Studies of Rapidly Quenched Superalloy Ribbons", Report 82CRD318, December, 1982, General Electric Corp. Res. and Development Center, Schenectady, NY.

14. H.A. Davies, N. Shohoji, and D.H. Warrington, "The Structures of Rapidly Quenched Nickel-Based Superalloy Ribbons Produced by Melt Spinning", in Rapid Solidification Processing; Principles and Technologies, Ed. R. Mehrabian, B.H. Kear, and M. Cohen, Claitor's Publ. Div., Baton Rouge, LA, 1980, pp. 153-164.

15. R.G. Rowe, unpublished research.

16. R. Amato, General Electric Company, Aircraft Engine Business Group, Evendale, Ohio, private communication.

17. D.G. Konitzer and H.L. Fraser, "The Production of Powders of Ti-Alloys with a Dispersion of Rare Earth Oxides by Laser Spin Atomization", Presentation, 1984 Powder Metallurgy Conference, Toronto, Canada, To be published.

18. S.W. Schwenker, D. Eylon and F.H. Froes, "Influence of Foreign Particles on Fatigue Behavior of Ti-6Al-4V Prealloyed Powder Compacts", Met. Trans. A, 17A, pp 271-280 (1986).

19. T.C. Peng, S.M.L. Sastry, and J.E. O'Neal, "Exploratory Study of Laser Processing of Titanium Alloys", in Lasers in Metallurgy, Ed. K. Mukherjee and J. Mazumber, Proc. Conf., Chicago, 1981, AIME, NY, 1981, pp. 279-292.

20. T.C. Peng, S.M.L. Sastry, J.E. O'Neal, and J.F. Tesson, "A Correlation Between Dendrite-Arm-Spacing and Cooling Rate for Laser-Melted Ti-15V-3Al-3Sn-3Cr", in Lasers in Materials Processing, Ed. E.A. Metzbower, Proc. Int. Conf., Los Angeles, CA, Jan. 1983, Amer. Soc. for Metals, Metals Park, OH, 1983, pp. 241-246.

21. T.W. Clyne, "Numerical Treatment of Rapid Solidification", Met. Trans. B, 15B, pp. 369-381, (1984).

22. L. Katgerman, "Computer Simulation of Cellular Growth During Rapid Solidification of Binary Aluminum Alloys", in Modeling of Casting and Welding Processes II, Ed. N.H. Henniker, Proc Conf., August 1983, Met. Soc. AIME, 1983, pp. 135-143.

23. D.R. Stull, JANAF Thermochemical Tables, Second Edition, Office of Std. Ref. Data, Nat. Bur. of Stds., Washington, DC, June, 1971, Report #NSRDS-NBS 37.

24. M. Hansen, Constitution of Binary Alloys, 2nd Edition, McGraw-Hill, New York, 1958.

25. S.M.L. Sastry, T.C. Peng, P.J. Meschter, and J.E. O'Neal, "Rapid Solidification Processing of Titanium Alloys", J. of Metals, September, 1983, pp. 21-28.

26. M.F.X. Gigliotti, R.G. Rowe, G.E. Wasielewski, G.K. Scarr, and J.C. Williams, "Tensile and Creep Properties of Rapidly Solidified Titanium Alloys Containing Complex Matrices and Fine Dispersions", in Rapidly Solidified Alloys and Their Mechanical and Magnetic Properties, Ed. B.C. Giessen, D.R. Polk, and A.I. Taub, Proc. Mat. Res. Soc. Symp., Boston, MA, Dec., 1985 To be published.

MECHANICAL EVALUATION OF

RAPIDLY SOLIDIFIED TITANIUM ALLOYS CONTAINING FINE DISPERSOIDS

M.F.X.Gigliotti*, G.E.Wasielewski**, and R.G.Rowe*

* General Electric Corporate Research and Development, Schenectady, NY
** General Electric Aircraft Engine Business Group, Evendale, OH

Abstract

Exploratory alpha titanium alloys containing various rare earth compound dispersoids were prepared by rapid solidification. The alloys were consolidated by HIP and extrusion and thermally exposed. The roles of matrix microstructure and dispersoid size and distribution were evaluated by comparing tensile and creep strengths of materials thermally exposed to produce different microstructures. Fundamental aspects of the role of matrix alloy chemistry were studied by comparing model alloys in which individual elements were systematically varied. Portions of this work were carried out under U.S. Air Force Contract No. F33615-83-C-5034.

Introduction

An opportunity for alloy development of new titanium alloys using rapid solidification processing arises from the ability to produce fine rare earth compound dispersions in titanium matrices. Initial work by Vordahl [1] showed that fine rare earth compound dispersoids could be achieved by rapid solidification, and the resulting properties were improved. The dispersoids evaluated were rare earth oxides, sulfides, and oxide-sulfide combinations.

Various binary and ternary titanium alloys have been evaluated to assess the role of rare earth compound dispersoids on strengthening [2,3] High levels of rare earth additions produce dispersoids which act to strengthen the alloy [4]. Er was found to be a preferred addition to titanium, because of a lower rate of coarsening than other rare earths [5-7], and that the dispersoids can be formed in Ti - Al alloys as well [8].

The dispersoid formed by erbium metal additions to titanium has been identified as erbium sesquioxide [9]. Additionally, heat treatments into the beta temperature range were demonstrated to coarsen rare earth oxide dispersoids. The greatly enhanced coarsening rates above the beta transus was explained in terms of the enhanced mobilities of oxygen and rare earth.

Titanium aluminide alloys containing rare earth oxide dispersions have been evaluated [10]. The consolidation step of HIP plus a subsequent extrusion acted to severely coarsen the dispersoid. This effect may be limited to the aluminides, since the consolidations of alpha titanium alloys cited above did not have coarsened dispersoids.

Model highly alloyed titanium compositions containing rare earths were rapidly solidified, consolidated, and mechanically evaluated. The strong restraint on grain growth by rare earth dispersoids is harmful to creep resistance at 650 C (1200 F) [11]. Annealing above the beta transus coarsened the rare earth dispersoids, but greatly improved creep resistance. The coarsened dispersoid appeared to have beneficial effects on post creep ductility.

It is the purpose of this work to describe fundamental aspects of the structure and mechanical properties of a series of dispersion containing titanium alloys as a function of alloying content.

Experimental Procedure

Alloy Chemistries.

The alloys for evaluation are listed in Table I. With respect to this table, alloy UT is an alpha titanium alloy with no additions that serves as a base for comparing properties. Alloy UV is the matrix alloy UT with an addition of Si. Alloy UW is the matrix alloy UT with an addition of Er. Alloys WE, WI, and WN are the matrix alloy UT with an Er addition plus additions of increasing levels of Al. Alloy WS is a modification to the original matrix UT in which Sn is substituted for Al. Alloy WY is the high tin alloy WS with an Er addition. Alloy WZ is the high tin alloy WS with Er and Si additions.

The alloys chemistries were chosen to evaluate the role of increasing Al contents with all other matrix elements held constant, and also to evaluate relative behaviors of Sn and Al on mechanical behaviors.

Table I. Alloy Compositions

Alloy	Ti	Al	Zr	Sn	Cb	Mo	Other	
				atom %				
UT	84.70	12.20	1.30	1.20	0.50	0.10		
UV	84.48	12.17	1.30	1.20	0.50	0.10	0.25Si	
UW	84.19	12.13	1.29	1.19	0.50	0.10	0.60Er	
WE	84.19	12.13	1.29	1.19	0.50	0.10	0.60Tb	
WI	81.21	15.11	1.29	1.19	0.50	0.10	0.60Er	
XL	80.57	15.35	1.29	1.19	0.50	0.10	1.00Er	
WN	78.42	17.89	1.29	1.20	0.50	0.10	0.60Er	
WS	85.90	6.20	1.30	6.00	0.50	0.10		
WY	85.39	6.16	1.29	5.96	0.50	0.10	0.60Er	
WZ	85.16	6.15	1.29	5.95	0.50	0.10	0.60Er	0.25Si

Alloy	Ti	Al	Zr	Sn	Cb	Mo	Other	
				weight %				
UT	86.26	7.00	2.52	3.03	0.99	0.20		
UV	86.12	6.99	2.52	3.03	0.99	0.20	0.15Si	
UW	84.46	6.85	2.46	2.96	0.97	0.20	2.10Er	
WE	84.54	6.86	2.47	2.96	0.97	0.20	2.00Tb	
XL	81.15	8.71	2.47	2.97	0.98	0.20	3.52Er	
WI	82.53	8.65	2.50	3.00	0.99	0.20	2.13Er	
WN	80.67	10.37	2.53	3.06	1.00	0.21	2.16Er	
WS	79.60	3.24	2.29	13.78	0.90	0.19		
WY	78.08	3.17	2.25	13.51	0.89	0.18	1.92Er	
WZ	77.96	3.17	2.25	13.50	0.89	0.18	1.92Er	0.13Si

Table II. Chemical Analysis of Melt Extracted Filament

Alloy		Al	Sn	Other	
		atom %			
UT	add	13.66	1.18		
	anal	10.63	1.20		
UV	add	13.63	1.18	0.25Si	
	anal	9.90	1.18	0.27Si	
UW	add	13.57	1.17	0.59Er	
	anal	11.14	1.19	0.39Er	
WE	add	13.57	1.17	0.59Tb	
	anal	13.27	1.12	0.46Tb	
WI	add	19.90	1.12	0.57Er	
	anal	16.93	1.04	0.34Er	
WN	add	23.89	1.11	0.56Er	
	anal	20.36	1.07	0.28Er	
WS	add	8.37	5.94		
	anal	7.17	5.91		
WY	add	8.29	5.91	0.60Er	
	anal	6.54	5.70	0.26Er	
WZ	add	8.28	5.89	0.59Er	0.24Si
	anal	6.73	5.78	0.31Er	0.17Si
XL	add	17.30	1.17	0.98Er	
	anal	14.72	1.11	0.60Er	

Rapid Solidification.

The alloys were prepared by non-consumable arc melting and drop casting 1.2 cm diameter by 5 cm long rods into copper chill molds. Next, the alloys were rapidly solidified by pendant drop melt extraction in vacuo. The melt extraction was accomplished by melting the top of a vertically oriented alloy rod with an electron beam and extracting filament with a chill wheel rotating in the horizontal plane. The melt extraction apparatus is illustrated in Figure 1.

Figure 1 - Melt extraction apparatus, showing horizontal wheel and metal brush. Melt extracted filament in background.

Consolidation and Thermomechanical Treatment.

The rapidly solidified alloy filaments were packed in steel cans and HIPped at 840 C, 210 MPa for 3 hours. The green densities of the packed alloy filaments were about 70% of theoretical. After HIPping, the steel jacket was machined off the titanium alloy cylinders. The alloy cylinders were then fit into steel jackets for extrusion. The diameter of the titanium cylinders was about 1.3 cm. The diameter of the steel extrusion jacket was 5.4 cm. The extrusion conditions were 840 C and an area reduction of 8:1. An extrusion assembly and extruded bar are shown in Figure 2. The extruded bars were X-rayed to precisely locate the titanium core, the bars cut into 2.5 cm lengths, and the steel machined off the titanium cores.

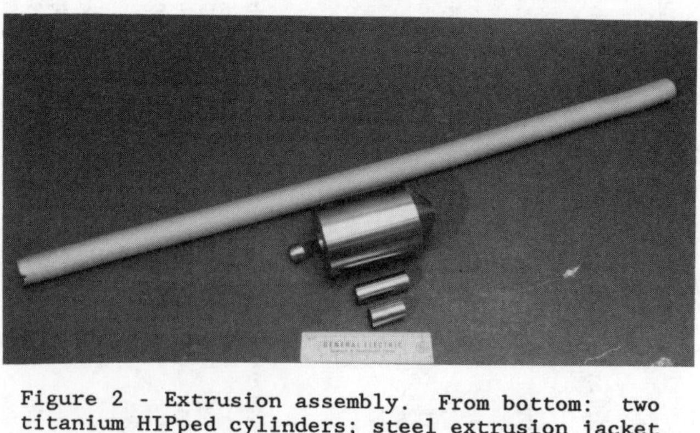

Figure 2 - Extrusion assembly. From bottom: two titanium HIPped cylinders; steel extrusion jacket and rear plug; extruded rod.

Thermal Exposures.

Thermal exposure samples were placed in titanium tubes welded closed in an argon glove box. Each titanium tube was then sealed in an evacuated fused silica tube. The alloys were thermally exposed above their beta transus, followed by a lower temperature age. Alloys UT, UV, UW, WE, WI, and WN were exposed 1200 C for 2 hours, then 600 C for 48 hours. Alloy XL was exposed 1130 C for 2 hours, then 760 C for 24 hours, then 600 C for 24 hours. Alloys WS, WY, and WZ were exposed 1100 C for 2 hours, then 600 C for 48 hours.

Mechanical Testing.

Tensile and creep tests were conducted using a specimen with a 1.3 cm (0.5 in.) gage length and 0.20 cm (0.80 in.) gage diameter. Final machining of gage and threaded ends was done after thermal exposure.

Tensile tests were conducted in air at room temperature and in vacuo at 650 C. Tensile tests were conducted at a constant crosshead speed of .05 cm/minute. A 0.2% offset yield strength was determined from load-time charts. Elongation to maximum load and elongation to failure were also calculated from load-time charts. Creep tests were conducted at 650 C and 138 MPa in argon. Creep data was electronically recorded from capacitance gages attached to the titanium alloy specimens.

Results and Discussion

Chemical Analysis.

The rapidly solidified melt extracted filaments were analyzed by wet chemistry techniques. The results are listed in Table II. The amounts of Zr, Cb, and Mo were found to be the amounts added to the original drop cast ingots. However, the amounts of Al, Sn and rare earth metal were lower than that added. The amounts of these elements lost in processing could be due to evaporation from the molten pool during melt extraction. Figure 3 plots the aluminum levels analyzed against the amount added; there appears to be a linear relationship with a slope slightly less than one. The rare earth metal analysis is similarly plotted in Figure 4. There is somewhat more scatter in the data for the rare earth metals, but the trend appears similar to that for aluminum. The Sn contents were slightly reduced for the high tin alloys WS, WY, and WZ.

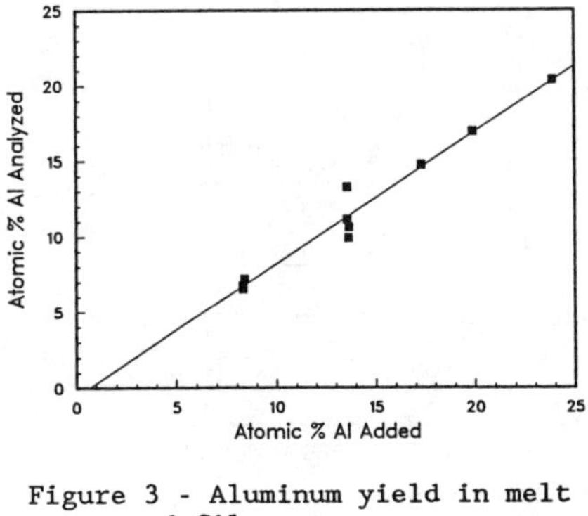

Figure 3 - Aluminum yield in melt extracted filament.

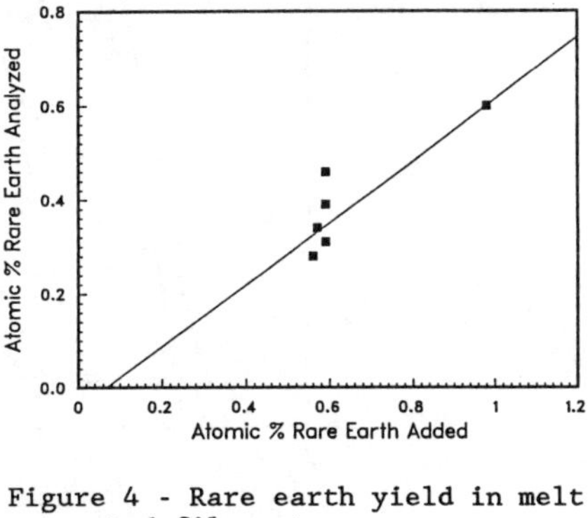

Figure 4 - Rare earth yield in melt extracted filament.

Oxygen analyses were conducted on the rapidly solidified filaments. The oxygen levels were between 670 and 1550 weight parts per million. There was no relationship between alloy content and oxygen analysis; the presence or absence of rare earth addition did not alter the oxygen level.

Microstructure of Consolidated Alloys.

The consolidated alloys were given the multistep thermal exposures described above. The steps comprised exposures above the beta transus followed by agings below the beta transus. These exposures resulted in coarsened rare earth particles in a transformed beta microstructure. The structure is illustrated in light and electron micrographs in Figure 5. The light microscopy reveals a plate like alpha microstructure. The transmission microscopy show that the particles of rare earth oxide have coarsened to around one half micro meter in diameter.

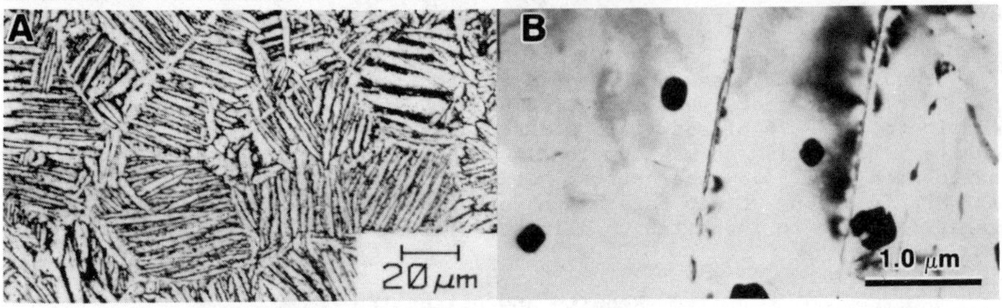

Figure 5 - Microstructure of consolidated and thermally exposed alloy UW. a) Light micrograph. b) Transmission electron micrograph.

The growth of dispersoid in model alloys was previously described [11]. Further analysis of the dispersoid growth is shown in Figure 6, in which the dispersoid size distribution is plotted for alloy UW in the as-HIP condition as well as after thermal exposures of two hours at 750 C, 875 C, or 1200 C, followed by 48 hour exposures at 600 C. The distributions were determined by counting particles on TEM micrographs within five size groupings appropriate for that anneal. Figure 6 shows that there is only slight coarsening associated with either the 750 C or 875 C anneals, both of which

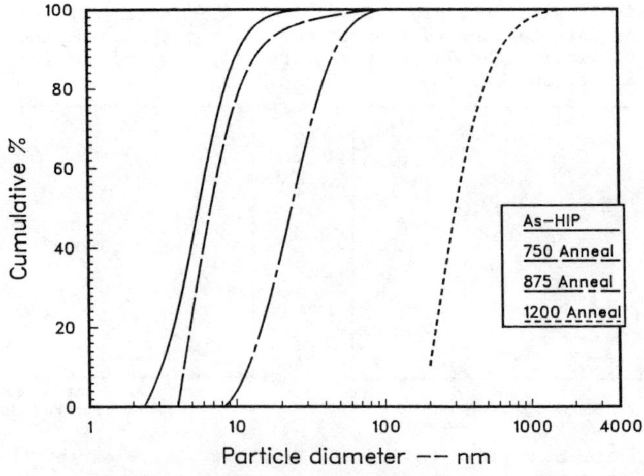

Figure 6 - Dispersoid size distribution in alloy UW after various thermal exposures.

are below the beta transus. However, the 1200 C anneal results in particle sizes typically about one half micro meter in diameter.

Tensile tests.

The results of the tensile tests are listed in Table III. The high Al alloys UT, UW, WE, XL, and WI behaved as a class differently from the high tin alloys WS, WY, and WZ.

The rare earth containing high Al alloys UW, WE, XL, and WI showed a tensile behavior that varied monotonically with analyzed Al content. Yield strength at 650 C increased with Al content; ultimate tensile strength increased to about 700 MPa for alloy XL and remained at this level for alloy WI. Ductility decreased monotonically with analyzed Al level. These trends are illustrated in Figure 7.

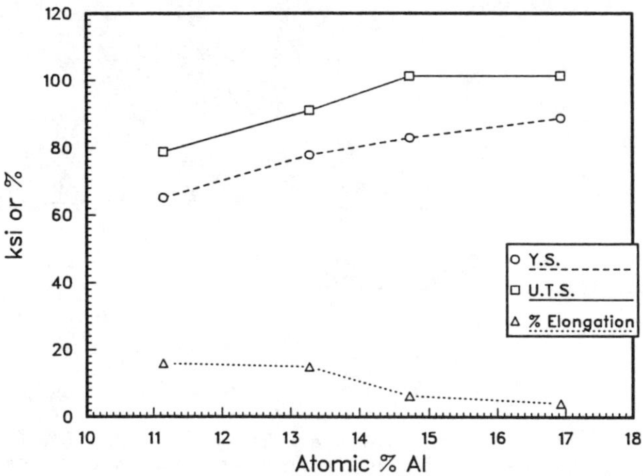

Figure 7 - Tensile behavior at 650 C as a function of analyzed aluminum content.

The addition of a rare earth to the high aluminum base can be compared in the tensile behaviors of alloy UT (Base) and alloy UW (+Er), where the erbium containing alloy is only slightly stronger than the base alloy, but has a much higher tensile ductility. The addition of erbium allows use of higher aluminum alloys while maintaining tensile ductility.

The high tin alloys WS, WY, and WZ were tested at both room temperature and at 650 C. At room temperature the erbia containing alloys WY and WZ had higher yield strengths and lower ductilities than the base alloy WS. In contrast, at 650 C, the alloys all have about the same strengths, but the erbia containing alloys are more ductile. These behaviors might be due to the combined effects of a smaller grain size of the rare earth containing alloy improving strength at low temperatures, while at elevated temperatures the strength is reduced due to grain boundary sliding.

It is possible to compare two alloy bases, UT and WS, and the erbia modifications, UW and WY, which have about the same sums of atomic % (Al+Sn). The substitution of tin for aluminum appears to raise the tensile strength at 650 C, but strongly lowers ductility.

Fracture surfaces of all tested alloys were evaluated by scanning electron microscopy. At both test conditions, the fracture surface of the rare earth containing alloy exhibited a dimpled rupture appearance. At the bottom of the dimples there were particles which analyzed high in rare earth metal, and are presumably the rare earth oxide dispersoid. The fracture appears to have occurred by the coalescence of voids forming around dispersoid particles. The fracture surfaces of the base alloys were flat and sharp, appearing to be separation along alpha plates. Figure 8 compares typical fracture surfaces for the base alloy WS and the erbium containing alloy WY.

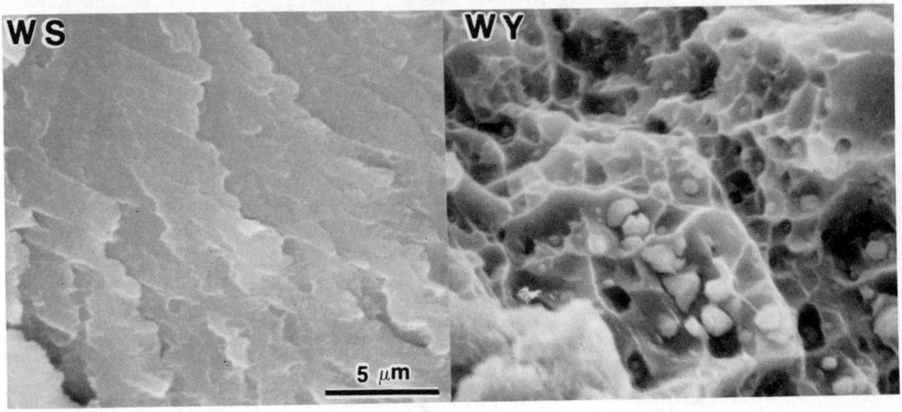

Figure 8 - Fracture surfaces of room temperature tensile specimens of base alloy WS and erbium containing alloy WY.

Creep tests.

The 650 C/138 MPa creep results are listed in Table IV. The creep resistances of the dispersoid containing alloys appear to track with (Al+Sn) content. Figure 9 plots time to 0.5% creep against analyzed (Al+Sn) content, and Figure 10 plots minimum creep rate against analyzed (Al+Sn) content for the alloys containing a rare earth addition. The high tin base alloy WS, and the silicon containing alloys UV (high Al base + Si) and WZ (high tin base + Er + Si) are not included in the comparison of Figures 9 and 10. With reference to the figures, a linear regression line through the data shows a general increase in creep resistance with increasing (Al+Sn) content.

Si additions increase creep resistance in the comparison of alloys WY (Er) with WZ (Er+Si). The alloy WZ has twice the creep life and one half the creep rate.

The overall effect of adding rare earth metals to these exploratory compositions is mixed. The addition of the rare earth metal increased 650 C tensile ductility, so that a 16 atomic % (Al+Sn) alloy with rare earth would have about the same ductility as a 12 atomic % (Al+Sn) alloy without a rare earth metal addition. This would permit the achievement of higher tensile and yield strengths while still maintaining some tensile ductility. On the other hand, creep strength is decreased by the rare earth metal additions for a fixed (Al+Sn) level. To achieve equivalent creep resistance as the 12% (Al+Sn) alloy, the alloy with the rare earth addition needs to have increased Al and Sn. The net result is that titanium base alloys containing a balance of better tensile strengths and equivalent creep strengths can be achieved by adding a rare earth metal plus increasing the amount of (Al+Sn).

Table III. Tensile Tests of Ti Alloys

Alloy	Code	Temp.	.2%Y.S. MPa	.2%Y.S. ksi	U.T.S. MPa	U.T.S. ksi	%El.ml	%El.f	%R.A.
UT	B	650C	437	63.4	518	75.1	5.9	10.2	38.6
UW	B	650C	449	65.1	543	78.8	6.0	15.9	24.4
WE	B	650C	537	77.9	629	91.2	5.9	14.9	29.7
WI	B	650C	613	88.9	700	101.5	3.4	4.1	12.1
XLE	Y	650C	573	83.1	699	101.4	5.5	6.3	8.6
WS	L	RT	900	130.5	974	141.2	2.4	3.1	9.0
WS	L	650C	538	78.0	576	83.5	0.5	0.7	5.9
WY	L	RT	976	141.5	1004	145.6	0.8	1.0	8.3
WY	L	650C	529	76.7	586	85.0	1.5	1.9	9.8
WZ	L	RT	972	141.0	1026	148.6	1.0	1.2	6.6
WZ	L	650C	507	73.6	585	84.8	3.8	4.5	9.9

Thermal Exposure Code:
B- 1200 C/600 C
L- 1100 C/600 C
Y- 1130 C/760/600

Table IV. Creep Test Results
650 C/138 MPa (1202 F/20 ksi)

Alloy	Code	Hours to % Creep 0.1	0.2	0.5	1.0	2.0
UV	B	3.0	9.0	53.5		
UW	B	0.9	2.3	22.8		
WE	B	2.0	7.5	43.9	134	375
WI	B	1.0	6.4	76	310	820
WN	B	3.00	12.3	71.0	231	602
WS	L	1.0	5.0	25.1	97	252
WY	L	0.45	2.2	14.8	60	164
WZ	L	0.95	4.2	34.2	128	366

Thermal Exposure Code:
B- 1200 C/600 C
L- 1100 C/600 C
Y- 1130 C/760 C/600 C

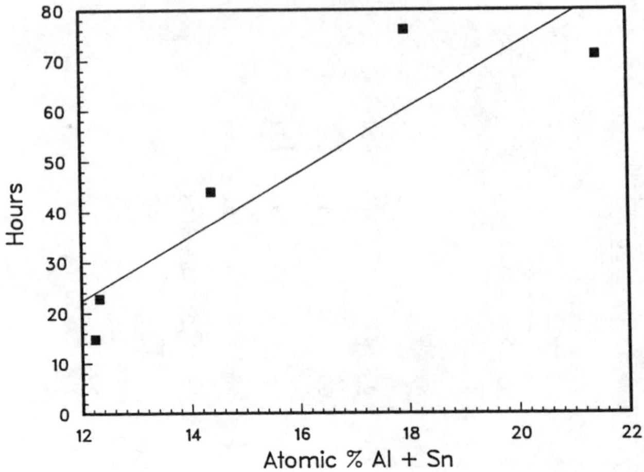

Figure 9 - Time to 0.5% creep at 650 C and 138 MPa versus analyzed aluminum plus tin content.

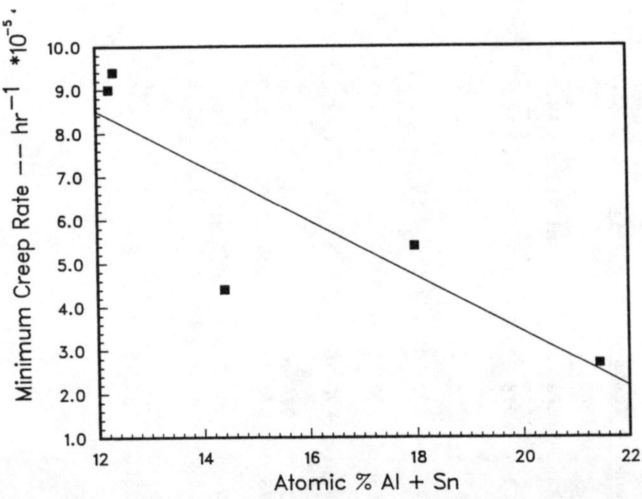

Figure 10 - Minimum creep rate at 650 C and 138 MPa versus analyzed aluminum plus tin content.

Summary

A series of model exploratory titanium alloys containing additions of erbium or terbium were prepared in consolidated form by melt extraction, HIP, and extrusion. Chemistry changes during rapid solidification of Al, rare earth metal, and tin appear due to evaporation from the melt. The rare earth dispersoid is severely coarsened by thermal exposures above the titanium alloy beta transus. Tensile and creep tests indicate a balance of better tensile strengths and equivalent creep strengths might be achieved by adding a rare earth metal modifying the matrix alloy chemistry.

Acknowledgements

Lee Perocchi was responsible for overseeing scheduling of the experiments and participated in most phases of material production, testing and characterization. Dave Burnham prepared the arc melted titanium alloys; Sid Young made the melt extracted filaments; Rick Nagy HIPped the alloys; John Hughes and Dick Laing extruded the alloys. Chris Canestraro and Paul Dupree conducted the mechanical tests. Curt Rodd did the optical metallography; Gary Scarr conducted TEM studies. Their care and diligence is gratefully acknowledged.

References

[1] Milton B. Vordahl, "Dispersoid Titanium and Titanium-Base Alloys", U.S.Patent 3,622,406, November 23, 1971.

[2] S.H. Whang, "Rapidly Solidified Ti Alloys Containing Novel Additives" Journal of Metals, 36 (1984), 34-40.

[3] S.M.L. Sastry, T.C. Peng, and J.E. O'Neal, "Superposition of Solid-Solution-, Precipitation-, Grain-Size-, and Dispersion- Strengthening in Ti-Al-X Alloys", Titanium Science and Technology, G. Lutjering, U. Zwicker, and W. Bunk, eds., Deutsche Gesellschaft fur Metallkunde E.V., 1984, 1811-1818.

[4] S.M.L. Sastry, J.E. O'Neal, R.J. Lederich, and B.B. Rath, "The effect of yttrium and erbium dispersoids on the deformation behavior of titanium" J. Materials Science, 14 (1979), 179-183.

[5] S.M.L. Sastry, T.C. Peng, P.J. Meschter, and J.E. O'Neal, "Dispersion Strengthened Powder-Metallurgy Titanium Alloys", AFWAL TR-83-4092, 1982.

[6] S.M.L. Sastry, P.J. Meschter, and J.E. O'Neal, "Structure and Properties of Rapidly Solidified Dispersion Strengthened Titanium Alloys: Part I. Characterization of Dispersoid Distribution, Structure, and Chemistry", Metallurgical Transactions, 15A (1984), 1451-1463.

[7] S.M.L. Sastry, T.C. Peng, and L.P. Beckerman, "Structure and Properties of Rapidly Solidified Dispersion Strengthened Titanium Alloys: Part II. Tensile and Creep Properties", Metallurgical Transactions, 15A (1984), 1465-1474.

[8] S.M.L. Sastry, T.C. Peng, P.J. Meschter, and J.E. O'Neal, "Rapid Solidification Processing of Titanium Alloys", J. Metals, 35 (9)(1983),21-28.

[9] D.G. Konitzer, B.C. Muddle, H.L. Fraser, and R. Kirchheim, "Refined Dispersions of Rare Earth Oxides in Ti-Alloys Produced by Rapid Solidification Processing", Titanium Science and Technology, G. Lutjering, U. Zwicker, and W. Bunk, eds., Deutsche Gesellschaft fur Metallkunde E.V., 1984, 405-410.

[10] J.A. Sutliff and R.G. Rowe, "Rare Earth Oxide Dispersoid Stability and Microstructural Effects In Rapidly Solidified Ti3Al and Ti3Al-Nb", Proceedings of the Materials Research Society Symposium on Rapid Solidification, B.C. Geissen, D.R. Polk, and A.I. Taub, eds., 1986.

[11] M.F.X. Gigliotti, R.G. Rowe, G.E. Wasiliewski, G.K. Scarr, and J.C. Williams, "Tensile and Creep Properties of Rapidly Solidified Titanium Alloys Containing Complex Matrices and Fine Dispersoids", Proceedings of the Materials Research Society Symposium on Rapid Solidification, B.C. Geissen, D.R. Polk, and A.I. Taub, eds., 1986.

RARE EARTH OXIDE DISPERSIONS

IN RAPIDLY SOLIDIFIED TITANIUM-ALUMINUM ALLOYS°°

David B. Snow and Anthony F. Giamei

United Technologies Research Center
Silver Lane, East Hartford, CT 06108

ABSTRACT

Significant potential exists for the improvement of creep strength by dispersion strengthening in rapidly solidified titanium alloys to which rare earth elements have been added. The relative effectiveness with which several different rare earth elements form oxide dispersions in splat quenched Ti-Al solid solution was examined, using Ti-Al-Er as a base. The introduction of oxygen during alloy preparation was controlled by using components of known composition, and continuously gettered argon during arc melting, splat quenching and annealing. The crystal structure and spatial distribution of the dispersed phases in both as-quenched and aged specimens was determined by analytical transmission electron microscopy. The structure of the intercellular precipitates in Ti-Al-Er was that of cubic Er_2O_3, while both cubic and monoclinic rare earth sesquioxides were observed at intercellular sites in Ti-Al-Gd and Ti-Er-Dy. Most of the precipitates formed in both the Gd-, Tb-, and Er-containing alloys during aging at 800°C for 15 min. were cubic "Type I": lattice parameter less than that of the equlibrium sesquioxide, semi-coherent, with $(11\bar{2})\parallel(1\bar{1}00)$ matrix and $[111]\parallel[0001]$ matrix. These rare earth suboxides (presumably) were faceted on (111) when viewed in the $[11\bar{2}]$ direction, and were rectangular in profile with a maximum dimension of 35 nm. A few "Type II" precipitates were also observed after aging, which had a lattice parameter equal to that of the equilibrium cubic sesquioxide, namely, 1.0547 nm, with $(11\bar{2})\parallel(1\bar{1}00)$ and $[1\bar{1}0]\parallel[0001]$.

°° Research supported in part by the Office of Naval Research, Contract N00014-85-C-0426

Introduction

The continuing effort to develop structural materials which have improved high temperature properties for aircraft gas turbine engine applications includes advanced alloys based on titanium. Currently, there is significant interest in the potential for increasing the yield, tensile and creep strength of titaium alloys by using rapid solidification to achieve refined second phase dispersions. Since the dispersed particles must precipitate readily upon aging and resist subsequent diffusional coarsening, they must also have a low solubility in the surrounding matrix. Oxides of the lanthanide group, and chemically similar metals, fit these criteria well, since they have large free energies of formation and are relatively insoluble in many alloys, including those of titanium. Such oxides have been successfully used for high temperature dispersion strengthening in alloys such as tungsten (1), nickel (2), and mechanically alloyed nickel base alloys (3,4). A key difference here, however, is that the dispersion is created by supersaturation (rapid solidification) followed by precipitation.

Early attempts to increase the high temperature strength of titanium and its alloys with oxide dispersions achieved limited success. Internally oxidized titanium-cerium alloys were reported to display an improvement in stress rupture life at temperatures as high as 650°C (5). Another study reported that the addition of cerium zirconate and thoria to titanium, followed by arc melting, caused the precipitation of oxide particles during solidification, which in turn increased the tensile strength at 538°C (6). A subsequent investigation (7) considered the potential of specific rare earth oxides for dispersion strengthening titanium, and concluded that Lu_2O_3, Nd_2O_3, and Dy_2O_3 should be particularly effective, since they did not appear to be reduced by titanium at 1066°C.

Efficient utilization of rare earth oxides for strengthening titanium alloys at elevated temperatures requires the creation of fine, homogeneous dispersions (8). Although mechanical alloying has been successfully used to produce oxide dispersion strengthened (ODS) nickel base superalloys (3,4), similar attempts to create Y_2O_3 dispersions in titanium base alloys did not significantly improve their mechanical properties (9,10). In the case of conventionally cast alloys, the beneficial effects which might be expected to derive from the relative insolubility of most rare earth elements in titanium (11), together with their strong affinity for oxygen (12), have frequently been negated by inherently slow solidification rates. This allows the rare earth oxides to precipitate as embrittling interdendritic segregates (13) and/or relatively coarse ($\simeq$ 260 nm) dispersions (6,12). It is noteworthy, however, that at low concentrations, their deoxidizing effect can result in ductility improvement (13,14). Even after mechanical processing, the presence of 20-70 nm yttrium- or erbium-containing particles in conventionally cast Ti-6Al-4V was less effective than expected in increasing strength at temperatures above 330°C (15). To avoid some of these difficulties, the consolidation of P/M alloys from blended rare earth oxide and titanium powders has been suggested (16,17), but it is not clear that a sufficiently refined or homogeneous dispersion of fine particles can be achieved by this procedure. Nevertheless, it has been recognized for some time that the affinity of the rare earth elements for oxygen in titanium-oxygen solutions, and the stability of the rare earth oxides thus formed, gave titanium-rare earth alloys significant potential for ODS (18). During the past several years the results of numerous investigations (8,19-29) have suggested that technologically significant dispersion strengthening can best be achieved in these alloys by several rapid solidification techniques.

Although early observations of rapidly solidified titanium-rare earth alloys revealed dispersed particles which contained a rare earth element (19), the identity of these phases remained uncertain. The morphology also required a detailed description. Subsequent research (8,20-22) used analytical transmission electron microscopy to positively identify the dispersed compounds in rapidly solidified Ti-Er and Ti-Al-Y alloys as equilibrium structure rare earth sesquioxides. Er_2O_3

dispersions have also been identified in rapidly solidified Ti-6242 and Ti Al (25-27). The precipitates in laser-processed Ti-6Al-4V-1.5Y, -3Nd or -1.5Er (wt%) were shown to have the equilibrium lattice structure of the corresponding rare earth sesquioxide (26). Recent investigations of other rapidly solidified titanium base alloys have reported the presence of lanthanum-tin intermetallics (28), and of cerium sulfides and oxysulfides (29). Dispersed precipitates of these or similar compounds within grain interiors have resisted coarsening sufficiently to remain 50 nm in size when aged below the beta transus of the titanium matrix (22,25,30). In addition, it has recently been demonstrated that rare earth compound dispersions are effective in increasing the creep strength of alpha titanium alloy matrices (29-32).

The results presented in this report are derived largely from a research program curently in progress, which concerns the characterization of rare earth oxide dispersions created by the rapid solidification and aging of a titanium--aluminum solid solution containing $\simeq 0.5$ at % of a rare earth. At present, the alloys which are being investigated each incorporate a lanthanide element with atomic number >63 (Gd, Tb, Dy, or Er).

Experimental Details

Particle dispersions were formed in rapidly solidified Ti-Al-rare earth alloys by the reaction of the rare earth components with elements, often regarded as impurities, which are interstitially dissolved in the titanium solid solution. The results of most of the investigations cited previously indicate that if tin or sulfur are not major alloy constituents, and the quantities of interstitial impurities nominal for commercially-pure titanium are present, the dispersed rare earth compounds which precipitate during or after rapid solidification will (structurally) be oxides. Consequently, knowledge and control of the alloy interstitial (particularly oxygen) content throughout processing becomes important. This has exemplified by the reported influence of an oxygen excess or deficiency in the Ti matrix on the coarsening rate of erbia (22,25). In the present investigation, attempts were made to monitor and control the interstitial content of the alloys at each stage of processing.

The elements selected for alloy consolidation were commercially-pure Ti (Grade 2), 99,999% Al, and, 99.99% (metallics), sublimed rare earth elements. The target composition in all cases was 11 Al and 0.5-0.6 Gd, Tb, Dy or Er (at%). The alloy components were arc-melted together under continuously-gettered argon in a modified commercial arc melter specifically designed to produce small quantities of alloys with very low levels of impurities (33). As an additional precaution, sacrificial Ti was melted to crete an oxygen sink prior to melting each experimental specimen. The arc was sustained by three sharpened tungsten electrodes, which slant toward the water-cooled copper hearth from the sides. Splat quenching was achieved by the hammer-and-anvil technique, using a flat copper hammer pneumatically driven vertically downward onto the molten alloy at the instant the arcs were shut off. The splat quenched specimens contained areas with smooth opposing surfaces, 75-150 μm in thickness, from which thin foils were prepared for examination by transmission electron microscopy. Splats selected for aging were wrapped in and supported on edge by Ta foil, then heated in a superalloy tube furnace at 800°C for 15 min. Oxidation protection was provided by flowing, continuously-gettered Ar, and by a container of Ti chips placed within the hot zone upstream of the specimens. Aging was terminated by removing the specimens from the hot zone, while keeping them within the protective atmosphere of the furnace.

Results and Discussion

Chemical Analysis

The interstitial content of the Grade 2 commercial purity titanium used for all alloys was determined by gas fusion analysis of 5 gram arc melted buttons. The carbon and nitrogen contents were each < 200 wt ppm in all samples analysed. the oxygen content was 2200 ± 200 wt ppm.

All of the rare earth elements used for the alloys described herein were stored in argon-filled sealed containers. As oxygen is usually the principal impurity in rare earth elements, the oxygen content of the Dy was analysed after several months of storage in argon. Five-gram samples cut from the Dy stock were determined by neutron activation analysis to contain 0.43 ± 0.02 wt% 0 (4.2 ± 0.16 at% 0). Thus the addition of ≃ 0.5 at% rare earth should not significantly increase the oxygen content of any of the alloys.

In order to evaluate the extent of alloy constituents loss during arc melting, the composition of an electropolished thin foil, splat quenched and aged Ti-11Al-0.5Gd (nominal at%), was determined by EDS with an electron microprobe. Quantitative analyses of spectra obtained from three 12 x 12 μm areas yielded values of 10.2 ± 0.3 at% Al and 0.42 ± 0.06 at% Gd, using previously analysed Ti-6Al-4V (wt%) and Gd_2O_3 specimens as standards.

As-Splat-Quenched Microstructures

The as-splat-quenched grain structure of the Ti-Al-Tb, -Dy and -Er alloys consisted predominately of acicular, alpha-prime HCP martensite; with some individual grains of relatively equiaxed alpha phase. It has been suggested that a similar observation of non-acicular alpha in as-splat-quenched Ti-6Al-4V (wt%) is indicative of cooling rates of > 10^5°C/s (34). An estimation of the prior-beta grain size could often be made by observing groups of martensite plates which had similar contrast. The only precipitates observed in these as-quenched alloys were those which formed intercellular networks; there was no evidence for solid-state precipitation at temperatures below the solidus during the quench, even in splats as thick as 175 μm (Fig. 1).

In contrast, as-splat quenched Ti-11Al-0.5Gd exhibited evidence of less rare earth supersaturation, even in specimens whose 75 μm thickness and greater proportion of relatively equiaxed alpha was indicative of a >10^5°C/s cooling rate at the solidus (34). Intracellular precipitation of 10 nm gadolinia, with a "Type I" structure and crystallographic orientation (8), occurred throughout the microstructure in a dispersion very similar to that observed in Ti-Al-Er and Ti-Al-Tb alloys after 800°C aging (subsequently discussed).

In agreement with previous research (20-22), many of the intercellular precipitates in all four alloys were identified by EDS to be rare earth compounds, and by selected area and convergent beam diffraction to have the structure of cubic rare earth sesquioxides. However, non-cubic intercellular precipitates were also observed in the Ti-Al-Gd and Ti-Al-Dy alloys (Fig. 2), and these were identified as the monoclinic, equilibrium high-temperature polymorphs of Gd_2O_3 and Dy_2O_3 (35). A similar observation of intercellular Nd_2O_3, which had retained its high-temperature equilibrium structure (hexagonal, A-type) was made in laser-processed Ti-6Al-4V-3Nd (wt%) (26).

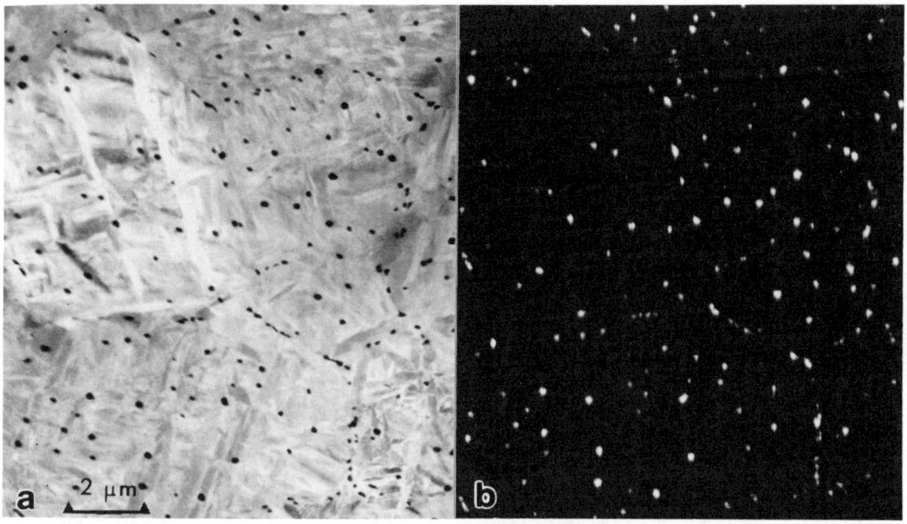

Fig. 1. Dy_2O_3 distribution in as-splat-quenched Ti-11Al-0.6Dy, 175 μm splat thickness. a) Annular dark field STEM. b) Backscattered electron image.

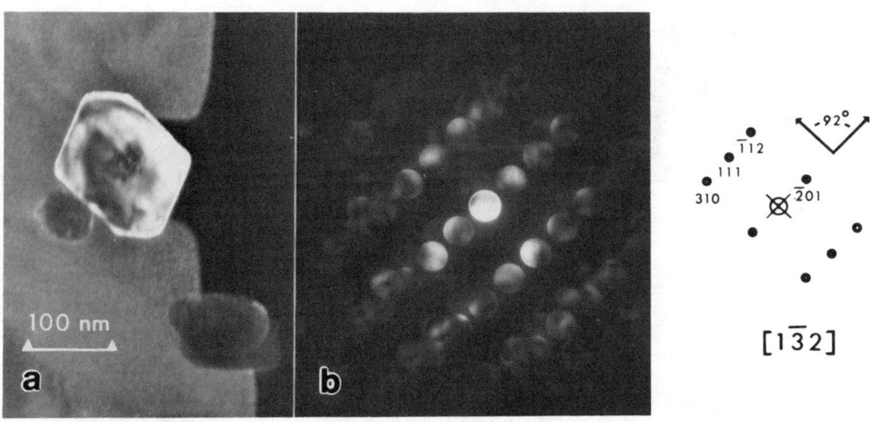

Fig. 2. Intercellular, monoclinic Dy_2O_3 in as-splat-quenched Ti-11Al-0.6Dy a) Dark field CTEM image, g = $\bar{4}02$. b) CBDP, 40 nm spot size, 575 mm camera length.

Microstructures of Specimens Aged at 800°C

Examination of microstructure of specimens splat-quenched and aged at 800°C for 15 min. has thus far been concentrated on the Ti-Al-Gd and Ti-Al-Er alloys. The grain structure and precipitate distribution in an aged specimen of the Ti-Al-Gd alloy (125 μm splat thickness) is shown in Fig. 3. This alloy was less martensitic (60-70%) after splat quenching, and exhibited much more evidence of grain boundary migration and recovery of the martensite laths after heat treatment than the other three. Most of the larger (100-150 nm) precipitates retained an intercellular pattern. Some of these were determined by selected area diffraction to have the equilibrium low-temperature (cubic) structure of Gd_2O_3, a = 1.081 nm (35), while others retained the high-temperature (monoclinic) structure observed in as-splat-quenched specimens. Many precipitates in this size range were surrounded by strain contrast in the matrix, and in some cases had generated punched-out dislocation loops at the particle-matrix interface. This was probably due to the difference in thermal expansion coefficients between gadolinia (36) and alpha titanium. Gadolinia particles of somewhat smaller size were distributed in a pattern indicative of precipitation on prior beta grain boundaries. Extensive precipitate-free zones and/or accelerated precipitate coarsening by grain boundary migration had occurred during aging. Similar, but less extensive, phenomina have been reported in specimens of melt-spun and aged Ti_3Al containing dispersed erbia (37).

A population of much smaller and more homogeneously distributed precipitates had formed in intercellular locations, without apparent correlation with the grain structure. These were predominately plate like in morphology when viewed in the $[1\bar{1}00]$ or $[11\bar{2}0]$ matrix directions, and were usually ≃ 30 nm in maximum dimension. They exhibited strong strain contrast under appropriate two-beam imaging conditions, indicative of semi-coherency (Fig. 4). Selected area diffraction patterns, calibrated from Ti-11at%Al matrix lattice parameters of a = 0.2925 m and c = 0.4675 nm (38), indicated that these small precipitates were cubic, with a lattice parameter of a $\simeq$ 1.04 nm. The crystallographic relationship of these precipitates to the HCP alpha matrix was the same as the "Type I" erbia precipitates observed in laser-processed and aged Ti-0.7at%Er (8,39): $(11\bar{2})$ ppt $[1\bar{1}00]$ matrix; and [111] ppt || [0001] matrix. Following (39), this crystallographic relationship, and a lattice parameter somewhat less than the equilibrium cubic sesquioxide, suggests that they may be suboxides.

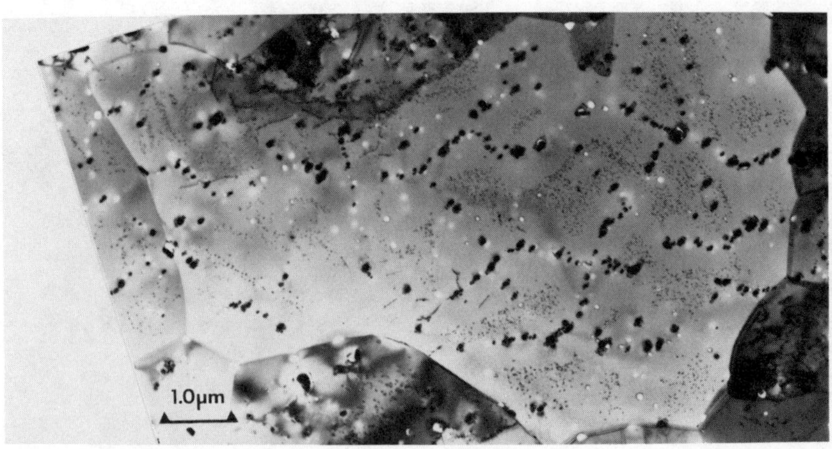

Fig. 3. Gadolinia distribution in Ti-11Al-0.5Gd. 125 μm thick splat quenched specimen, aged at 800°C for 15 min.

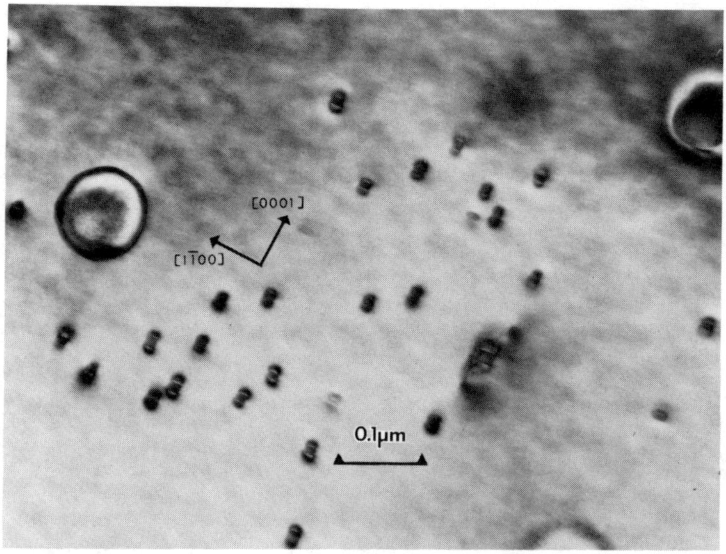

Fig. 4. Semi-Coherent cubic gadolinia precipitates in Ti-11Al-0.5Gd. 125 μm thick splat quenched specimen, aged at 800°C for 15 min.

The microstructure of Ti-11Al-0.5Er, splat-quenched to 125 μm thickness and aged at 800°C for 15 min., was distinctively different from that of Ti-Al-Gd (Fig. 5). The grain structure present after aging was not markedly different from that of the as-quenched specimens, in that it was almost completely martensitic with a few relatively equiaxed alpha grains. Some areas of precipitate-free zones and grain boundary migration could be observed, but were few in number. The formation of intercellular patterns by the larger precipitates was less in evidence of precipitation on martensite plate boundaries and dislocations, and there was a larger population of fine, semi-coherent precipitates than was the case in a larger population of fine, semi-coherent precipitates than was the case in aged Ti-Al-Gd. This suggests a greater degree of Er supersaturation, and consequently a greater potential for dispersion strengthening.

The distribution of erbia precipitates, viewed in a direction near the [11$\bar{2}$] ppt and [1$\bar{1}$00] matrix axes, is shown at higher magnification in Fig. 6. Most of the smaller precipitates have the same "Type I" crystallography and shape described previously in the case of gadolinia in aged Ti-Al-Gd; [11$\bar{2}$] ppt || [1$\bar{1}$00] matrix and (111) ppt || (0001) matrix. The distinctive contrast immediately adjacent to these precipitates reveals the coherency strain associated with the (111) erbia / (0001) Ti-Al matrix interface. Analysis of selected area diffraction patterns of the [0001], [11$\bar{2}$0], [1$\bar{1}$00] and [1$\bar{1}$02] matrix zone axes which contained "Type I" precipitate patterns confirmed that it was cubic with a latice parameter close to 1.015 nm, in agreement with previous observations (8,39). A smaller population of "Type II" erbia precipitates, with the 1.0547 nm lattice parameter of cubic Er_2O_3, was also observed (Fig. 7). These had the crystallographic orientation [11$\bar{2}$] ppt || [1$\bar{1}$00] matrix and (1$\bar{1}$0) ppt || (0001) matrix, again in agreement with the observations of previous repors (8,39).

Preliminary examination of 100 μm splat quenched and aged specimens of Ti-11Al-0.5Tb show a grain structure and precipitate distribution quite similar to that of Ti-11Al-0.5Er. There were however, many more precipitate free zones associated with prior beta grain boundaries in the Tb-containing alloy. Again, the majority of the $\simeq$ 35 nm precipitates formed by aging at 800°C are of the semi-coherent, "Type I" variety, despite the multiplicity of equilibrium oxide stoichiometries in the Tb-O system.

Fig. 5. Erbia distribution in Ti-11Al-0.5Er. 125 μm thick splat quenched specimen, aged at 800°C for 15 min.

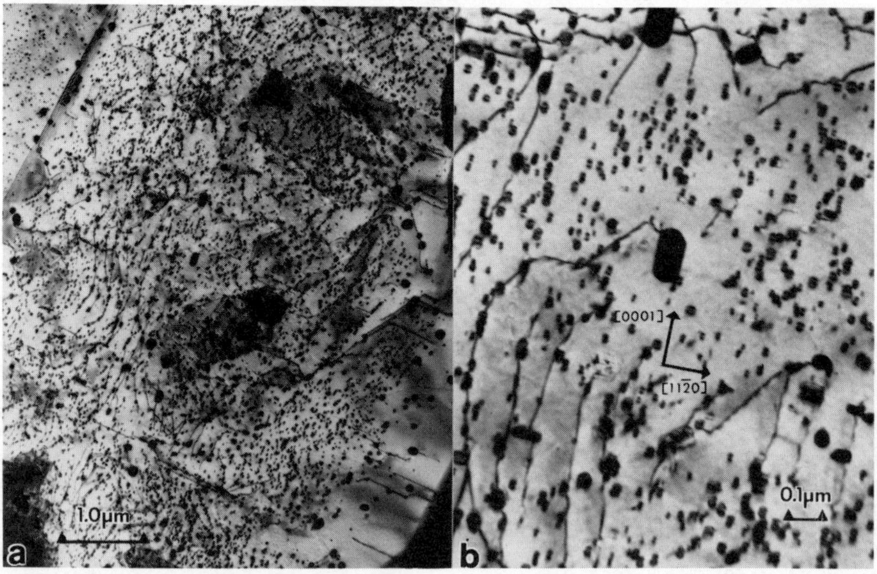

Fig. 6. Erbia distribution precipitated in Ti-11Al-0.5Er. 125 μm thick splat quenched specimen, aged at 800°C for 15 min. Center of a) shown in b) at higher magnification.

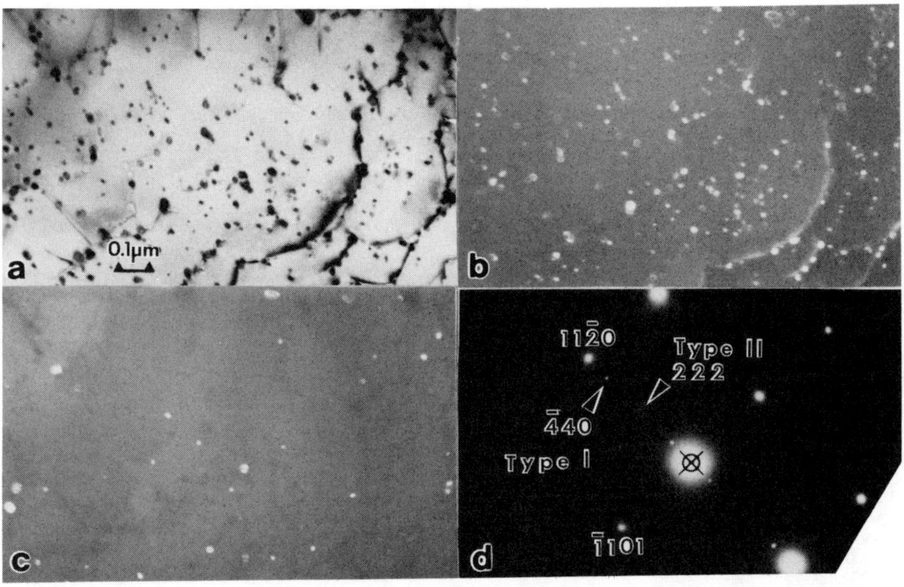

Fig. 7. Erbia distribution in laged Ti-11Al-0.5Er. a) Bright field. b) "Type I" distribution. c) "Type II" distribution. d) Associated [1$\bar{1}$02] matrix SADP with erbia spots used for b) and c) marked.

Conclusions

1. Rapid solidification of titanium alloys by arc melting and splat quenching in high purity environment can produce specimens with controlled and reproducable solidification rates and compositions.

2. At constant cooling rate and alloy composition, less gadolinium is retained in metastable Ti-Al solid solution than terbium or erbium.

3. The intercellular/dendritic precipitate spacing (1-2 μm) and prior-beta grain size (3-6 μm) of all four alloys are consistent with a cooling rate of $\simeq(10)^5$ °C/s at 1650°C.

4. Both the cubic and monoclinic equilibrium structures of Gd_2O_3 and Dy_2O_3 precipitates were observed in as-splat-quenched alloys.

5. ~ 35 nm semi-coherent oxides were precipitated in splat quenched Ti-11Al-0.5Gd, Tb and Er after aging at 800°C for 15 min. In all cases, these precipitates induced strong strain contrast in the adjacent matix.

6. In the alloys which contained gadolinium or erbium, the structure of the majority of the precipitates formed by 800°C aging was consistent with that of cubic gadolinia and erbia with lattice parameters less than those of the equilibrium cubic sesquioxides. The crystallographic relationship of these cubic precipitates to the HCP matrix was [11$\bar{2}$] ppt || [1$\bar{1}$00] matrix and (111) ppt || (0001) matrix.

7. Precipitate-free zones were observed in all specimens aged at 800°C, associated with prior-beta grain boundary sites and with grain boundary migration.

References

1. G. W. King and H. G. Sell, "The Effect of Thoria on the Elevated-Temperature Properties of Recrystallized High-Purity Tungsten," Trans. TMS-AIME, 233 (1965) 1104-1113.

2. B. A. Wilcox and A. H. Clauer, "Creep of Thoriated Nickel above and below 0.5 Tm," Trans. TMS-AIME, 236 (1966) 570-580.

3. J., S. Benjamin, "Dispersion Strengthened Superalloys by Mechanical Alloying," Met. Trans., 1 (1970) 2943-2951.

4. J. S. Benjamin and M. J. Bomford, "Effect of Oxide Volume Fraction and Particle Size on Elevated Temperature Strength of a Dispersion Strengthened Alloy," Met. Trans., 5 (1974) 615-621.

5. R. H. Hiltz, Jr. and N. J. Grant, "The Elevated Temperature Characteristics of Internally Oxidized Titanium-Cerium Alloys," Trans. AIME, 212 (1958) 383-387.

6. C. F. Dixon and H. M. Skelly, "Dispersion Strengthening Titanium with Refrac tory Oxides," Canadian Metall Quart., 11 (1972) 491-495.

7. R. C. Waugh, "Suitable Oxides for Dispersion Strengthening of Titanium Alloys," Int. J. Powder Met. & Powder Tech., 12 (1976) 85-89.

8. D. G. Konitzer et al., "The Production of Ultrafine Dispersions of Rare Earth Oxides in Ti Alloys Using Rapid Solidification," Rapidly Quenched Metals, vol. 1, ed. S. Steeb and H. Warlimont (London: Elsevier Science Publishers B.V., 1985), 953-956.

9. M. J. Buczek, G. S. Hall, S. R. Seagle, and H. B. Bomberger, "Grain Refinement of Titanium Alloys," (Final Technical Report AFML-TR-74-255, RMI Co., November, 1974.

10. I. G. Wright and A. H. Clauser, "Study of Intermetallic Compounds, Task A: Dispersion-Hardened TiAl," (Report AFML-TR-76-107, Battelle Columbus Laboratories, July, 1976).

11. O. Kubaschewski-vonGoldbeck, "Part II: Phase Diagrams," Titanium: Physico-Chemical Properties of its Compounds and Alloys, ed. K. L. Komarek (Vienna: International Atomic Energy Agency, 1983) 75-197.

12. K. A. Gschneidner, Jr., N. A. Kippenhan, and O. D. McMasters, "Thermochemistry of Rare Earths," (Rare Earth Information Center, Institute for Atomic Research, Iowa State University, 1973).

13. R. P. Simpson, "Controlled Weld-Pool Solidification Structure and Resultant Properties with Yttrium Inoculation of Ti-6AL-6-2Sn Welds," Welding J. 56 (1977) 67s-77s.

14. R. E. Shalin, S. G. Glasunov, and A. I. Khorev, "Advanced Trends in Developing High-Strength Titanium Alloys," Titanium '80, Science and Technology, vol. 2, ed. H. Kimura and O. Izumi, (Warrendale, PA: The Metallurgical Society, 1980), 1265-1275.

15. B. B. Rath et al., "The Influence of Erbium and Yttrium Additions on the Microstructure and Mechanical Properties of Titanium Alloys," 1185-1196, ibid.

16. J. A. Miller and G. Brodi, "Consolidation of Blended Elemental Ti-6Al-4V Powder to Near Net Shapes," (AFML-TR-4028, March, 1979).

17. C. A. Kelto et al., "Titanium-Powder Metallurgy - A Perspective," Powder Metallurgy of Titanium Alloys, ed. F. H. Froes and J. E. Smugeresky, (Warrendale, PA: The Metallurgical Society, 1980), 1-19.

18. J. C. Williams, "Precipitation in Titanium-Base Alloys," Precipitation Processes in Solids, ed. K. C. Russell and H. I. Aaronson, Warrendale, PA: The Metallurgical Society, 1978), 191-221.

19. B. B. Rath, R. J. Lederich, and J. E. O'Neal, "Effect of Rare Earth Additions on the Grain Refinement in Titanium" Grain Boundaries in Engineering Materials, ed. J. L. Walter, J. H. Westbrook and D. A. Woodford (Baton Rouge, LA: Claitor's Publ. Div., 1975), 39-42.

20. D. G. Konitzer and H. L. Fraser, "An Example of the Use of Combined Techniques of Analytical Transmission Electron Microscopy for Phase Identification," Microbeam Analysis - 1984, ed. K. F. J. Heinrich (San Francisco, CA: San Francisco Press, 1982), 393-398.

21. D. G. Konitzer, B. C. Muddle, and H. L. Fraser, "A Comparison of the Microstructures of As-Cast and Laser Surface Melted Ti-8Al-4Y," Met. Trans. A, 14A (1983) 1979-1988.

22. D. G. Konitzer, B. C. Muddle, and H. L. Fraser, "Formation and Thermal Stability of an Oxide Dispersion in a Rapidly Solidified Ti-Er Alloy," Scripta Met., 17 (1983) 963-966.

23. S. H. Wang, "Rapidly Solidified Ti Alloys Containing Novel Additives," J. Metals, 36(4) (1984) 34-40.

24. S. M. L. Sastry, P. J. Meschter, and J. E. O'Neal, "Structure and Properties of Rapidly Solidified Dispersion-Strengthened Titanium Alloys: Part I. Characterization of Dispersoid Distribution, Structure and Chemistry," Met. Trans. A, 15A (1984) 1451-1463.

25. D. G. Konitzer and H. L. Fraser, "The Production and Thermal Stability of a Refined Dispersion of Er_2O_3 in Ti_3Al Using Rapid Solidification Processing," High-Temperature Ordered Intermetallic Alloys, ed. C. C. Koch, C. T. Liu, and N. S. Stoloff (Pittsburgh, PA: Materials Research Society, 1985), 437-442.

26. D. B. Snow, "Structure and Mechanical Properties of Laser-Consolidated Ti-6Al-4V and Ti-6Al-2Sn-4Zr-6Mo with Rare Earth Element Additions," Laser Processing of Material, ed. K. Mukherjee and J. Mazumder (Warrendale, PA: The Metalurgical Society, 1985), 83-98.

27. R. G. Rowe et al., "Microstructural Study of Rapidly Solidified Titanium Containing Erbium and Boron," Rapidly Solidified Materials, ed. P. W. Lee and R. S. Carbonara (Metals Park, OH: American Society for Metals, 1985), 107-114.

28. C. S. Chi and S. H. Wang, "Microstructures and Mechanical Property of Rapidly-Solidified Ti5Al-4.5La and Ti-5Al-5.4Er Alloys," Mechanical Behavior of Rapidly Solidified Materials, ed. S. M. L. Sastry and B. A. MacDonald (Warrendale, PA: The Metallurgical Society, 1986), 231-245.

29. R. G. Rowe and E. F. Koch, "Rapidly Solidified Titanium Alloys Containing Cerium Sulfide and Oxysulfide Dispersions," Rapidly Solidified Materials, ed. P. W. Lee and R. S. Carbonara (Metals Park, OH: American Society for Metals, 1985), 115-127.

30. S. M. L. Sastry, T. C. Peng, and L. P. Beckerman, "Structure and Properties of Rapidly Solidified Dispersion Strengthened Titanium Alloys: Part II. Tensile and Creep Properties," Met. Trans. A, 15A (1984) 1465-1474.

31. S. F. X. Gigliotti et al., "Tensile and Creep Properties of Rapidly Solidified Titanium Alloys Containing Complex Matrices and Fine Dispersions," Rapidly Solidified Alloys and Their Mechanical and Magnetic Properties, ed. B. C. Giessen, D. R. Polk and A. I. Taub (Boston, MA: Materials Research Society, 1986), to be published.

32. M. F. X. Gigliotti, "Mechanical Evaluation of Rapidly Solidified Titanium Alloys Containing Fine Dispersions," this publication.

33. T. B. Reed and E. R. Pollard, "Tri-Arc Furnace for Czochralski Growth with a cold Crucible," J. Crystal Growth, 2 (1968) 243-247.

34. T. F. Broderick et al., "The Effect of Cooling Conditions on the Microstructure of Rapidly Solidified Ti-6Al-4V," Met. Trans. A, 16A (1985) 1951-1959.

35. F. A. Shunk, Constitution of Binary Alloys, Second Supplement (New York, NY: McGraw-Hill Book Co., 1969), 305-306, 379.

36. K. A. Gscheidner, Jr., Rare Earth Alloys (New York, NY: D. Van Nostrand Co., 1961) 245.

37. J. A. Sutliff and R. G. Rowe, "Rare Earth Oxide Dispersiod Stability and Microstructural Effects in Rapidly Solidified Ti_3Al and Ti_3Al-Nb," Rapidly Solidified Alloys and Their Mechanical and Magnetic Properties, ed. B. C. Giessen, D. R. Polk and A. I. Taub (Boston, MA: Materials Research Society, 1986), to be published.

38. P. Rogl, "Part III: Crystal Structure and Density Data," Titanium: Physico-Chemical Properties of its Compounds and Alloys, ed. K. L. Komarek (Vienna: International Atomic Energy Agency, 1983) 201-369.

39. J. T. Stanley et al., "A Comparison Between the Microstructures of Laser Surface Melted Ti Alloys Containing Either Er or La," this publication.

LASER SURFACE MELTING OF Ti ALLOYS CONTAINING Er OR La

S.A. Court, J.T. Stanley, D.G. Konitzer*, M.H. Loretto** and H.L. Fraser

Department of Materials Science, University of Illinois, Urbana, IL 61801
*Alcoa Technical Center, Alcoa Center, PA 15069
**Department of Metallurgy and Materials, University of Birmingham
Birmingham B15 2TT, United Kingdom

Abstract

The microstructures of rapidly solidified (RS) Ti alloys containing additions of Er or La have been fully characterized and compared. RS was effected by a laser surface melting (LSM) technique. RS Ti-Er (0.7at%) samples consist of a supersaturated solid solution based on α'-Ti. Subsequent aging at 700°C produces a refined dispersion (8-20 nm diameter) of Er_2O_3 precipitates. The oxides are of two types, exhibiting different orientation relationships and slightly different lattice parameters. These differences are explained on the basis of the mechanisms of precipitation of the particles. RS Ti and Ti-Al (8.9at%) alloys containing additions of La (0.9 at%) do not form a supersaturated single phase, but exhibit a moderately refined dispersion of La particles (~ 30-200 nm diameter). On subsequent aging the structure changes from La to La_2O_3, presumably by internal oxidation. The mechanism of formation of microstructure in the La-containing alloys is thought to involve a monotectic reaction.

Introduction

Recent work has shown that fine dispersions of rare earth oxides can be introduced into Ti-based alloys by rapid solidification processing (RSP) (1-3), and that significant improvements in mechanical properties can then be achieved (4). The nature of the dispersoid is of interest in terms of a detailed understanding both of the elevated temperature stability and of the strengthening effects, and the present work has been aimed at assessing the chemical composition and crystal structure of the dispersoids together with the orientation relationships which exist between the matrix and second phase.

Experimental Procedure

Small ingots of Ti-0.7at%Er, Ti-0.9at%La and Ti-8.9at%Al-0.9at%La were prepared by argon arc melting and portions of these ingots were laser surface melted using the techniques described elsewhere (5). Subsequent heat treatments were carried out in sealed quartz capsules under an argon atmosphere. Thin foil samples and extraction replicas were prepared for examination by analytical electron microscopy using standard techniques.

Results

Ti-Er Alloy

A typical microstructure observed in the LSM samples is shown in Fig. 1. It is clear that only a low density of precipitates is present and the microstructure consists essentially of single phase α'-Ti supersaturated in Er, but as can be seen from Fig. 2 a very much higher density of precipitates is observed after heat treating for 20 hrs at 700°C. It is apparent from micrographs taken with g = 0002 and g = $11\bar{2}0$ (Figs. 3a and b) that there are two different types of precipitate; firstly, those that show strong (strain) contrast with g = 0002 and only weak contrast with g = $11\bar{2}0$ (referred to as Type I) and secondly, those that show the opposite behavior in that strong contrast is observed with g = $11\bar{2}0$ (referred to as Type II). These precipitates may be tentatively identified as oxides since precipitate reflections in selected area diffraction (SAD) patterns can be fairly consistently indexed if the precipitates are assumed to be cubic Er_2O_3

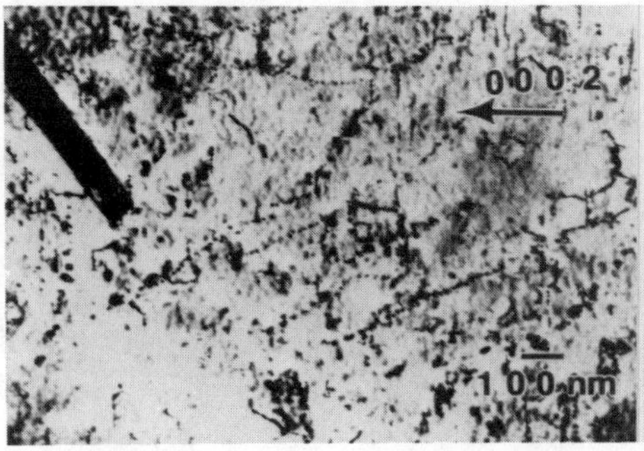

Figure 1 - Transmission electron micrograph of LSM Ti-0.7E (at%).

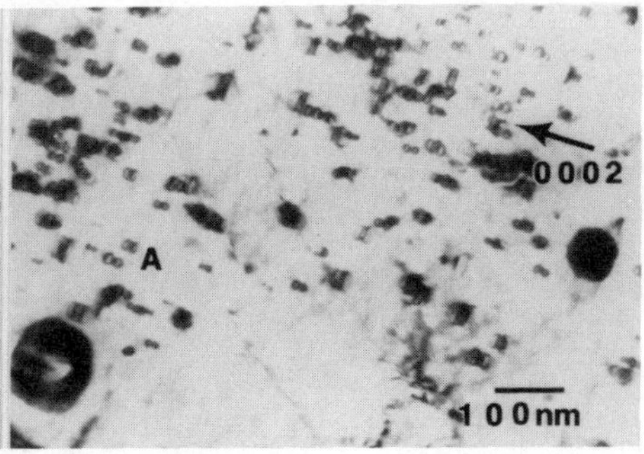

Figure 2 - Transmission electron micrograph taken from a sample of Ti-0.7Er (at%) heat treated for 20 hrs at 700°C after laser treating.

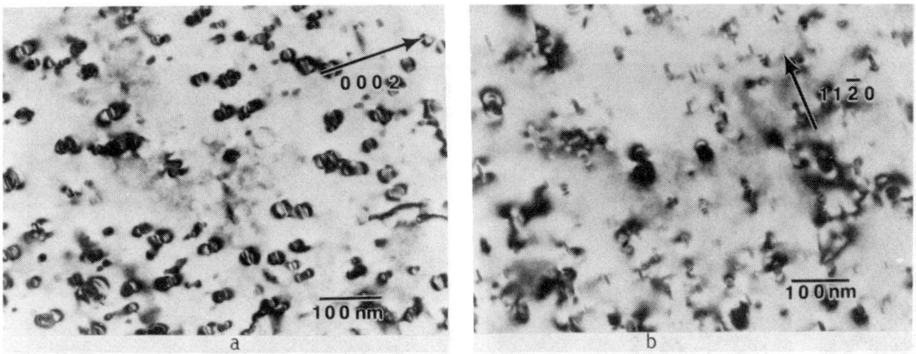

Figure 3 - Transmission electron micrographs taken from a sample of Ti-0.7Er (at%) heat treated for 20 hrs at 700°C after laser treating. (a) imaged with $\underline{g}$ = 0002 in Ti; (b) imaged with $\underline{g}$ = $11\bar{2}0$ in Ti. $\underline{B}$ near $[10\bar{1}0]$.

(erbia). Furthermore, similar electron energy loss (EEL) spectra, containing an O-absorption edge, are obtained from both types of particle, an example being given in Fig. 4. However, it appears from detailed SAD analysis that

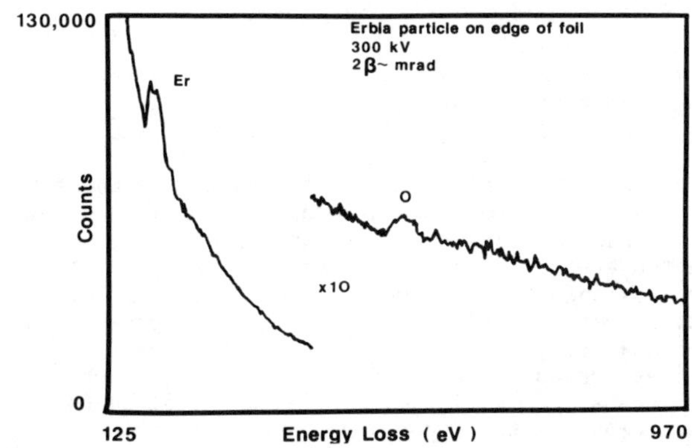

Figure 4 - Electron energy loss spectrum taken from a particle situated on the edge of a perforated thin foil.

there are two different orientation relationships between the precipitates and the Ti-matrix and that the two different families of precipitates, corresponding to the two types, I and II, have slightly different lattice parameters. Microdiffraction patterns taken from individual particles, together with computer-generated patterns, enable a clear distinction to be drawn between the lattice parameters of the two habits, and these results are described in turn.

Microdiffraction from Type I Precipitates

Microdiffraction patterns taken from individual Type I precipitates are shown in Fig. 5 with $\underline{B}$ = [0001] and $[10\bar{1}0]$ for Ti. From these and other patterns corresponding to different electron beam directions it can be shown that the precipitate is indeed cubic with an orientation relation with the matrix which can be expressed as:

$$(111) \parallel (0001)_{Ti}$$
$$\langle 110\rangle \parallel \langle 11\bar{2}0\rangle_{Ti}$$

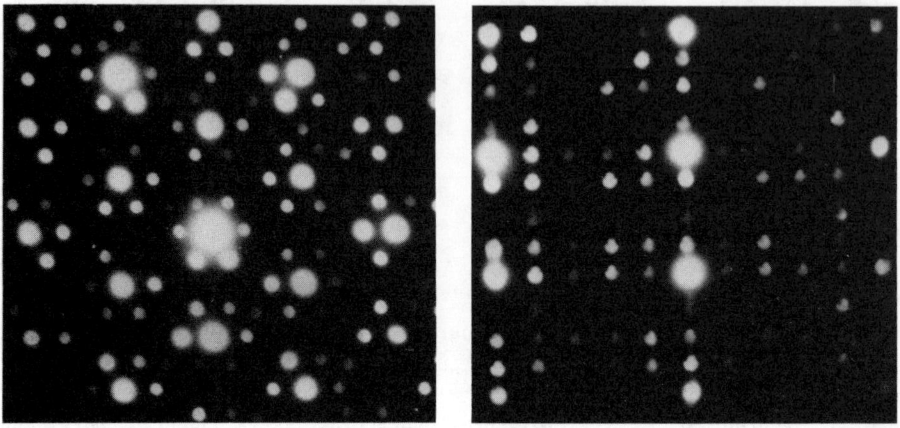

a b

Figure 5 - Microdiffraction patterns taken from an individual Type I Er_2O_3 particle found in a sample of Ti-0.7Er (at%), heat treated for 20 hrs at 700°C after being laser treated. (a) $\underline{B}$ = [0001]; (b) $\underline{B}$ = [10$\bar{1}$0].

It is not possible to obtain an accurate measure of the lattice parameter of these precipitates by direct measurement of the sizes of diffracting vectors from these diffraction patterns although measurements made on these patterns and on selected area diffraction patterns suggest that the lattice parameter is slightly less than that of Er_2O_3. There is, however, a significant amount of information which suggests that despite the apparent lack of agreement with the lattice parameter, these precipitates are Er_2O_3. Thus, the very strong 440 reflections are characteristic of rare earth oxides which have the same structure as Er_2O_3. In addition, earlier work (6) on the coarsening behavior of these particles and the dependence of coarsening rate on oxygen concentration is consistent with the view that the precipitates are Er_2O_3 and the indication that the lattice parameter is smaller than the known value for Er_2O_3 is surprising.

A measure of the lattice parameter of the precipitates can be obtained by using computed diffraction patterns to simulate the experimental observations. Thus, Fig. 6(a) shows a portion of the pattern shown in Fig. 5(a) together with the corresponding computer simulated pattern for Ti and Er_2O_3 with the lattice parameter for Er_2O_3 taken as 10.15 Å and 10.547 Å, respectively, in Figs. 6(b) and (c). It is apparent that the groups of four reflections indicated make up a rectangle in Fig. 6(b), similar to those in the experimental pattern, but approximate to a square in Fig. 6(c). A similar comparison between experimental and simulated patterns has been made for the case when the incident beam direction is parallel to [10$\bar{1}$0]. Thus, a portion of Fig. 5(b) and simulated patterns, with Er_2O_3 lattice parameters given again by 10.15 Å and 10.547 Å, are shown in Fig. 7(a,b,c). The measured ratios of the distances AC to AD in Fig. 7(b) are 0.81 (experimental), 0.80 (a_o = 10. 15 Å)and 0.76 (a_o = 10.547 Å). There is excellent agreement between the experimental pattern and that simulated with a_o = 10.15 Å. It may therefore be concluded that the lattice parameter of the Type I precipitates is given approximately by 10.15 Å, rather than the value expected for Er_2O_3.

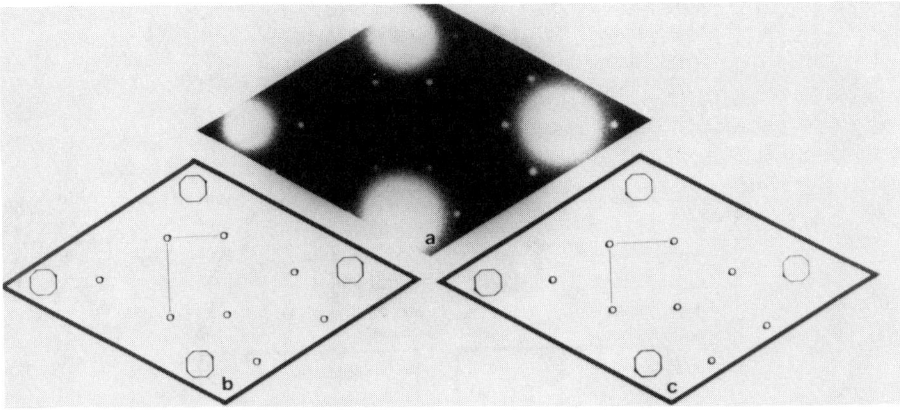

Figure 6 - Comparison of (a) experimental and (b) and (c) computed diffraction patterns. The experimental pattern in (a) is a portion of the microdiffraction pattern from a Type I precipitate shown in Fig. 5(a), and the computed patterns correspond to patterns predicted for $[0001]_{Ti}$ and $[111]_{Er_2O_3}$ using a lattice parameter for Er_2O_3 of 10.15 Å and 10.547 Å for (b) and (c), respectively. The various symbols on the computed patterns have the following meanings: ◯ Ti reflections; • reflection from double diffraction between matrix and precipitate.

Microdiffraction from Type II Precipitates

Microdiffraction patterns taken from individual Type II precipitates are shown in Fig. 8 with $\underline{B}$ = [0001] and $[1\bar{2}10]$ for Ti. From the patterns shown in Fig. 8 it can be shown that the precipitate is cubic with an orientation relationship with the matrix which can be expressed as following:

$$(\bar{1}10) \parallel (0001)_{Ti}$$
$$\langle 111\rangle \parallel \langle 1\bar{2}10\rangle_{Ti}$$

In contrast to the observations made on the Type I precipitates, measurements made on patterns such as that shown in Fig. 8 are entirely consistent with the precipitates being Er_2O_3 with a lattice parameter of 10.547 Å. This is again best seen by comparing the experimental and computed diffraction patterns. Thus Fig. 9(a) shows a part of a diffraction pattern obtained from a Type II precipitate with $\underline{B} = [1\bar{2}10]_{Ti}$ and Fig. 9(b) and (c) show computed patterns assuming a_o = 10.15 Å and a_o = 10.547 Å for Er_2O_3, respectively. The line shown parallel to that which joins two of the diffraction maxima in Fig. 9(a) clearly slopes to the left as does the corresponding line on Fig. 9(c) which is computed for a lattice parameter of 10.547 Å. Additional confirmation concerning the lattice parameter of the Type II oxides arises from comparison of experimental and computed patterns for the case where the electron beam is parallel to $\langle 0001\rangle_{Ti}$. Fig. 10(a) shows a part of the experimental pattern, whereas the computed patterns with the oxide lattice parameter given by 10.15 Å and 10.547 Å are given in Fig. 10(b) and (c), respectively. As can be seen, in the case of the experimental pattern and that computed with a_o = 10.547 Å, (Fig. 10(c)), the double diffracted reflections marked A and B are positioned significantly and equivalently away from a line (dashed in Fig. 10(c)) along the $01\bar{1}0$ systematic row. However, the same reflections in the computed pattern corresponding to an oxide lattice parameter of 10.15 Å are positioned almost on that line (Fig. 10(b)). Therefore, it may be concluded that the lattice parameter of Type II oxides is 10.547 Å.

The observation that there are apparently two forms of Er_2O_3 which differ in lattice parameter and which form with different orientation relationships

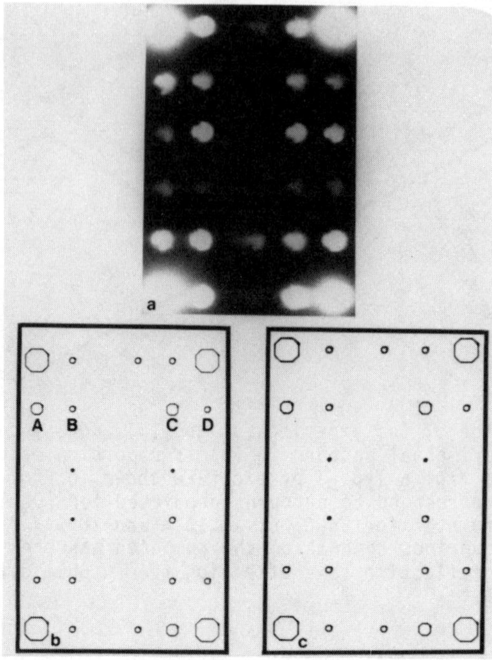

Figure 7 - Comparison of (a) experimental and (b) computed diffraction patterns. The experimental pattern in (a) is a portion of the micro-diffraction pattern from a Type I precipitate shown in Fig. 5(b), and the computed patterns correspond to patterns predicted for $[10\bar{1}0]_{Ti}$ and $[11\bar{2}]_{Er_2O_3}$ using a lattice parameter for Er_2O_3 of 10.15 Å and 10.547 Å for (b) and (c), respectively. The various symbols on the computed patterns have the following meanings: ◯ Ti reflection; ○ strong precipitation reflection; ◦ weak precipitate reflection or strong reflection from double diffraction between matrix and precipitate; • reflection from weak double diffraction between matrix and precipitate.

with Ti is unexpected and an attempt is now made to rationalize these observations.

Consider first the two orientation relationships; the expected orientation relation between Er_2O_3 and Ti would be that observed for the Type II precipitates with the closest packed planes, $\{110\}_{Er_2O_3}$ being parallel to $(0001)_{Ti}$ and the $<111>_{Er_2O_3}$ being parallel to $<11\bar{2}0>_{Ti}$. It is possible to explain the orientation relationship for Type I precipitates on the basis that these precipitates initially form as Er and are oxidized subsequently by diffusion of oxygen from the Ti matrix. This is consistent with the observation that Type II precipitates are far more common in oxygen-excess samples (5) after high temperature annealing where internal oxidation of any Type I precipitates could go to completion during the particle coarsening that occurs. This observation also suggests that the oxygen-deficient Type I precipitates are less stable against coarsening during high temperature annealing than the stoichiometric Type II precipitates. Since Er is hexagonal it would be expected to precipitate with (0001) Er, parallel with $(0001)_{Ti}$ and $<11\bar{2}0>_{Er}$ parallel to $<11\bar{2}0>_{Ti}$. Subsequent oxidation of the Er would be expected to transform $(0001)_{Er}$ to $\{111\}_{Er_2O_3}$ so that the observed orientation

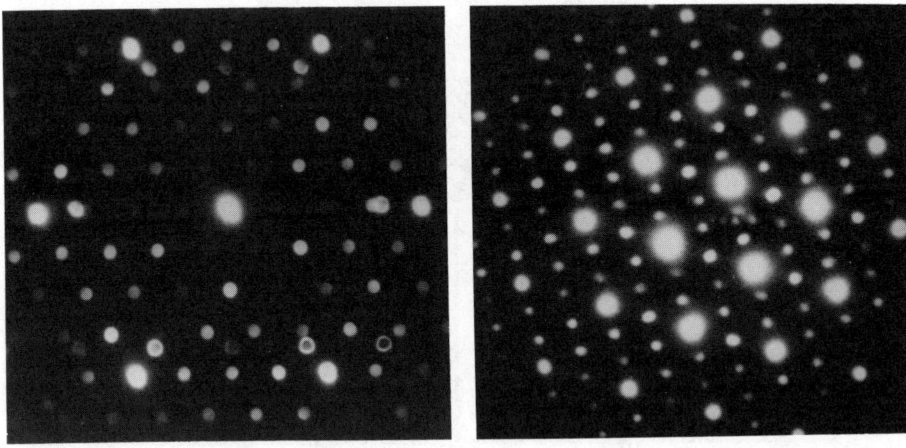

a b

Figure 8 - Microdiffraction patterns taken from an individual Type II Er_2O_3 particle found in a sample of Ti-0.7Er (at%), heat treated for 20 hrs at 700°C after being laser treated. (a) Beam direction, $\underline{B} = [11\bar{2}0]_{Ti}$; (b) $\underline{B} = [0001]_{Ti}$.

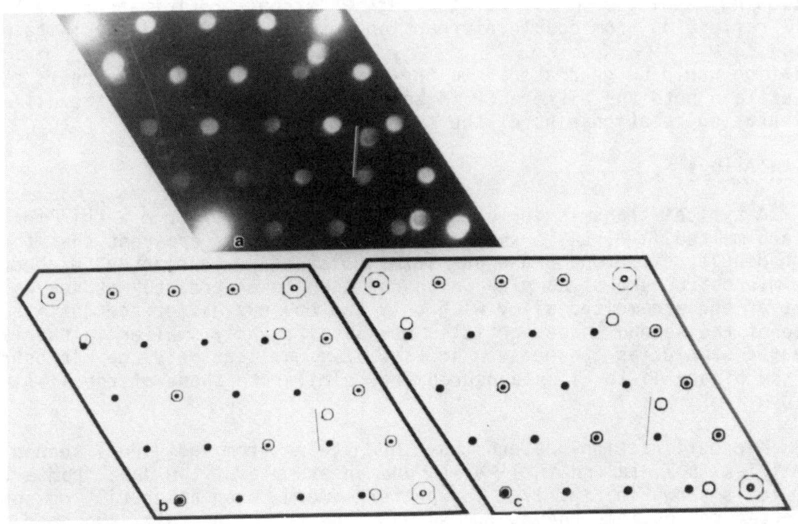

Figure 9 - Comparison of (a) experimental and (b) and (c) computed diffraction patterns. The experimental pattern in (a) is a portion of the micro-diffraction pattern from a Type II precipitate shown in Fig. 8(a), and the computed patterns correspond to patterns predicted for $[1\bar{2}10]_{Ti}$ and $[111]_{Er_2O_3}$ using a lattice parameter for Er_2O_3 of 10.15 Å and 10.547 Å for (b) and (c), respectively. The various symbols on the computed patterns have the following meanings: ◯ <4$\bar{4}$0> reflection in Er_2O_3; ○ Ti matrix and strong precipitate reflections; • reflection from double diffraction between precipitate reflections.

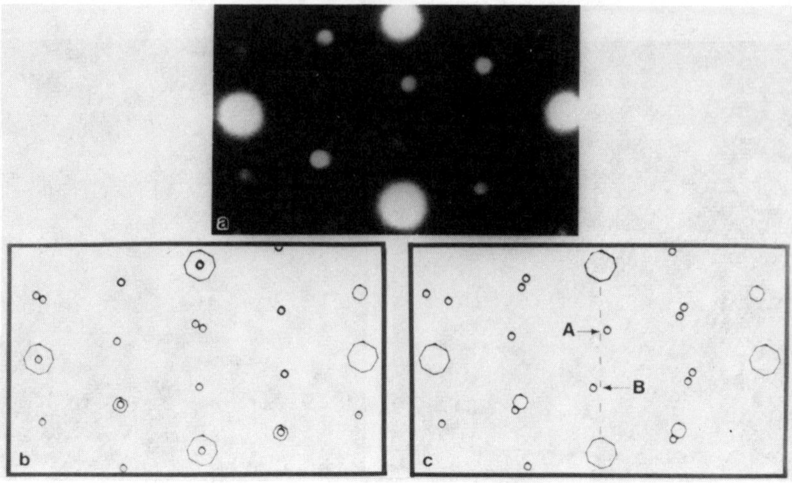

Figure 10 - Comparison of (a) experimental and (b) and (c) computed diffraction patterns. The experimental pattern in (a) is a portion of the microdiffraction pattern from Type II precipitate shown in Fig. 8(b), and the computed patterns correspond to patterns predicted for $[0001]_{Ti}$ and $[\bar{1}10]_{Er_2O_3}$ using a lattice parameter for Er_2O_3 of 10.15 Å and 10.547 Å for (b) and (c), respectively. The various symbols on the computed patterns have the following meanings: ◯ Ti reflection; ○ strong precipitate reflections; ∘ reflection from double diffraction between matrix and precipitate.

relation would be generated. On the basis of the above, it appears possible to explain both the difference in the lattice parameters and the different orientation relationships of the two types of precipitates.

Ti-La Alloys

A typical transmission electron micrograph taken from a thin foil sample of arc-melted Ti-Al-La is shown in Fig. 11 and it is apparent that there is a high density of second phase particles which range in size up to about 1 μm. The microstructure of samples which had been laser treated was very similar to that of the arc-melted alloy with only the obvious difference that the average size of the second phase particles was significantly smaller in the laser treated sample, as can be seen in Fig. 12. Qualitatively the microstructures of the binary Ti-La alloy appeared very similar to those of the Ti-Al-La alloy.

Microdiffraction patterns have been taken from individual second phase particles in Ti-La and in Ti-Al-La and an example of the data from a Ti-Al-La sample is shown in Fig. 13. The patterns were taken after tilting about selected reflections through measured tilt angles and this, together with the observed symmetries shows that the particles are fcc and have a lattice parameter of about 5.6 Å. Identical results were obtained from the Ti-La alloy. On this basis alone, it appears that the precipitates are effectively pure La, and from SAD patterns it can be shown that the precipitates possess an orientation relationship with the Ti-matrix which can be expressed as:

$$\{111\} \parallel (0001)_{Ti}$$
$$\langle 110\rangle \parallel \langle 1\bar{2}10\rangle_{Ti}$$

The presence of a second phase is a rather surprising result in view of previous results involving Er and Y, where a supersaturation of rare earth

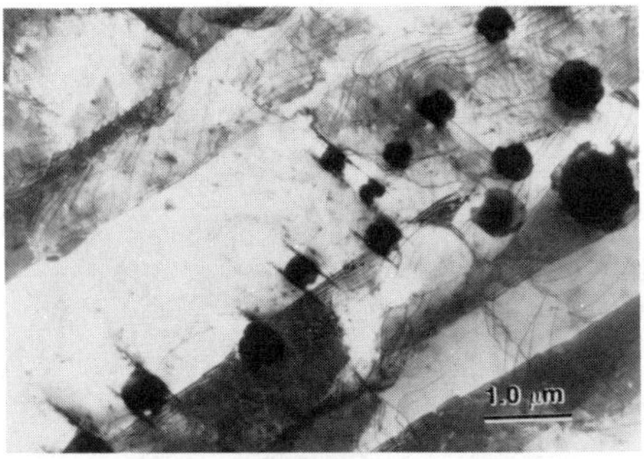

Figure 11 - Bright field micrograph of arc-melted Ti-8.9Al-0.9La (at%).

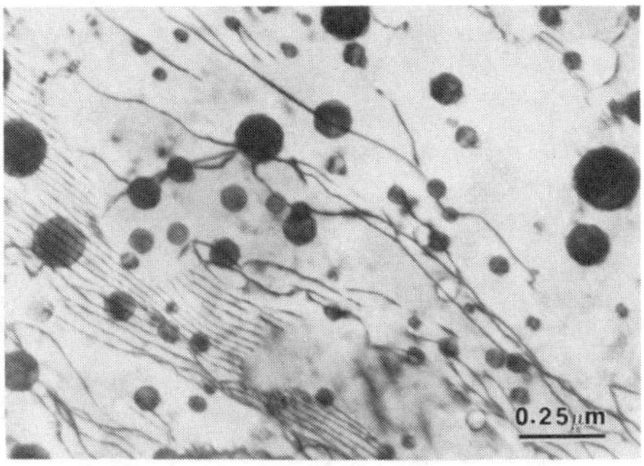

Figure 12 - Bright field micrograph from a laser treated Ti-8.9Al-0.9La (at%) sample.

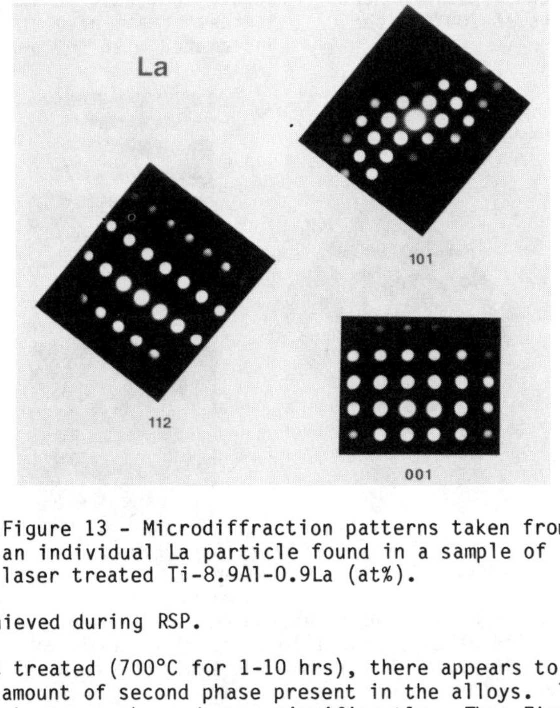

Figure 13 - Microdiffraction patterns taken from an individual La particle found in a sample of laser treated Ti-8.9Al-0.9La (at%).

element is achieved during RSP.

When heat treated (700°C for 1-10 hrs), there appears to be little or no change in the amount of second phase present in the alloys. The microstructure has, however, changed very significantly. Thus Fig. 14 shows that the particles now show strong strain contrast. Individual particles have been imaged with several g-vectors and in each case strain contrast is observed, indicating that the strain field is radial in nature, i.e. an hydrostatic state of stress. Dark field images of precipitates near the foil surface show that the vector joining white to black contrast features is in the direction of +g (Fig. 15), indicating that the strain fields are interstitial in nature.

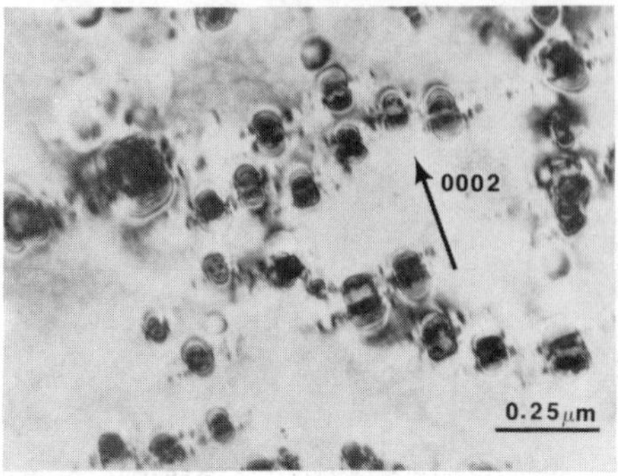

Figure 14 - Transmission electron micrograph taken from a Ti-8.9Al-0.9La electron (at%) specimen heat treated for 10 hrs at 700°C after laser treating.

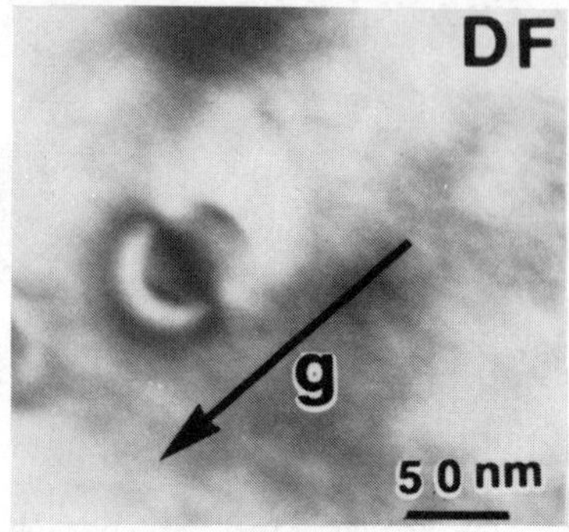

Figure 15 - Dark field transmission electron micrograph showing anomalous strain contrast, indicating the "interstitial" nature of the strain associated with the particles.

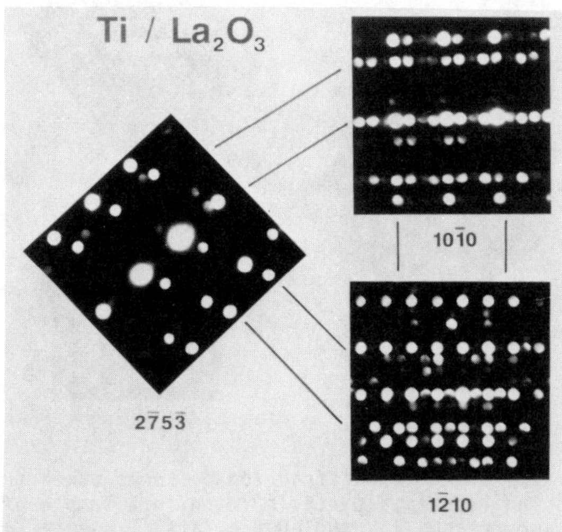

Figure 16 - Microdiffraction patterns taken from an individual La_2O_3 particle found in a sample of Ti-8.9Al-0.9La (at%) heat treated for 1 hr at 700°C after being laser treated.

Microdiffraction analysis (Fig. 16) shows that the particles are no longer fcc but are hcp and are consistent with the particles now being La_2O_3. Again an orientation relationship exists between the precipitates and matrix and this can be expressed as:

$$(0001) \parallel (0001)_{Ti}$$
$$\langle 11\bar{2}0\rangle \parallel \langle 11\bar{2}0\rangle_{Ti}$$

Further confirmatory evidence for the particles being La_2O_3 is obtained by the observation of the O-edge in the EELS spectra recorded with the electron probe incident on a particle extracted from the Ti-Al-La alloy in the form of a replica (Fig. 17).

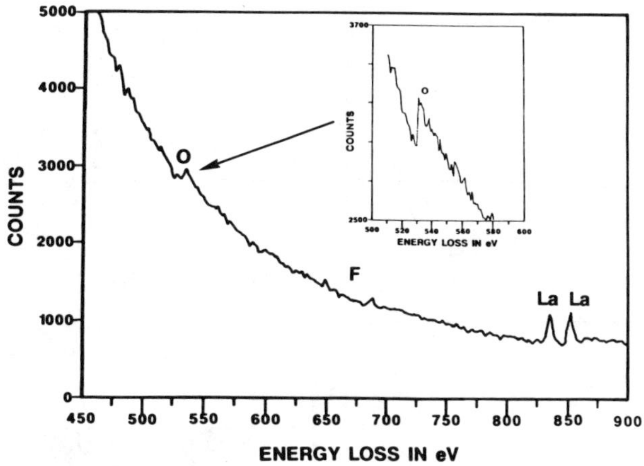

Figure 17 - EELS spectrum taken from a particle extracted from a sample of Ti-8.9Al-0.9La (at%) heat treated for 1 hr at 700°C after being laser treated. Accelerating voltage 300 kV.

Discussion

It is clear that significant differences exist between the behavior exhibited by the La-containing alloys and those involving either Y or Er. The main differences appear to be the inability of rapid solidification to effect a significant supersaturation of solute in the case of the La-containing alloys. In addition, in both arc-melted and rapidly solidified conditions, the La-containing alloys exhibited particles of elemental La, whereas in the Er-containing alloy, the particles observed appear to be invariably oxides. The difference in behavior of these two alloys may be explained on the basis of their respective phase diagrams. Thus, in the cases of Er, the diagram consists of a simple binary eutectic, Fig. 18(a). In this case, it is necessary only to extract heat from the sample at such a rate that solidification occurs at a temperature below that corresponding to T_o (defined for a given composition by the temperature where the free energies of solid and liquid are equal). In the case of the Ti-La system, the phase diagram contains a liquid miscibility gap, as shown in Fig. 18(b) and the liquid phase separation may be seen in arc-melted samples. Consequently, the material being rapidly solidified has a slightly hypo-monotectic composition and during rapid solidification undercooling occurs such that liquid phase separation occurs again, as depicted schematically on the phase diagram (Fig. 18(b)). In this way, both arc-melted and rapidly solidified samples contain a distribution of La particles. In the case of rapidly solidified samples, it is expected that La particles from the pro-monotectic reaction would be present, in addition to those produced during the monotectic reaction itself (7), as appears consistent with the micrographs (e.g. see Fig. 13). Upon heat treatment, the La particles, constrained by the Ti matrix, are internally oxidized producing an increase in volume, and the elastic strains, revealed by electron microscopy.

Acknowledgements

This research was supported in part by the US DoE, Division of Materials Sciences, under contract DE AC02 76ER01198. The electron microscopy was

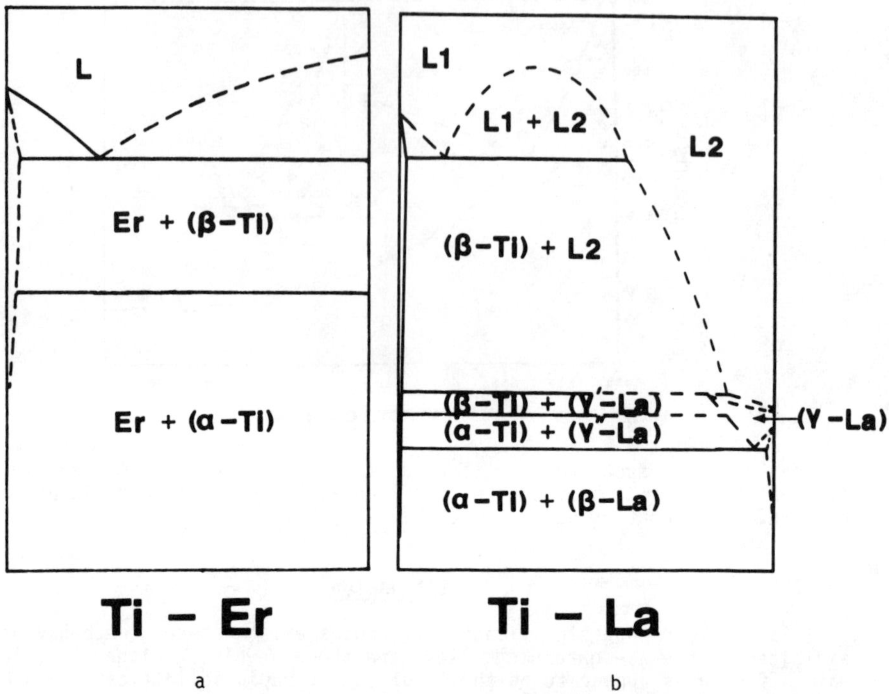

Figure 18 - Phase equilibrium diagrams for the systems (a) Ti-Er and (b) Ti-La.

performed using the facilities of the Center for the Microanalysis of Materials, operated by the Materials Research Laboratory of the University of Illinois.

References

1. K.K. Sankaran, S.M.L. Sastry and P.S. Pao, Metall. Trans. A 11A (1980) 196-198.

2. D.G. Konitzer, B.C. Muddle and H.L. Fraser, Metall. Trans. A 14A (1983) 1979-1988.

3. D.G. Konitzer, B.C. Muddle and H.L. Fraser, Scripta Met. 17 (1978) 963-966.

4. S.M.L. Sastry, T.C. Peng and L.P. Beckerman, Metall. Trans. A 15A (1984) 1465-1474.

5. D.G. Konitzer, J.T. Stanley, M.H. Loretto and H.L. Fraser, accepted for publication in Acta Met.

6. D.G Konitzer, J.P.A. Lofvander, S.A. Court, R. Kirchheim and H.L. Fraser, submitted to Acta Met.

7. D.C. Van Aken and H.L. Fraser, Acta Met. 33 (1985) 963-974.

EFFECT OF Er, Si, AND W ADDITIONS ON POWDER METALLURGY

HIGH TEMPERATURE TITANIUM ALLOYS

R. G. Vogt[a], D. Eylon[b], and F. H. Froes[a]

[a]Air Force Wright Aeronautical Laboratories
Materials Laboratory, AFWAL/MLLS
Wright-Patterson AFB, OH 45433

[b]Metcut-Materials Research Group
now with the University of Dayton
Graduate Materials Engineering
Dayton, OH 45469

Abstract

In an effort to improve the high temperature capabilities of titanium alloys, Ti-6Al-2Sn-4Zr-2Mo-0.08Si (Ti-6242) and Ti-6Al-2Sn-4Zr-6Mo (Ti-6246) were modified by Er dispersoids, Si precipitates, and through the addition of W and prepared using the plasma rotating electrode process (PREP). Results of creep tests at 480°C (900°F)/245MPa (35ksi) and 565°C (1050°F)/245MPa (35ksi) of heat treated conditions are presented and compared to base alloys. The modified Ti-6246 showed little improvement over baseline material, however the Si and W alloying additions significantly enhanced the high temperature performance of the Ti-6242 base alloy. Dispersoid coarsening characteristics at 925°C (1700°F) and 1040°C (1900°F) are also shown.

Introduction

The maximum level of alloy additions in high temperature titanium alloys is in some cases kept low due to the detrimental effects of slow cooling associated with conventional titanium alloy ingot metallurgy practices [1]. Among those effects are segregation and coarsening of second phase particles such as silicides. Also, the ability to add dispersoid formers, such as the rare earths, is limited due to segregation particularly at grain boundaries during ingot solidification and subsequent solid state cooling [1].

Rapid solidification technology (RST) of titanium alloys has advanced to a point where solid solubility limits can be extended and fine precipitates and dispersions can be developed and kept uniform and fine. This technique then offers the potential for increasing the alloying levels in high temperature titanium alloys. Examples of titanium RST product forms include splats, ribbons, wires, particulate (staple fibers), and spherical powder. The range of cooling rates for these processes range from 10^3°C/s for gas atomized titanium powder up to 10^7°C/s for splat quenched material [2]. Unfortunately, the amount of material currently produced by these methods is small making it difficult to undertake a full scale investigation of mechanical properties. Because of this difficulty semi-commercial spherical plasma rotating electrode powder (PREP) was used in the present program. Existing equipment allows production of clean powders in hundred of pound quantities with a calculated cooling rate of 10^4°C/s for the 150μm average powder size [3].

The effect of erbium on titanium alloy properties has been investigated fairly extensively and is well documented [4-6]. Sastry et al [6] demonstrated that the addition of a fine erbia dispersion produced an order of magnitude improvement in creep rate. Cooling rates on the order of 10^3-10^5°C/s^{-1} must be achieved to produce a beneficial fine dispersion [5]. The effect of Er additions was, therefore, investigated in this work.

The segregation prone eutectoid formers like Si and W are normally difficult to add to titanium in large quantities. However, RST minimizes segregation and therefore enhances the strengthening effects due to these elements. These elements were therefore chosen as secondary additions in the present investigation.

The base alloys selected for this study were Ti-6Al-2Sn-4Zr-2Mo-0.08Si (Ti-6242) and Ti-6Al-2Sn-4Zr-6Mo (Ti-6246), with additions of Er, Si, W or combinations thereof, a total of seven conditions.

The modified alloy compositions and baseline compositions were evaluated in the extruded condition with emphasis on high temperature creep behavior. The dispersoids coarsening behavior of the Ti-6242 alloys was also studied.

Experimental Procedure

Powder Production

Five 10 kg (22 lb) titanium alloy ingots with the compositions PRST 2-6 as listed in Table I were produced by TIMET. The material was produced from titanium sponge and master alloys with erbium added to form an erbia (Er_2O_3) dispersion by combining with internal oxygen. Ingots were double vacuum arc melted and subsequently forged at 1065°C (1950°F) into 75 mm (3 in.) diameter bars. After sand blasting and acid pickling the surface, bars were turned down to 100 mm (2.5 in.) electrodes which were converted into spherical powder by the PREP process [1] at 15,000 rpm.

Table I. Experimental Alloys

Billet No.	Nominal Composition	Designation
PRST-0	Ti-6Al-2Sn-4Zr-2Mo-0.08Si	Ti-6242
PRST-00	Ti-6Al-2Sn-4Zr-6Mo	Ti-6246
PRST-2	Ti-6Al-2Sn-4Zr-2Mo-0.08Si+2Er	Ti-6242S+2Er
PRST-3	Ti-6Al-2Sn-4Zr-6Mo+2Er	Ti-6246+2Er
PRST-4	Ti-6Al-2Sn-4Zr-2Mo+0.4Si	Ti-6242+0.4Si
PRST-5	Ti-6Al-2Sn-4Zr-2Mo+0.4Si+2Er	Ti-6242+0.4Si+2Er
PRST-6	Ti-6Al-2Sn-4Zr-2Mo-0.08Si+3W	Ti-6242S+3W

In addition, two base alloy powders were evaluated: Ti-6242 and Ti-6246. As can be seen from Table I, the Ti-6242 base composition was modified in this work with all three alloying additions (Er, Si, and W), while the Ti-6246 base composition was only modified with Er.

Figure 1 shows the microstructures of the as-received powder. As expected from a process with a cooling rate on the order of 10^4°C/s, a dendritic structure is observed.

Compaction

Direct powder extrusion (DPE) was selected to compact the powders into a form suitable for this program. It was found that full density could be achieved by DPE at 760°C (1400°F) with a 3 hr preheat. It is interesting to note that in a parallel work, a full density was achieved by vacuum hot pressing (VHP) these alloys only at 870°C (1600°F)/172MPa (25ksi)/4 hr.

Powders were filled into a commercially pure (CP) titanium extrusion can which was subsequently evacuated and sealed after heating at 390°C (750°F) to remove moisture. The cans were extruded, after 3 hr preheat, at a ratio of 20:1 into 20 mm (0.75 in.) diameter rods. This rod was then machined into cylindrical blanks 75 mm (3 in.) long x 12.5 mm (0.5 in.) diameter.

From the optical and SEM photomicrographs (Figures 2 and 3, respectively), it can be seen that the microstructure of all as-extruded compacts is an ultrafine equiaxed alpha in a beta matrix. It can be further seen from the STEM micrographs (Figure 4), that the average grain size in these extrusions is less than 1 micron. The fine dispersion of Er_2O_3 particles can also be seen both intra-granularly and at the grain boundaries in Figure 4a. The average dispersoid size in the as-compacted material ranged from 70Å for smaller particles to 400Å for larger ones.

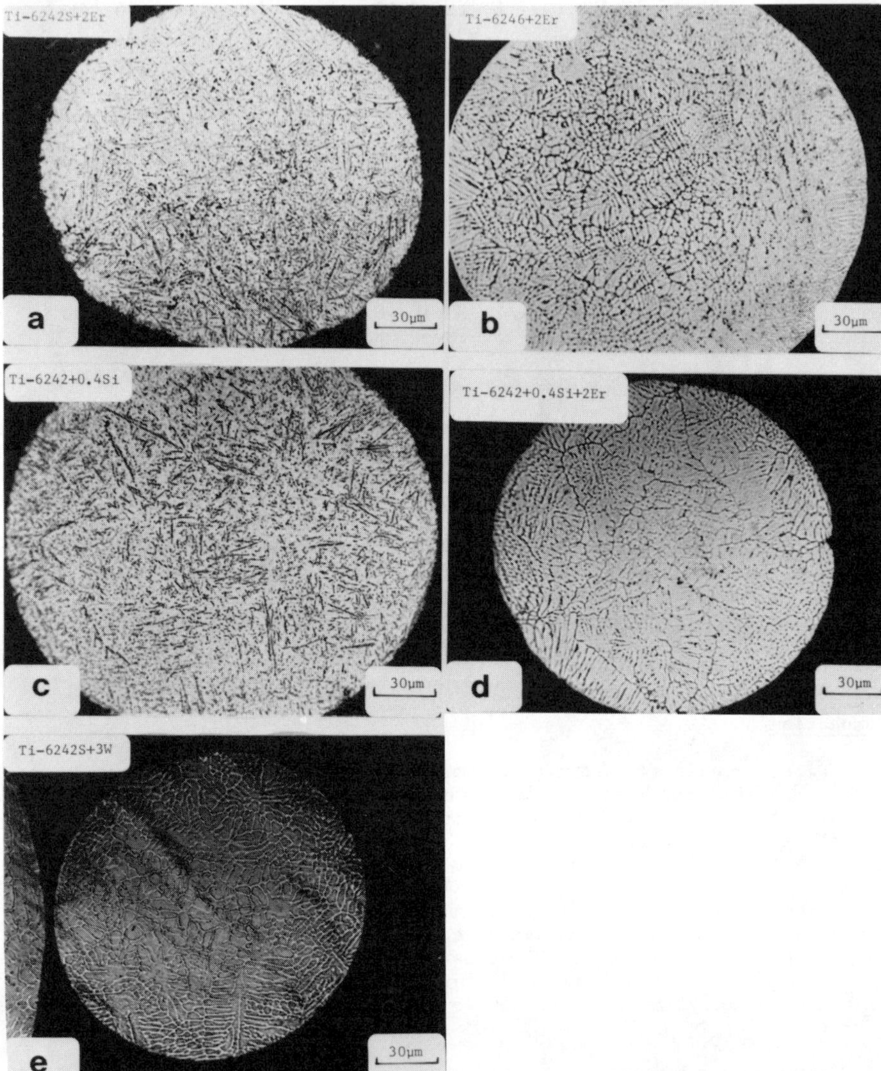

Figure 1 - As-solidified PREP powder particles. (a) Ti-6242+2Er (PRST-2); (b) Ti-6246+2Er (PRST-3); (c) Ti-6242+0.4Si (PRST-4); (d) Ti-6242+0.4Si+2Er (PRST-5); and (e) Ti-6242+3W (PRST-6).

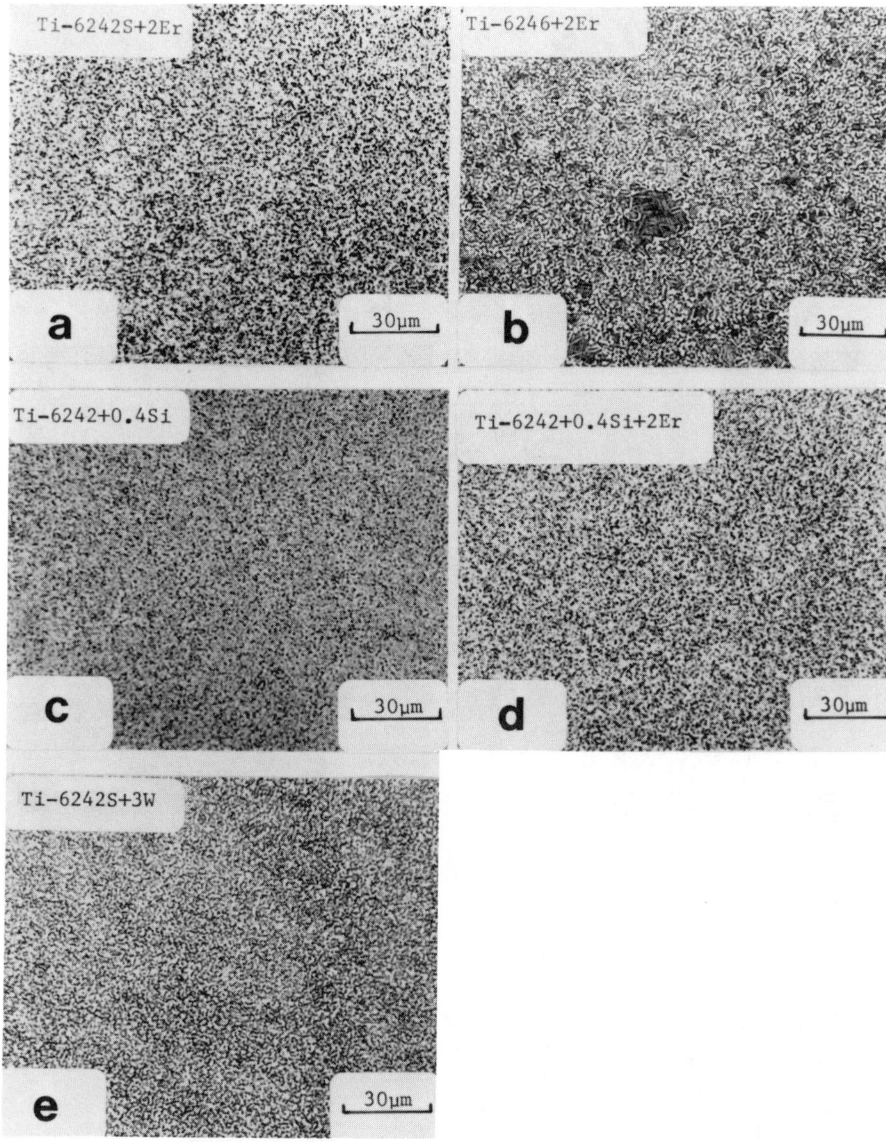

Figure 2 - As-compacted microstructure (optical). (a) Ti-6242+2Er (PRST-2); (b) Ti-6246+2Er (PRST-3); (c) Ti-6242+0.4Si (PRST-4); (d) Ti-6242+0.4Si+2Er (PRST-5); and (e) Ti-6242+3W (PRST-6).

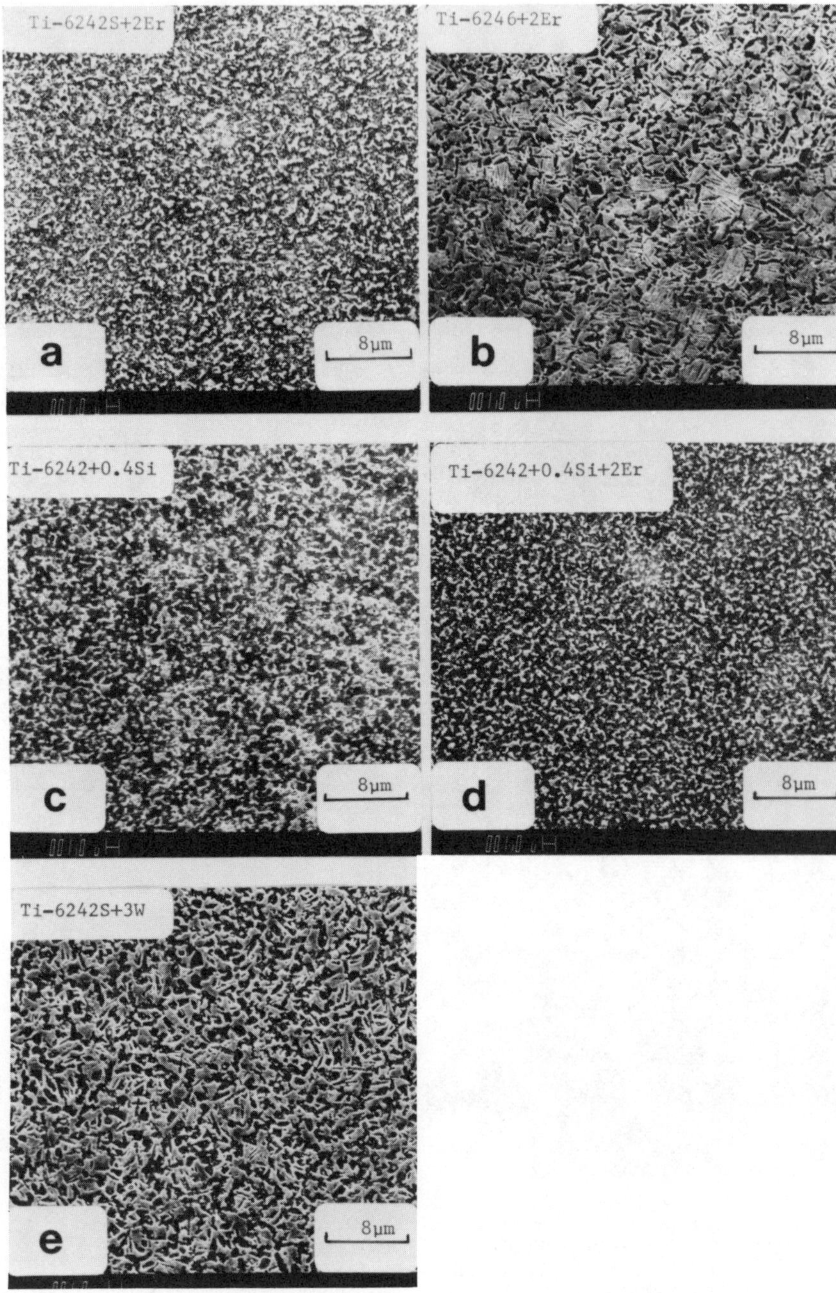

Figure 3 - As-compacted microstructure (SEM).
(a) Ti-6242+2Er (PRST-2); (b) Ti-6246+2Er (PRST-3); (c) Ti-6242+0.4Si (PRST-4); (d) Ti-6242+0.4Si+2Er (PRST-5); and (e) Ti-6242+3W (PRST-6).

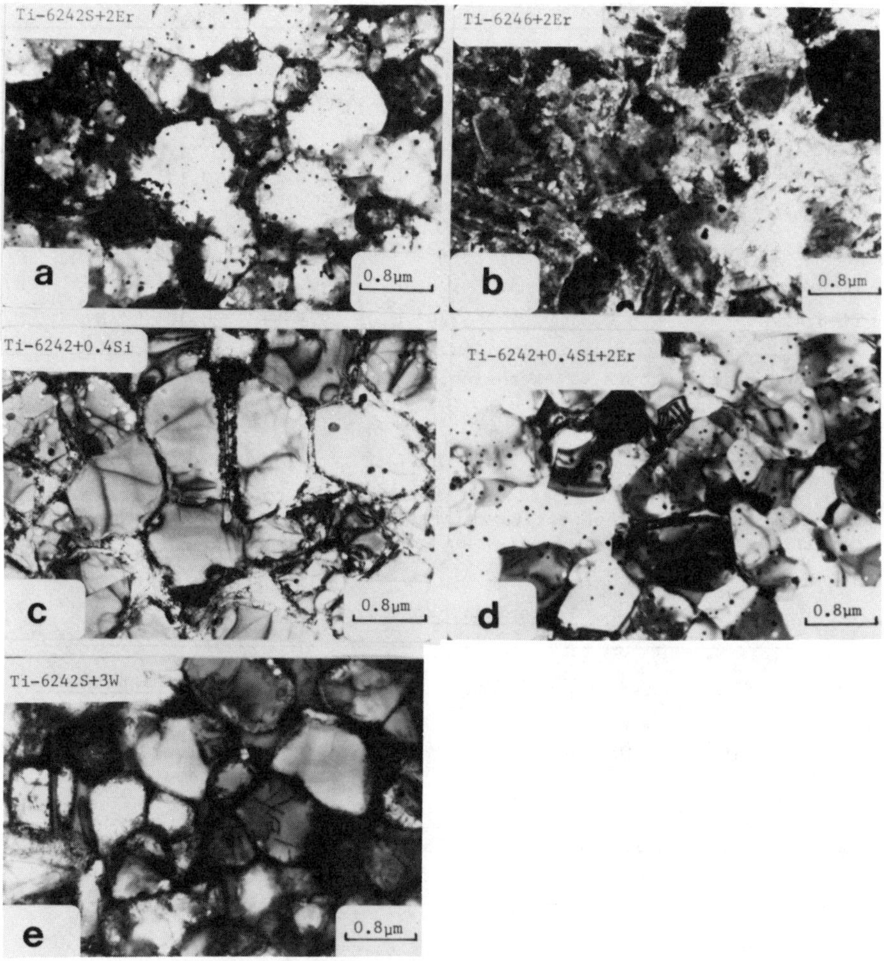

Figure 4 - As-compacted microstructure (STEM).
(a) Ti-6242+2Er (PRST-2); (b) Ti-6246+2Er (PRST-3);
(c) Ti-6242+0.4Si (PRST-4); (d) Ti-6242+0.4Si+2Er
(PRST-5); and (e) Ti-6242+3W (PRST-6).

Chemical Analysis

Wet chemical analysis was performed on PRST 2-6 compositions in the extruded form. The baseline compositions (PRST-0 and PRST-00) were analyzed in the powder form. The results are summarized in Table II. All compositions were well within the nominal levels (Table I). The oxygen levels of PRST 2-6 were low (1000-1250 ppm), the result of careful melting, conversion, and clean handling of the powders.

Table II. Chemical Composition[a] in Wt % (Ti = Balance)

Alloy	Al	Sn	Zr	Mo	Si	W	Er	O	H	N
Ti-6242[a]	6.0	2.1	3.9	2.0	0.01	-	-	0.148	0.0039	0.006
Ti-6246[a]	6.1	2.3	4.1	5.8	-	-	-	0.103	0.0021	0.012
Ti-6242+2Er	5.9	2.0	3.8	1.9	0.06	-	1.8	0.124	N/A	N/A
Ti-6246+2Er	5.8	1.9	4.0	5.7	-	-	2.0	0.121	N/A	N/A
Ti-6242+0.4Si	6.0	2.0	4.0	1.9	0.34	-	-	0.108	N/A	N/A
Ti-6242+0.4Si+2Er	5.9	2.0	3.9	1.8	0.29	-	2.1	0.101	N/A	N/A
Ti-6242+3W	6.0	1.9	4.2	2.0	0.08	2.9	-	0.104	N/A	N/A

[a]Of powder (all others of extrusion).

Heat Treatments

The heat treatments used in this study for both alloys are summarized in Table III. Representative microstructures of the solution treated conditions of both alloys are shown in Figures 5a and 5b.

Table III. Heat Treatments

Alloy	Heat Treatments, °C (°F)/hr	Designation
Ti-6246	595 (1100)/8/AC	Direct age
Ti-6246	910 (1675)/1/AC+595 (1100)/8/AC	Solution treated and aged
Ti-6242	595 (1100)/8/AC	Direct age
Ti-6242	1025 (1880)/½/AC+595 (1100)/8/AC	Solution treated and aged

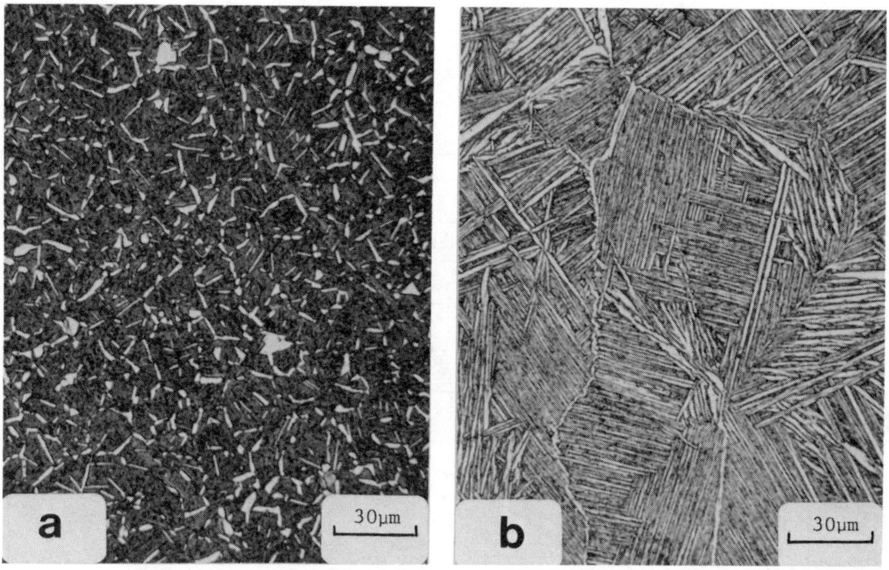

Figure 5 - Photomicrographs of heat treated conditions of (a) Ti-6246 at 910°C (1675°F)/1 hr/AC + 595°C (1100°F)/ 8 hr/AC and (b) Ti-6242 at 1025°C (1880°F)/½ hr/AC + 595°C (1100°F)/8 hr/AC.

Creep Testing

Two creep tests of each heat treatment were preformed (a total of 4 tests per alloy). The Ti-6246 based alloys were tested at 480°C (900°F)/240MPa (35ksi) while the Ti-6242 based alloys were tested at 565°C (1050°F)/240MPa (35ksi). Test limits were 0.2% creep or 150 hr, whichever occurred first.

Results and Discussion

As can be seen from the creep curves of the Ti-6246 based alloys in the directly aged condition (Figure 6), and the solution treated condition (Figure 7) the creep of both the baseline and erbia modified alloys are the same. This indicates that the erbia dispersion did not provide any beneficial strengthening for this alloy in the heat treatment condition tested.

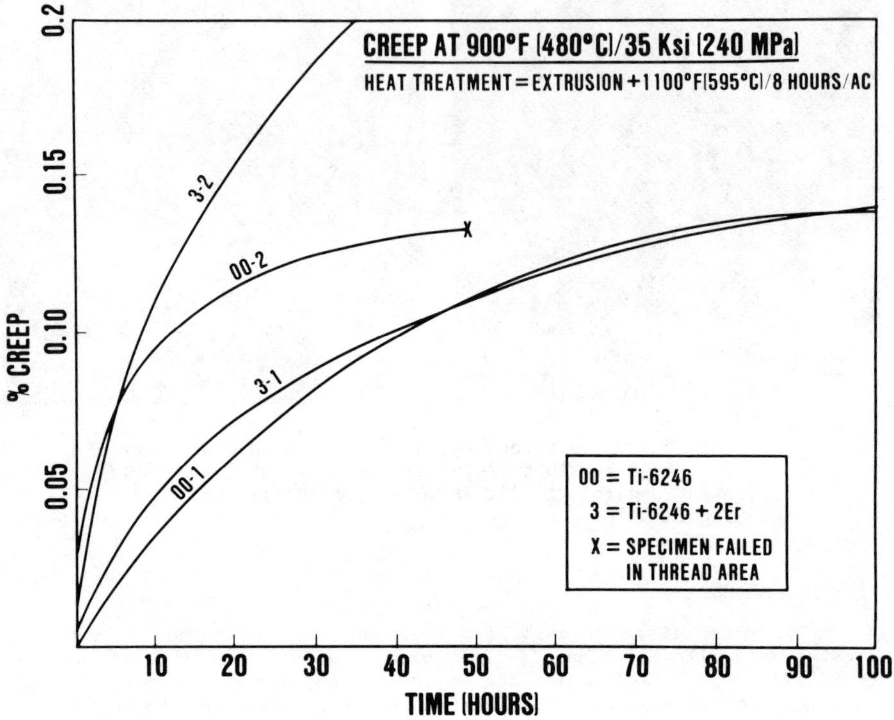

Figure 6 - Creep curves of directly aged Ti-6246 without and with Er additions (PRST-00 and PRST-2).

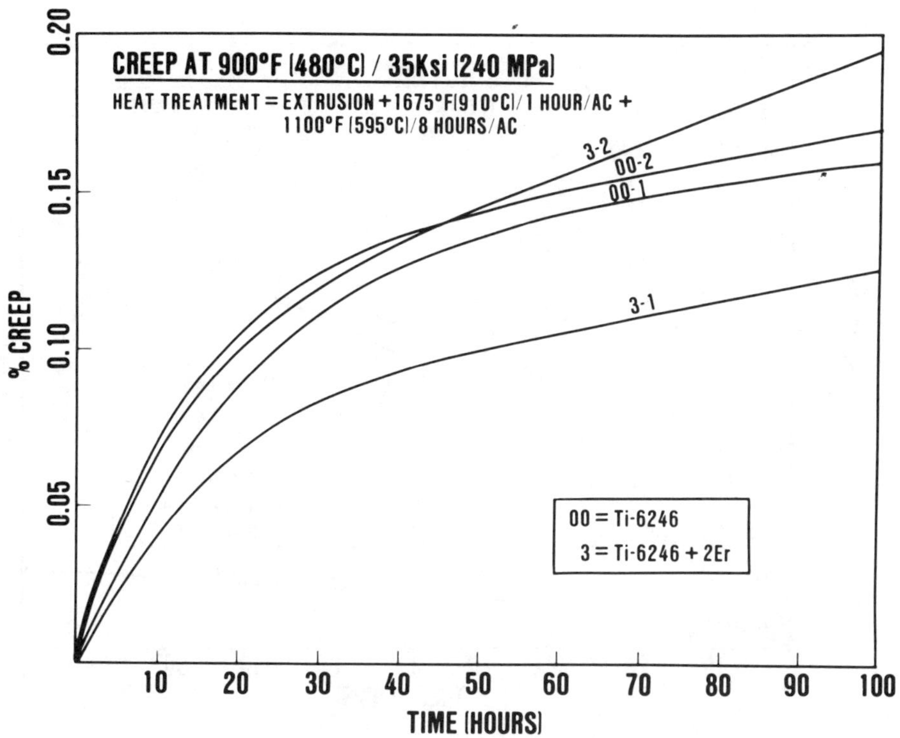

Figure 7 - Creep curves of solution treated Ti-6246 without and with Er additions (PRST-00 and PRST-2).

Optimization of the creep resistant microstructures in the Ti-6242 based alloys proved to be much more successful. As might have been expected, the creep tests conducted on the direct aged material resulted in lower creep resistance. This is certainly due to the very fine grain size (<1μm) found after the aging treatment which is similar to the as-compacted microstructures shown in Figure 3. None of the alloy additions strengthened the material enough to provide adequate creep strengthening. The beta solution treated and aged material, on the other hand, provided some very promising data. The alloys modified with silicon (PRST-4), silicon and erbia (PRST-5), and tungsten (PRST-6) demonstrated good creep properties as compared to the base material (Figure 8). Unfortunately, due to equipment malfunction no data for the erbia modified alloy was obtained, but additional tests are being performed.

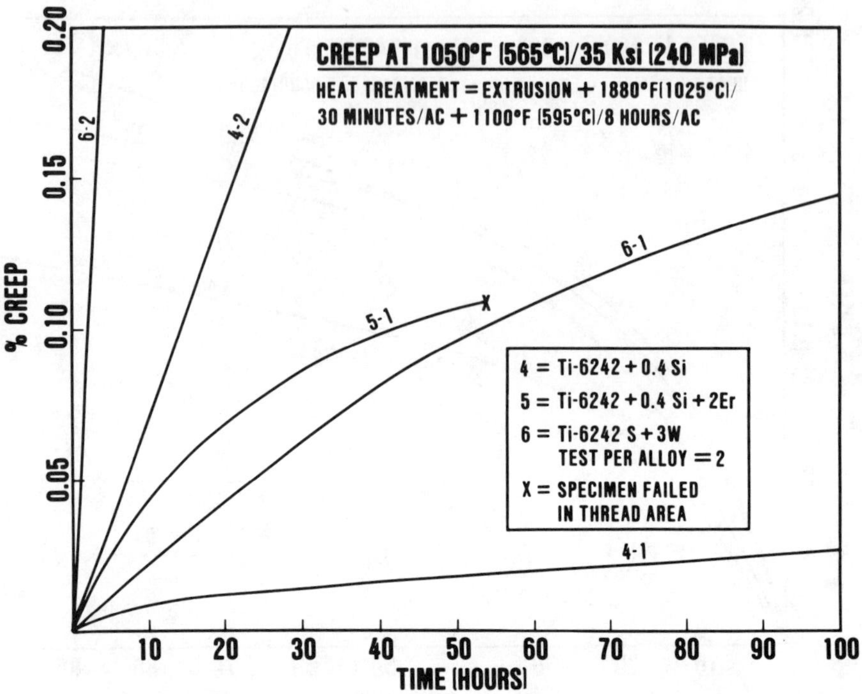

Figure 8 - Creep curves of solution treated Ti-6242 with Si, Er, and W additions (PRST-4, PRST-5, and PRST-6).

Figure 9 shows a comparison of the modified alloys and ingot metallurgy Ti-6242 heat treated to the same condition. All three alloys demonstrated superior creep properties when compared to the ingot material.

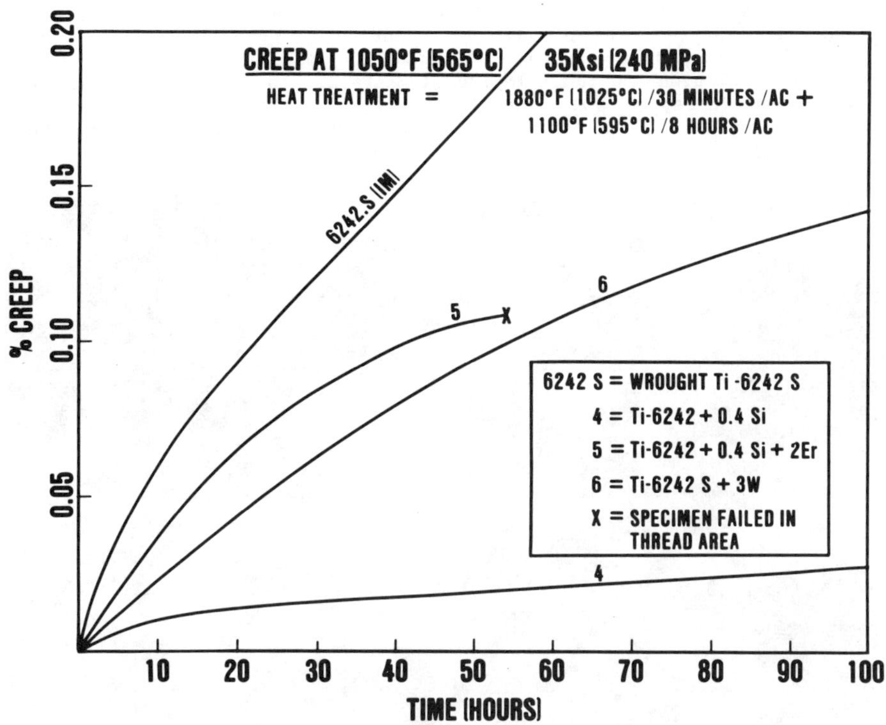

Figure 9 - Comparison of creep curves of Ti-6242 (PRST-4, PRST-5, and PRST-6) with IM baseline material.

Dispersoid Coarsening Measurements

While the ability to produce very fine and stable dispersoids in a powder product is important, the degree of property improvement will be dependent upon the dispersion in the final heat treated compacted product. The degree of Orowan strengthening achieved will be dictated by final particle size and the final interparticle spacing of the erbia particles [7]. Particle coarsening is most likely to occur during compaction, dictating a low processing temperature. On the other hand, the compaction temperature must be high enough to give a high integrity product. Thus a balance must be struck. Preliminary investigations of the rate of coarsening was done on the erbia containing alloy PRST-3 (Ti-6242+2Er) at 930°C (1700°F) and 1040°C (1900°F) for ½, 2, 8, and 24 hr (8 total conditions). The two coarsening temperatures evaluated, 930°C (1700°F) and 1040°C (1900°F) represented temperatures above and below the beta transus of the material (990°C [1815°F]) [8]. After each treatment, TEM foils were prepared and analyzed in a JEOL 2000FX STEM. The size of representative groups of dispersoids were measured to give an estimate of the coarsening characteristics of the erbia particles.

A bimodal distribution of coarsened particles was measured. The dispersoids were divided into two groups for measurement; greater than 200nm and below 200nm. As can be seen in Figures 10 and 11, coarsening occurred very rapidly, being quite pronounced after 30 minute aging at the lower temperature. The bimodal distribution of particles can be seen in these figures. It appears that some of the smaller particles are going back into solution and the alloying elements are then being absorbed by the larger particles, allowing further growth. This is supported by the growing denuded regions around the larger particles at longer aging times.

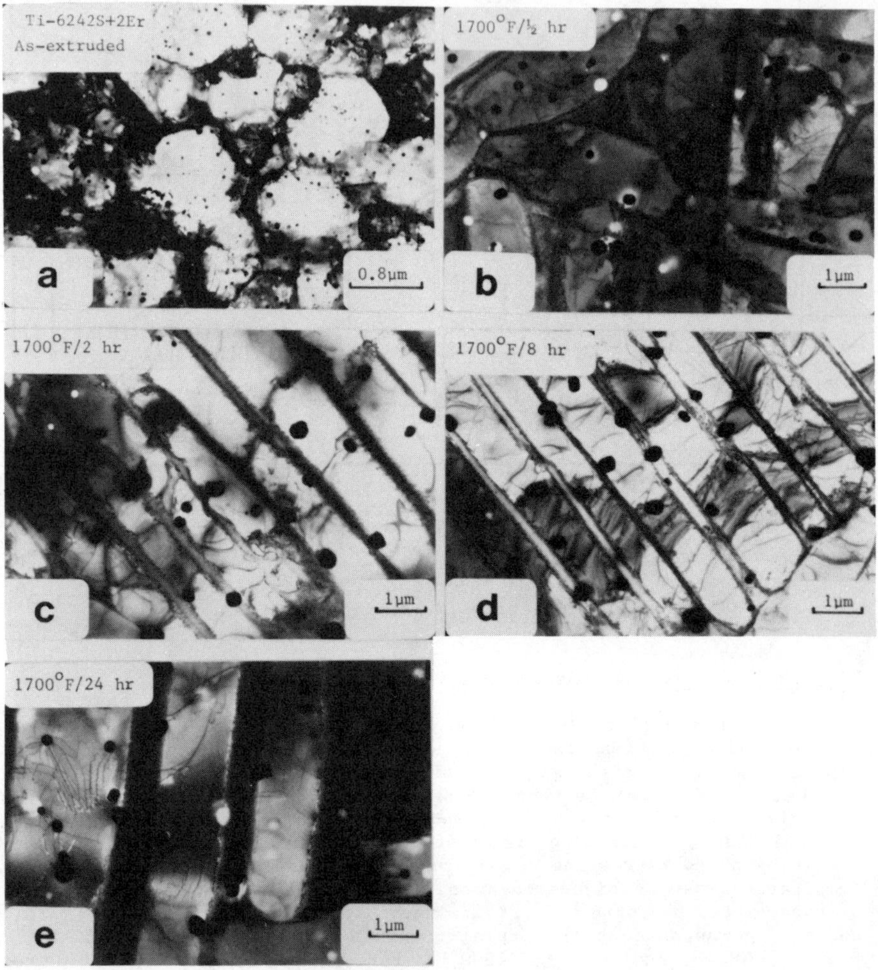

Figure 10 - Ti-6242+2Er (PRST-3) aged below the beta transus temperature at 930°C (1700°F). (a) as-extruded; (b) ½ hr aged; (c) 2 hr aged; (d) 8 hr aged; and (e) 24 hr aged.

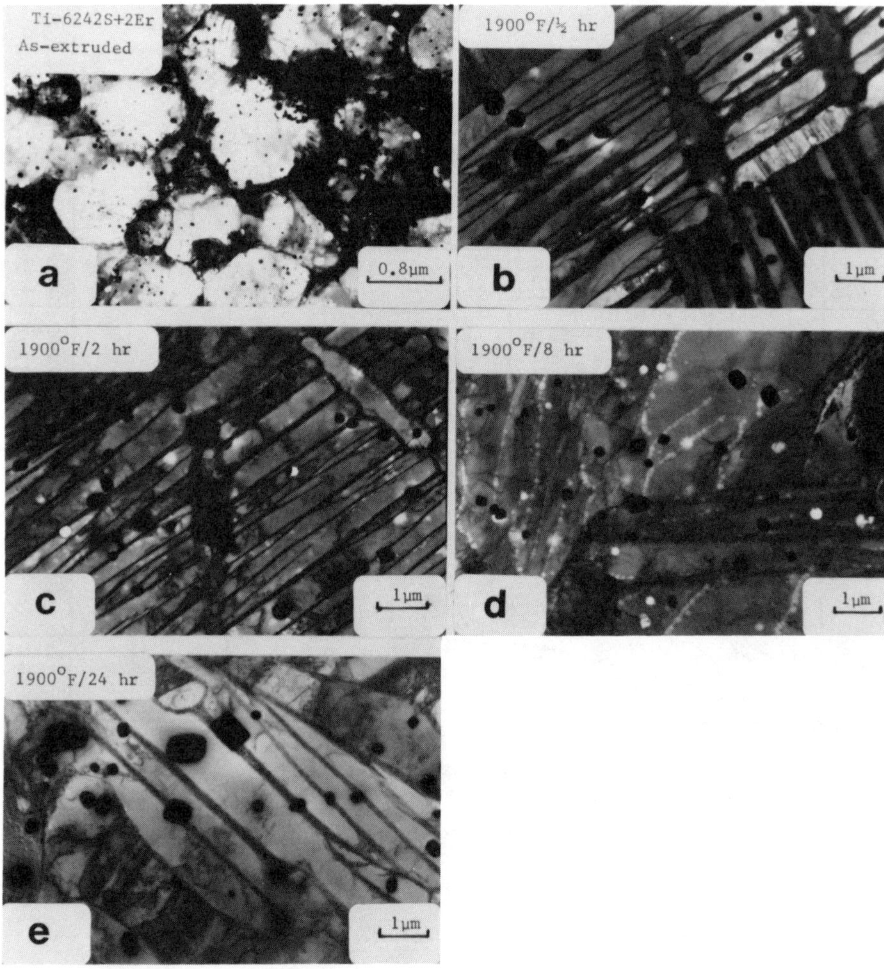

Figure 11 - Ti-6242+2Er (PRST-2) aged above the beta transus temperature at 1040°C (1900°F). (a) as-extruded; (b) ½ hr aged; (c) 2 hr aged; (d) 8 hr aged; and (e) 24 hr aged.

The coarsening rates of the erbia dispersoids are plotted in Figures 12 and 13. As expected, the coarsening rate of the large dispersoids is higher at 1040°C (1900°F) than at 930°C (1700°F). A very significant increase in coarsening rate is observed after 8 hr for both temperatures.

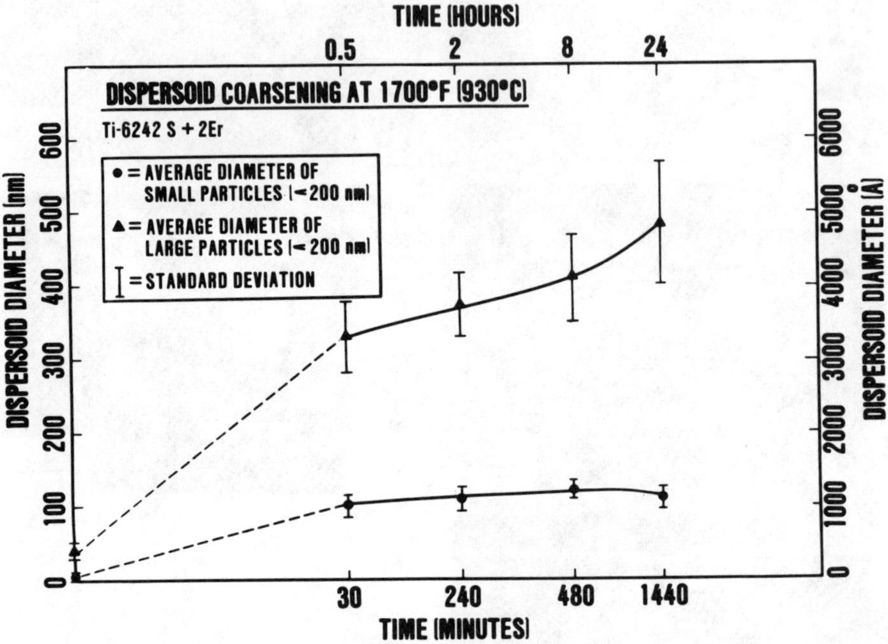

Figure 12 - Coarsening of erbia dispersoids in Ti-6242+2Er (PRST-2) alloy aged at 930°C (1700°F).

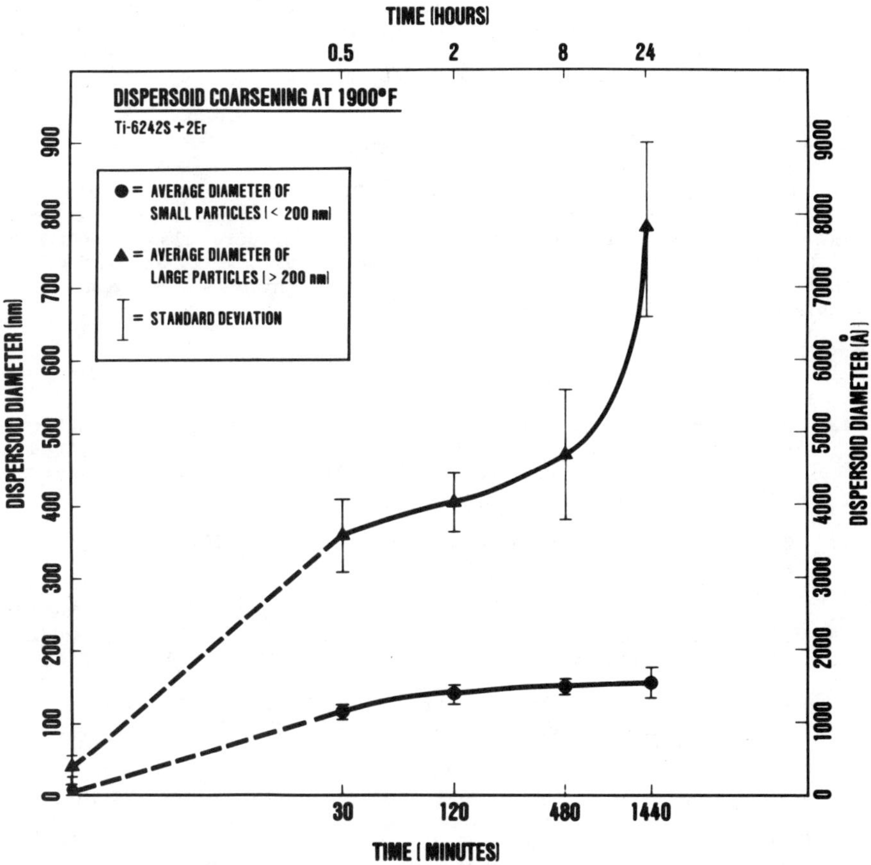

Figure 13 - Coarsening of erbia dispersoids in Ti-6242+2Er (PRST-2) alloy aged at 1040°C (1900°F).

Conclusions

1. Spherical prealloyed powder of Er, Si, and W modified Ti-6Al-2Sn-4Zr-2Mo-0.08Si (Ti-6242) and Ti-6Al-2Sn-4Zr-6Mo (Ti-6246) base compositions can be produced by the plasma rotating electrode process (PREP).

2. In the heat treatment conditions studied the erbia dispersion appears to offer no improvement in creep properties over the baseline materials.

3. When compared to the creep properties of wrought Ti-6242S in the beta solution treated and aged condition, the Ti-6242+0.4Si and the Ti-6242S+3W show significantly better creep properties.

4. A bimodal erbia dispersoid population was observed in the Er-containing alloys. Coarsening experiments showed that at temperatures below and above the beta transus temperature the large dispersoids further coarsen at the expense of the finer dispersoids, with the supra transus rate being considerably higher then that below the transus.

Acknowledgments

The authors would like to thank Dr. P. Bania of TIMET for his help in melting the alloy ingots. Ms. E. Klee of NMI is acknowledged for her help in conversion of the alloys to powder. The authors also wish to thank R. E. Omlor, R. D. Brodecki, W. A. Houston, N. G. Lovell, and S. D. Apt for their help with experiment work, and Ms. Karen A. Sitzman for her assistance with the manuscript.

References

1. F. H. Froes and D. Eylon, "Powder Metallurgy of Titanium Alloys - A Review," in Titanium, Science and Technology, vol. 1, ed. G. Lutjering, U. Zwicker, and W. Bunk (Oberursel, West Germany: DGM, 1985), 267-286.

2. S. J. Savage and F. H. Froes, "Production of Rapidly Solidified Metals and Alloys," Journal of Metals, 36(4)(1984), 20-33.

3. T. F. Broderick, A. G. Jackson, H. Jones, and F. H. Froes, "The Effect of Cooling Conditions on the Microstructure of Rapidly Solidified Ti-6Al-4V," Met. Trans. A, (16A)(11)(1985), 1951-1959.

4. S. M. L. Sastry, T. C. Peng, P. J. Meschter, and J. E. O'Neal, "Rapid Solidification Processing of Titanium Alloys," Journal of Metals, (35)(9)(1983), 21-28.

5. P. J. Meschter, J. E. O'Neal, and S. M. L. Sastry, "Structure and Properties of Rapidly Solidified Dispersion-Strengthened Titanium Alloys, Part 1: Characterization of Dispersoid Distribution, Structure, and Chemistry," Met. Trans. A, (15A)(7)(1984), 1451-1463.

6. S. M. L. Sastry, T. C. Peng, and L. P. Beckerman, "Structure and Properties of Rapidly Solidified Dispersion-Strengthened Titanium Alloys, Part 2: Tensile and Creep Properties," Met. Trans. A, (15A)(7)(1984), 1465-1484.

7. E. Orowan, Symposium on Internal Stresses in Metals (London: Institute of Metals, 1948), 451.

8. R. A. Wood, Titanium Alloy Handbook (Columbus, OH: Metals and Ceramics Information Center, Battelle, 1972), Publication No. MCIC-HB-02.

PRODUCTION OF HIGH STRENGTH BETA TITANIUM

ALLOYS THROUGH POWDER METALLURGY

R. G. Vogt[a], D. Eylon[b], R. R. Boyer[c], and F. H. Froes[a]

[a]Air Force Wright Aeronautical Laboratories
Materials Laboratory, AFWAL/MLLS
Wright-Patterson AFB, OH 45433

[b]Metcut-Materials Research Group
now with the University of Dayton
Graduate Materials Engineering
Dayton, OH 45469

[c]Boeing Commercial Airplane Company
P.O. Box 3707, Mail Stop 73-43
Seattle, WA 98124

Abstract

In an effort to produce a high strength segregation free beta titanium alloy, a Ti-8V-6Fe-1.5Al was double melted, converted to plasma rotating electrode process (PREP) powder, consolidated by extrusion, and subsequently heat treated. An attractive combination of strength (>200 ksi) and ductility (8% elongation) suggests that further exploration of high strength alloys using the rapid solidification/powder metallurgy (RS/PM) approach is warranted.

Introduction

Beta titanium alloys are an attractive class of alloys due to desirable combinations of high strength, ductility, workability, fracture resistance, and deep hardenability [1, 2]. However, as the amount of beta stabilizers is increased (specifically eutectoid formers), segregation can occur [3]. One method to circumvent this problem is through rapid solidification/powder metallurgy (RS/PM) [4]. To enable a meaningful evaluation of mechanical properties, relatively large amounts of material must be produced. The only RS method which can, presently, produce quantities large enough of titanium alloy clean powder for mechanical property evaluation is the plasma rotating electrode process (PREP). Cooling rates of $\sim 10^4$°C/s were calculated for a 150μm average diameter PREP particle [5]. It was the purpose of this reported work to evaluate the potential of this process to produce a compact of the Ti-8V-6Fe-1.5Al alloy which would exhibit attractive properties in the high strength condition (>200 ksi YS).

Experimental Procedure

The alloy selected for evaluation was Ti-8V-6Fe-1.5Al which is based on the Mallory-Sharon alloy MSM 185 (Ti-8V-5Fe-1.3Al) evaluated 30 years ago [6]. A base alloy ingot, double melted by TIMET, was forged at 1065°C (1950°F) into 3 in. diameter round bars. After sand blasting and acid pickling of the surface, the bars were turned down to 2.6 in. bars to be used as PREP electrodes.

The electrodes were converted into 6 lb of powder by the PREP process [7]. Chemical analysis of the electrode alloy before conversion is shown in Table I.

Table I. Chemical Composition in Wt % (Ti = Balance)

Nominal Composition	Chemical Analysis			
	V	Fe	Al	O
Ti-8V-6Fe-1.5Al	7.73	5.62	1.40	.110

It is important to note the low oxygen level (0.11 wt%), since previous work on the MSM 185 alloy [6] showed an oxygen level in excess of 0.35 wt% which could have been partially responsible for the reported high strength level. Previous experience with the PREP process has shown that the oxygen pickup during powder conversion and subsequent HIP'ing is no more than 0.02 wt% [7].

Two lots of this powder were extruded at 595°C (1100°F) and 760°C (1400°F) using a 3 hr preheat and commercially pure titanium extrusion cans. The 20:1 extrusion ratio yielded rods of 0.75 in. diameter. The extruded material was evaluated metallographically and was found to be fully dense.

Various heat treatments were performed including solution treatments at 675°C (1250°F) and 705°C (1300°F) followed by aging at 400°C (750°F), 440°C (825°F), 496°C (925°F), and 550°C (1025°F). The solution treatments were for 30 minutes at 675°C (1250°F) and for 15 minutes at 705°C (1300°F) followed by water quench. The aging treatment times were for 1, 2, 4, and 8 hr followed by air cooling. All the heat treated conditions were evaluated by Rockwell hardness (R_c) testing. The treatments showing the highest R_c values were then tested in tensile. The treatments selected for tensile testing are listed in Table II.

Table II. Heat Treatments Selected for Tensile Testing

Heat Treatment Number	Heat Treatment
1	675°C (1250°F)/30 min/WQ + 440°C (825°F)/4 hr/AC
2	705°C (1300°F)/15 min/WQ + 400°C (750°F)/4 hr/AC
3	705°C (1300°F)/15 min/WQ + 400°C (750°F)/16 hr/AC
4	705°C (1300°F)/15 min/WQ + 440°C (825°F)/4 hr/AC
5	705°C (1300°F)/15 min/WQ + 496°C (925°F)/2 hr/AC

A beta transus determination was also performed. First, the material was annealed at 510°C (950°F)/24 hr. This insured that the volume fraction of alpha present was maximized. This was followed by separate 15 minute solution treatments with ice brine quench (IBQ) starting at 650°C (1200°F) and increasing in 15°F increments. The transus was determined by finding the temperature at which a minimum amount of alpha was detected in the quenched structure and the first temperature at which no alpha remained. This is illustrated in Figure 1 as small amounts of alpha still appear in the 690°C (1275°F) treated material (Figure 1a), but is not present in the 705°C (1300°F) treated material (Figure 1b). The transus, therefore, was determined to be 700°C (1290°F).

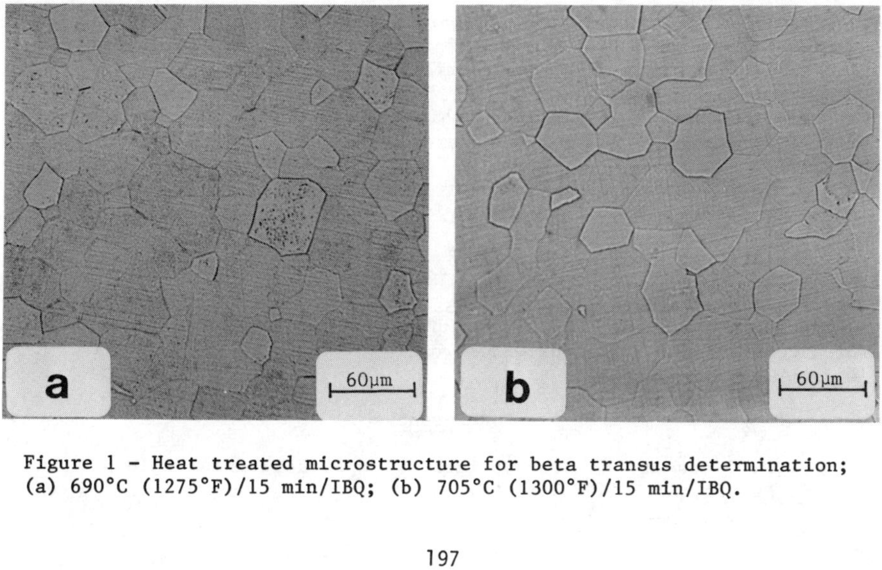

Figure 1 - Heat treated microstructure for beta transus determination; (a) 690°C (1275°F)/15 min/IBQ; (b) 705°C (1300°F)/15 min/IBQ.

Results and Discussion

All heat treatments listed in Table II were applied only to the beta extruded material (760°C [1400°F]). Heat treatments 2 and 3 were applied also to the alpha+beta extruded material (595°C [1100°F]). The tensile results are listed in Table III.

Table III. Mechanical Property Data

Heat Treatment Number	Type Extrusion	0.2% YS, MPa	ksi	UTS, MPa	ksi	EL., %	RA, %
1	Beta	1317	191	1407	203	9	12
2	Alpha+Beta	1158	168	1207	175	3	10
	Beta	1103	160	1151	167	9	31
3	Alpha+Beta	1448	210	1517	220	1	3
	Beta	1427	207	1510	219	4	11
4	Beta	1392	202	1482	215	8	7
5	Beta	1262	183	1351	196	14	24

The 705°C (1300°F)/15 min/WQ + 440°C (825°F)/4 hr/AC heat treatment resulted in an excellent combination of high strength and ductility. No significant difference between material extruded above or below the beta transus was apparent possibly because adiabatic heating which occurs during extrusion actually caused a temperature excursion over the beta transus even when extrusion is nominally in the alpha+beta region.

Conclusions

1. Use of a rapid solidification/powder metallurgy (RS/PM) approach allowed strengths in excess of 200 ksi to be obtained in the Ti-8V-6Fe-1.5Al alloy with good ductility (8% elongation).

2. Further investigation of high strength titanium alloys using the RS/PM approach is warranted.

Acknowledgments

The authors wish to thank P. Bania of TIMET for his help in melting the alloy. The assistance of W. A. Houston and N. G. Lovell in metallography and mechanical testing and Ms. Karen A. Sitzman in typing the manuscript is appreciated. Parts of this work were done under U.S. Air Force Contract No. F33615-82-C-5078.

References

1. F. H. Froes and H. B. Bomberger, "The Beta Titanium Alloys," Journal of Metals, 37(7)(1985), 28-37.

2. R. R. Boyer and H. W. Rosenberg, ed., Beta Titanium Alloys in the 1980's (Warrendale, PA: TMS Publications, 1984).

3. D. B. Hunter, "Screening and Selection of Candidate Alloys, Part II" (AFML Technical Report WAL-TR-405/2-15, Part II, TIMET Development, 1966).

4. H. Jones, Rapid Solidification of Metals and Alloys, Monograph No. 8 (London: The Institute of Metallurgists, 1982).

5. T. F. Broderick, A. G. Jackson, H. Jones, and F. H. Froes, "The Effect of Cooling Conditions on the Microstructure of Rapidly Solidified Ti-6Al-4V," Met. Trans. A, 16A (1985), 1951-1959.

6. J. R. Gross and S. Abkowitz, "An Investigation of the Mechanical Properties of Ti-8V-5Fe-1.3Al" (Report No. 1000 R164, Mallory-Sharon Research, 1956).

7. F. H. Froes and D. Eylon, "Powder Metallurgy of Titanium Alloys - A Review," in Titanium, Science and Technology, vol. 1, ed. G. Lutjering, U. Zwicker, and W. Bunk (Oberursel, West Germany: DGM, 1985), 267-286.

COARSENING OF YTTRIUM TIN PARTICLES

IN THE β PHASE OF TI-5SN-3Y

Y.D. Hahn, D. Vujic and S.H. Whang

Department of Metallurgy and Materials Science
Polytechnic University
333 Jay St.
Brooklyn, N.Y. 11201

Ostwald ripening of yttrium tin dispersion (Y_5Sn_3; D88) in Ti-5Sn-3Y alloy was studied at the temperature range of 900-1000°C. The results show that the coarsening obeys the cube law, which indicates that volume diffusion is dominant in this process. The coarsening rate of this temperature range is 4.1×10^{-27} m^3/s at 900°C; 9.6×10^{-27} m^3/s at 950°C; 2.7×10^{-26} m^3/s at 1000°C. The activation energy for the diffusion of Y in Ti at this temperature range estimated from the coarsening data was found to be 247 KJ/mol.

Introduction

In the recent years, much effort has been focused on research of rare earth dispersion strengthened Ti alloys since coarsening resistance of rare earth dispersoids at high temperatures, namely above 600°C is excellent[1,2]. Such coarsening resistant behavior of rare earth dispersoids in Ti generates enough scientific curiosity and implication for potential applications.

Experimental results on some of rare earth dispersoids such as La_2Sn[3] and Y_5Sn_3[4] in α-Ti have been reported. The results indicate that in both cases, volume diffusion is a dominant mechanism for the particle coarsening. Furthermore, the low coarsening rate in these alloys is primarily a result of the low diffusion coefficient and the low solubility of the rare earth metals in Ti.

All of the Ostwald ripening experiments reported have been performed in the α phase regime (700-850°C) of Ti alloy. However, it is interesting to know the coarsening behavior at the high temperature regime (β phase), i.e., above 900°C. In this paper, the coarsening behavior of Y_5Sn_3 particle in the β phase regime is reported.

Theoretical Background

A systematic description of particle coarsening, Ostwald ripening, was first made by Lifshitz and Slyozov[5], and Wagner[6] independently in 1961. This theory predicts that 1) average particle size proportional to time and materials constants; 2) the distribution of particle size reaches a steady state after a long period and remains constant independent of time. This model is based on an assumption that the particle diameter is much smaller than the interparticle spacing.

In the experiments, a broader size distribution than that predicted by LSW has always been found. As a result, much effort has been rendered to the modification of the LSW model by many authors [7-12] to bring the theory and the experiments together.

In the following, LSW and modified LSW (MLSW) equations are given:

LSW MODEL

$$\bar{r}^3 - \bar{r}_o^3 = f(T)\cdot(t - t_o) \qquad (1)$$

with

$$f(T) = \frac{8}{9} \cdot \frac{C_s(\infty)\cdot D\cdot \gamma\cdot V_m}{R\cdot T} \qquad (1\text{-}1)$$

where $f(T)$ = measured slop of $\bar{r}^3$ in plot of $\bar{r}^3$ vs t

MLSW MODEL

$$f'(K,T) = f(T)\cdot K(C,\phi) \qquad (1\text{-}2)$$

where

$\bar{r}_o$, $\bar{r}$ = average particle radii for the onset and final states
K = volume fraction factor
$C_s(\infty)$ = equilibrium solubility of solute in the matrix with particles of infinite size at a given temperature T
D = diffusion coefficient of solute, cm^2/sec
γ = interfacial free energy of particle, J/m^2
V_m = gram-molar volume of precipitate, cm^3/g-mole

t_o, t = onset and final annealing time, sec
R = gas constant, J/mole-k
T = absolute annealing temperature, K
C = concentration of solute in the matrix
ϕ = volume fraction of second phase particle

The approaches taken by the authors to attack the problem may be divided into two categories: one is to modify the diffusional flow pattern resulting from a new situation, in which the large volume fraction of particles decreases the distance among particles and thereby increases the concentration gradient around the particles and the diffusion rate as well [7,8,9,11,12]; the alternative is a particle coalescence mechanism which better describes the very broad distribution of particle size[10].

Nevertheless, it is very difficult to determine which model best represents the coarsening of second phase particles. In other words, no universal model to explain all experimental results is available at this moment.

Experimental

In order to prepare nearly oxygen free alloys, metals of high purity were used: electron beam melted Ti 99.99% with less than 140ppm oxygen; pure Sn 99.999% with less than 20ppm oxygen; and pure Y 99.99% with 83ppm oxygen. Alloy melting was performed in an arc furnace under argon atmosphere. The alloy button was broken into small pieces, which were subsequently splat quenched into foils by the hammer and anvil quench technique. These foils were annealed in a vacuum tube for various temperatures and durations. The annealed foils were prepared into thin films by twin jet polishers for TEM study. Particle coarsening analysis based on the TEM micrographs was carried out using an image analyzer.

Results

An examination of TEM micrographs from the as-quenched samples shows that the particle size distribution occurred in a broad range during solidification as shown in Fig. 1a. Also, dislocation networks are seen in the matrix, which presumably formed during rapid quenching. The annealed sample at 900°C for 25h reveals that particle coalescence is under way (Fig. 1b). At this stage, the dislocations were already annealed out. Further annealing of the alloy sample for 75h leads to significant coarsening of the particles (Fig. 1c). Also, it should be noted that the particles remain spherical and incoherent throughout the entire process.

Neverthless, the particle size distribution deviates significantly from the model proposed by LSW[5,6]. Fig. 2a and b show typical large deviations from $\rho = r/\bar{r}$ value of 1.5 which is a cut-off point predicted by LSW where $\rho^2 h(\rho)$ is the theoretical distribution function[6]. A plot of the particle radius vs. isothermal annealing time is shown in Fig. 3. Also, the same data results in a linear correlation in the plot of the cube of the particle radius vs. time as shown in Fig. 4.

The slope of these linear lines are given in Table I.

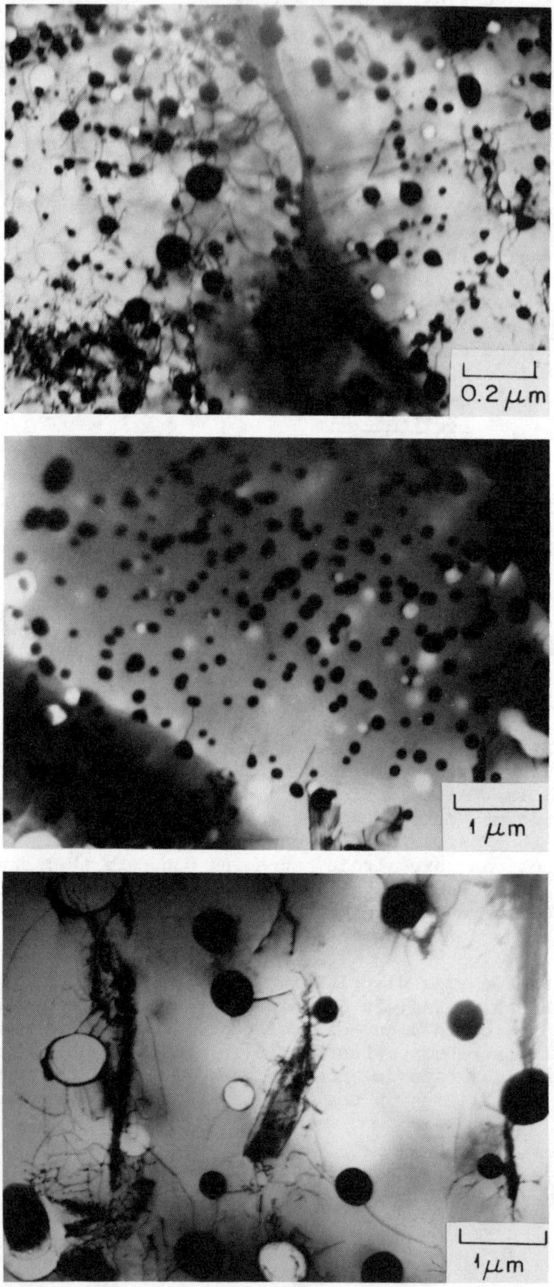

a) As-quenched State

b) Annealed at 900°C for 25h

c) Annealed at 900°C for 75h

Fig. 1 - Ti-5Sn-3Y Alloy Foil

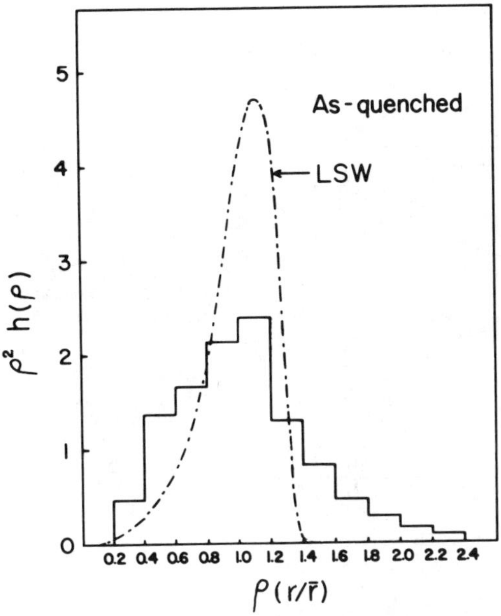

a) As-quenched State

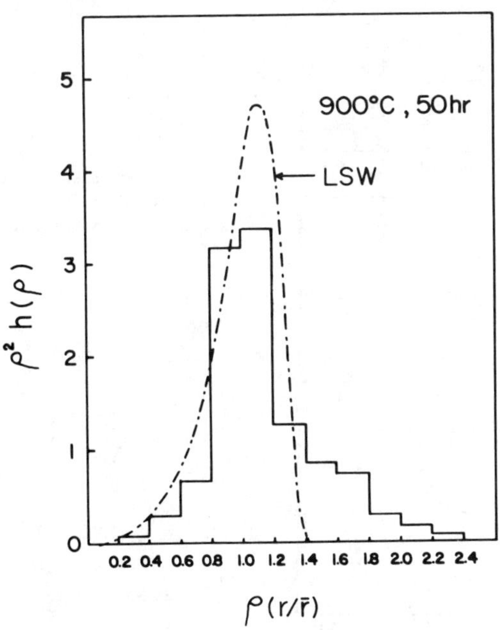

b) Annealed at 900°C for 50h

Fig. 2 - Particle Size Distribution

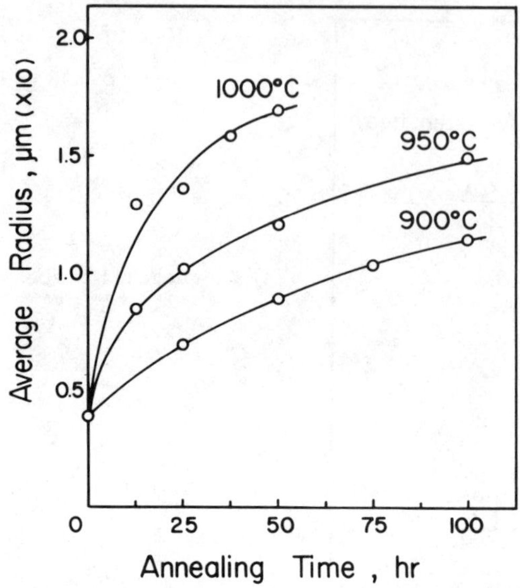

Fig. 3 - Plot of Average Particle Size, $\bar{r}$ vs. Annealing Time, t

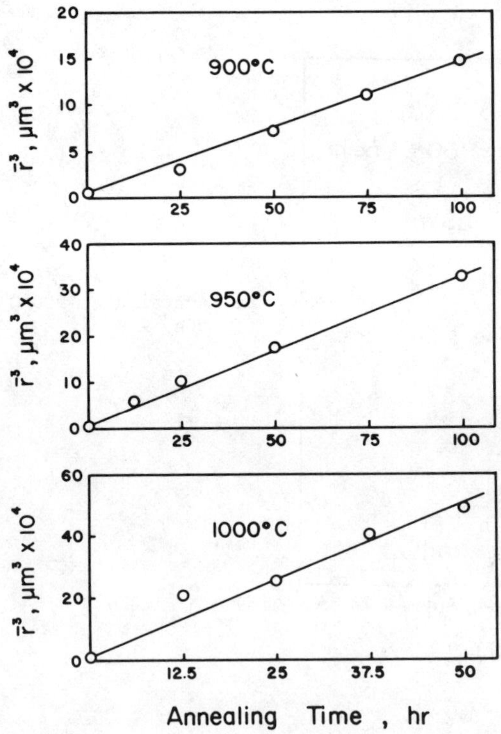

Fig. 4 - Plot of the Cube of Average Particle Radius vs. Annealing Time

Table I Ostwald Ripening of Y_5Sn_3 Particle in Ti-5Sn-3Y

Alloy	Temperature °C	C_s[13]	Voume[4] Fraction %	K [10]	f′(K,T) m^3/s
Ti-5Sn-3Y	950	0.0088	3.4	1.2	4.1×10^{-27}
	950	0.0088	3.4	1.2	9.6×10^{-27}
	1000	0.0088	3.4	1.2	2.7×10^{-26}

Discussion

In order to obtain the diffusion coefficient from Eq. 1-2, the following is considered.

The solubility of Y in Ti is nearly constant in this range of 900-1350°C [7]. The additon of 2.5 at % Sn to Ti is assumed not to affect the solubility of Y in Ti. The volume fraction factor, K is determined by one of the theoretical models[10]. This is done for two reasons: 1) in these experiments, particle coalescence, in fact, is active as shown in Fig. 1b; 2) this model provides as much broader distribution of particle size as observed in these experiments. Since the interfacial free energy of yttrium particles in Ti is not known, the value is arbitrarily assumed to be 1 J/m^2, which is an intermediate value among the values for incoherent particles, i.e., 0.5 J/m^2 for Zn-Cu[14] to 2.0 J/m^2 for ThO_2 in Ni[5].

The diffusion coefficients can be obtained by inserting all these values into Eq. 1-2. These values are listed in Table II along with diffusion coefficients for other rare earth elements and Th in various transition metals.

Table II Diffusion Coefficient and Activation Energy for Rare Earth Metals in Transition Metals

Alloy	Temperature °C	Phase	Diffusion Coefficient(D) cm^2/s	Activation Energy (Q) KJ/mole
Ti-5Sn-3Y (Y in Ti)	760	α	1.9×10^{-13}[4]	272[4]
	800	α	5.6×10^{-13}[4]	
	840	α	1.9×10^{-12}[4]	
	900	β	2.5×10^{-12}	247[this work]
	950	β	6.2×10^{-12}	
	1000	β	1.8×10^{-11}	
Sn in Ti	900	β	4.9×10^{-10}	132[16]
	950	β	8.6×10^{-10}	
	1000	β	1.4×10^{-9}	
Ce in Ni	900		2.46×10^{-12}	254[17]
	1000	fcc	2.4×10^{-11}	
Nd in Ni	900		2.43×10^{-12}	250[17]
	1000	fcc	2.19×10^{-11}	

Ce in Zr	900 1000	β	9.99×10^{-11} 1.44×10^{-8}	130[18]
Nd in Zr	900 1020	β	6.61×10^{-11} 1.78×10^{-8}	134[18]

The obtained diffusion coefficients are regarded as those of Y in Ti. The coarsening of Y_5Sn_3 particles requires transport of Y and Sn atoms from the smaller particles to the larger particles. The measured diffusion coefficients of Sn shown in Table II are much larger than those measured in these experiments. In addition, the solubility of Y in Ti is much smaller than that of Sn in Ti. Therefore, the coarsening process is controlled by the diffusion of Y in Ti. Hence, it is concluded that these diffusion coefficients are those of Y in Ti.

The apparent diffusion coefficient value in this temperature range increases rapidly by an order of magnitude for every 100°C increment. The increment in the apparent diffusion coefficient is very large compared with the increment in the solubility and the interfacial free energy, which are slowly changing functions. Therefore, it is mainly attributed to the increase in the diffusion coefficient of Y, and not to the increase in either the solubility or the interfacial free energy.

Furthermore, with an additional assumption that the diffusion of Y against reciprocal temperature can be described by an Arrhenius relation in this temperature range, the activation energy for the diffusion may be obtained from the following relation.

From Eq. (1-2), $f'(K,T) = Y(K)\cdot 1/T$

$$\frac{d\ln Y}{d(1/T)} = \frac{d\ln Do}{d(1/T)} + \frac{d}{d(1/T)} \cdot \frac{-Q}{R} \cdot \frac{1}{T} + \frac{d}{d(1/T)} \ln \frac{8K\cdot C_s\gamma\cdot V}{9R} \qquad (2)$$

The first and third terms on the right hand side vanish due to the given assumption. Hence,

$$\frac{d\ln Y}{d(1/T)} = \frac{d\ln D}{d(1/T)} = \frac{-Q}{R} \qquad (2\text{-}1)$$

where Q is activation energy for diffusion

From the plot of lnD and 1/T (Fig. 5), the activation energy Q for diffusion may be obtained. The determined value is 247 KJ/mole, which is similar to the value for the α phase, 272 KJ/mole[4]. These activation energy values include some uncertainty since the last term in Eq. 2 in a strict sense is not a constant independent of temperature. In other words, the concentration and the interfacial free energy[19] are not independent of temperature, though the change may be small. Another source of uncertainty may be that the curve deviates from an Arrhenius type equation at this temperature range. This means that the D_o value becomes a function of temperature. Nevertheless, the combined uncertainty is much less than an order of magnitude.

The diffusion coefficient and the activation energy for diffusion for various rare earth elements in transition metals are also given in Table II.

From the given information, the diffusion coefficient of rare earth elements in the transition metals is smaller than that of Sn in Ti or the self diffusion coefficient of Ti[20].

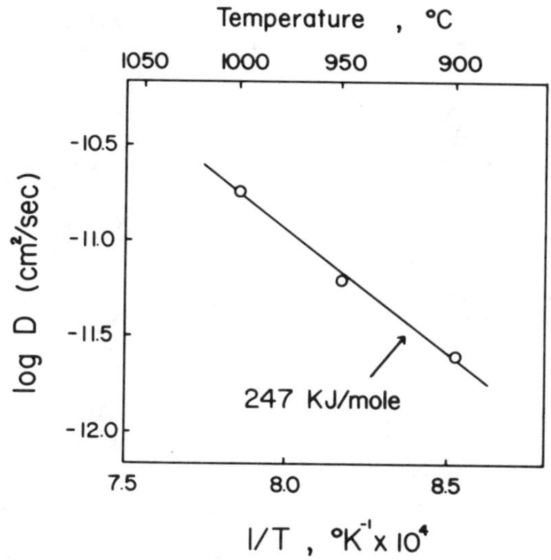

Fig. 5 - Plot of Log D vs. 1/T in Ti-5Sn-3Y Alloy

Summary

1. Ostwald ripening of yttrium dispersoids (Y5Sn3) in the β phase of Ti-5Sn-3Y shows a similar trend to that shown in the α phase.
2. No major jump in the particle coarsening is observed due to the transition from α phase to β in this alloy.
3. The estimated diffusion coefficient of Y in β-Ti is very small as has been observed for other rare earth elements in transition metals.
4. The activation energy for Y diffusion in β-Ti is similar to Ce and Nd in Ni at this temperature range.

Acknowledgement

The authors gratefully acknowledge the support of the Office of Naval Research, VA for this work (Contract No. N00014-85-K-0787).

References

1. S.H. Whang: "Rapidly Solidified Ti Alloys for High Temperature Applications", to appear in J. Mat. Sci., July 1986.
2. F.H. Froes and R.G. Rowe: "Rapidly Solidified Titanium", Proc. Sym. on Rapidly Solidified Alloys and Their Mechanical and Magnetic Properties, Dec. 2-6, 1985. Proceedings to be published in 1987.
3. Y.Z. Lu, C.S. Chi and S.H. Whang: "Second Phase Coarsening in Rapidly Solidified Ti-5Sn-4.5La System", Proc. Fifth Int. Conf. on Rapidly Quenched Metals, (S. Steeb and H. Warlimont, Eds., Elsevier Science Pub. B.V. 1985), pp. 949-952.

4. Y.Z. Lu, B.C. Giessen and S.H. Whang: "Particle Coarsening of Rare Earth Dispersoids in Ti-5Sn-R.E. Systems", submitted to Proc. Sym. on Rapidly Solidified Alloys, (B.C. Giessen, D.E. Polk and A.I. Taub, Eds., 1986).

5. I.M. Lifshitz and V.V. Slyozov: "The Kinetics of Precipitation from Supersaturated Solid Solutions", J. Phys. Chem. Solids, 19 (1961) pp. 35-50.

6. C. Wagner: "Theorie der Alterung von Neiderschlagen durch Umlosen", Z. Electrochem, 65 (1961), pp. 581-591.

7. R. Asimow: "Clustering Kinetics in Binary Alloys", Acta Met., 11 (1963), pp. 72-73.

8. A.J. Ardell: "The Effect of Volume Fraction on Particle Coarsening: Theoritical Considerations", Acta Met., 20 (1972), pp. 61-71.

9. A.D. Brailsford and P. Wynblatt: "The Dependence of Ostwald Ripening Kinetics on Particle Volume Fraction", Acta Met., 27 (1979), pp. 489-497.

10. C.K.L. Davies et al: "The Effect of Volume Fraction of Precipitate on Ostwald Ripening", Acta Met., 28 (1980), pp. 179-189.

11. K. Tsumaraya and Y. Miyata: "Coarsening Models Incorporating Both Diffusion Geometry and Volume Fraction of Particles", Acta Met., 31 (1983), pp. 437-452.

12. P.W. Voorhees and M.E. Glicksman: "Ostwald Ripening During Liquid Phase Sintering - Effect of Volume Fraction on Coarsening Kinetics", Met. Trans A, 15A (1984), pp. 1081-1088.

13. E.M. Savitskii and G.S. Burkhanov: "Phasen Diagramme von Titan - Legierungen mit Seltenen Erdmetallen", J. Less Comm. Metals, 4 (1962), pp. 301-314.

14. G.R. Pardy: "Widmanstatten Precipitation from Non-Ideal Solid Solution: α in β - CuZn", Metal Science Journal, 5 (1971), pp. 81-85.

15. C.B. Alcock and P.B. Brown: "Physicochemical Factors in the Dissolution of Thoria in Solid Nickel", Metal Science Journal, 3 (1969), pp. 116-120.

16. J. Askill and G.B. Gibbs: "Tracer Diffusion in β-Titanium", Physica Status Solidi, 11 (1965), pp. 557-565.

17. A.R. Paul and RP. Agarwala: "Lattice and Grain Boundary Diffusion at Cerium and Neodymium in Nickel", Met Trans., 2 (1971), pp. 2691-2695.

18. A.R. Paul and R.P. Agarwala: "Transport of Fission Products in Zircaloy -2", Int. Conf. on Thermodynamics of Nuclear Materials, Vol. 1, International Atomic Energy Agency, Vienna 1975.

19. L.E. Murr: Interfacial Phenomena in Metals and Alloys (Massachusetts, Addison-Wesley Pub. Co., 1975), pp. 87-164.

20. J.F. Murdock et al: "Diffusion of Ti^{44} and V^{48} in Titanium", Acta. Met., 12 (1964), pp. 1033-1039.

RAPID SOLIDIFICATION STUDIES IN EUTECTOID FORMING Ti ALLOYS

L.S. Chumbley, M.A. Ohls and H.L. Fraser

Department of Materials Science,
University of Illinois
Urbana, IL 61801

Abstract

Laser surface melting (LSM) and melt spinning (MS) have been used to produce rapidly solidified (RS) Ti-Si alloys, containing 4-10 wt.%Si. The microstructures characteristic of each technique have been found to differ considerably. In this respect, the role of oxygen contamination in microstructural development is thought to be of particular importance. In addition, the nature of the precipitation in a series of heat treated, LSM Ti-Cu-Er alloys containing 6-10 wt.% Cu and 1.5 wt.% Er was investigated.

Introduction

Several Ti-based alloys are currently being used in the aircraft industry for both structural airframe and gas turbine components, where a high strength to weight ratio is desirable. Most of these commercial alloys exhibit a two phase α (hcp) + β (bcc) microstructure. The major drawback of these alloys is that they are limited to temperatures of operation below ≈ 500°C. Attempts are thus being made to increase the strength of Ti alloys for elevated temperature service. These efforts are aimed at producing a fine dispersion of second phase particles by means of rapid solidification and subsequent heat treatment, so that strengthening may be achieved by the Orowan mechanism. In this respect, the rare earth oxides have received much attention (e.g. refs. 1,2) as have the borides (e.g. ref. 3) and sulfides (e.g. ref. 4), although to a lesser extent. A further class of dispersion-former involves a number of elements which form intermetallic compounds in Ti. This class includes Si, Cr, Ni, Fe and Cu. In the present work, the dispersion-forming potential of two of these systems has been investigated, namely Ti-Si and Ti-Cu (-Er). The Ti-Si system forms the equilibrium intermetallic Ti_5Si_3 and exhibits an eutectic at approximately 8.5 wt.%Si. The Ti-Cu system forms the intermetallic Ti_2Cu, and shows an eutectoid at 7.1 wt.%Cu.

Experimental Procedure

Small ingots of the Ti-Si alloys (containing 4, 6, 8.5 and 10 wt.%Si) and the Ti-Cu-Er alloys (containing 6, 8 and 10 wt.%Cu and 1.5 wt.%Er) were prepared by argon arc melting and portions of these ingots were laser surface melted using the techniques described elsewhere (2). Subsequent heat treatments were carried out in sealed quartz capsules under an argon atmosphere. Melt spun ribbon of the Ti-Si alloys was produced by induction melting small, pre-alloyed ingots in a quartz crucible, followed by the ejection of the molten material onto a copper wheel rotating with a surface speed of ≈ 35 ms^{-1}. The ribbon produced was roughly 3 mm in width and ≈ 25 μm in thickness.

Thin foil samples were prepared for analytical electron microscopy using standard techniques.

Results

Ti-Si Alloys

Four binary alloys were examined, namely Ti-4wt.%Si and Ti-6wt.%Si (hypoeutectic compositions), Ti-8.5wt.%Si (eutectic) and Ti-10wt.%Si (hypereutectic). The hypoeutectic alloys exhibited similar RS microstructures consisting of α'-Ti with an eutectic of α-Ti and the intermetallic Ti_5Si_3 forming in the interdendritic regions. The microstructure of the Ti-6wt.%Si alloy is shown in Fig. 1. All phases were identified using convergent beam electron diffraction (CBED). Energy dispersive x-ray (EDX) analysis revealed very little Si in solution in the α'-Ti matrix. Typical microstructures observed in the LSM Ti-8.5wt.%Si and Ti-10wt.%Si exhibited an eutectic of α-Ti and Ti_5Si_3 (rods), and an example from the 10 wt.%Si alloy is shown in Fig. 2.

The microstructures produced by melt spinning varied greatly from those which evolved during laser surface melting. However, it was also discovered that identical alloys melt spun in quartz crucibles under similar conditions might exhibit totally different microstructures. These variations were attributed to oxygen contamination from the quartz crucible during the MS process. The microstructure of a melt spun Ti-4wt.%Si alloy, thought to

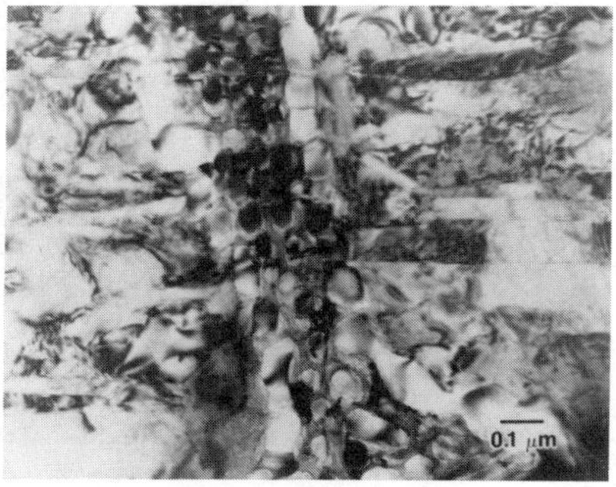

Figure 1 - Transmission electron micrograph from a laser treated Ti-6wt.%Si sample.

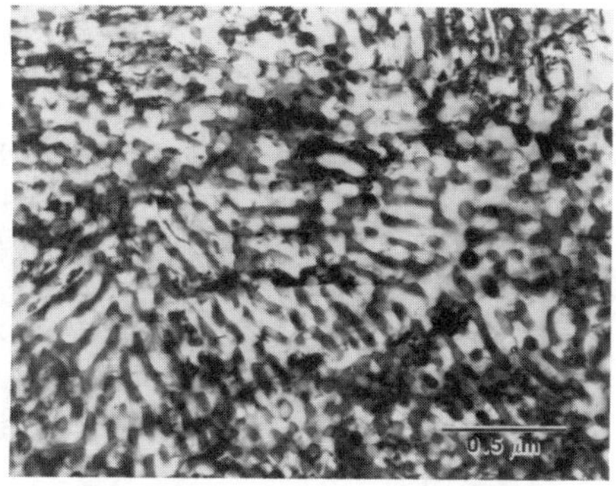

Figure 2 - Transmission electron micrograph from a laser treated Ti-10wt.%Si sample.

contain a relatively low oxygen concentration, is shown in Fig. 3, and can be seen to consist entirely of fine martensitic α' plates. EDX analysis showed that all the Si was in solution and no silicide particles were observed in any of the thin foils examined. The MS microstructure is seen to change significantly as the Si content is increased. For alloys containing 6 wt.%Si and 8.5 wt.%Si, the structure appears to be single phase β, although streaking seen in selected area diffraction (SAD) patterns is indicative of the ω phase of Ti (see Fig. 4). This phase is often present in rapidly quenched β-Ti and could be seen in dark field when the heaviest streaks were used to image the specimen. Melt spinning of the Ti-10wt.%Si alloy yields a predominantly amorphous structure as shown in Fig. 5. X-ray diffraction studies revealed no crystalline peaks, although a few small β crystallites were observed in thin foil specimens.

If the material to be melt spun was held in the quartz crucible too long (i.e. the oxygen pickup was increased), the microstructures produced during MS for all the Ti-Si alloys studied were seen to be similar, being cellular in nature and consisting chiefly of α-Ti and Ti_5Si_3. A third phase appeared as very small, roughly spherical particles only within the Ti_5Si_3 regions. The

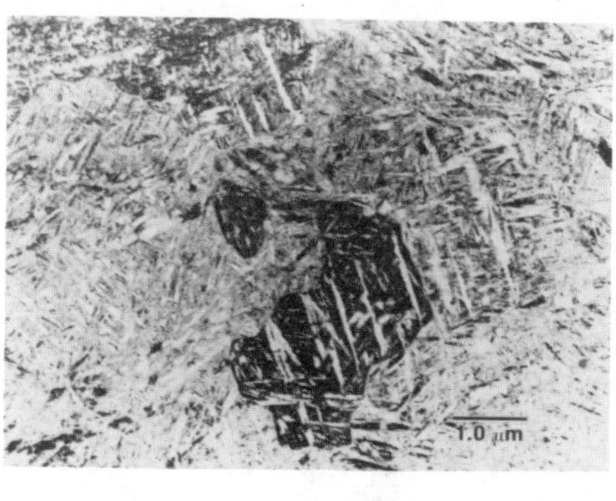

Figure 3 - Transmission electron micrograph from a melt spun Ti-4wt.%Si (low oxygen) sample.

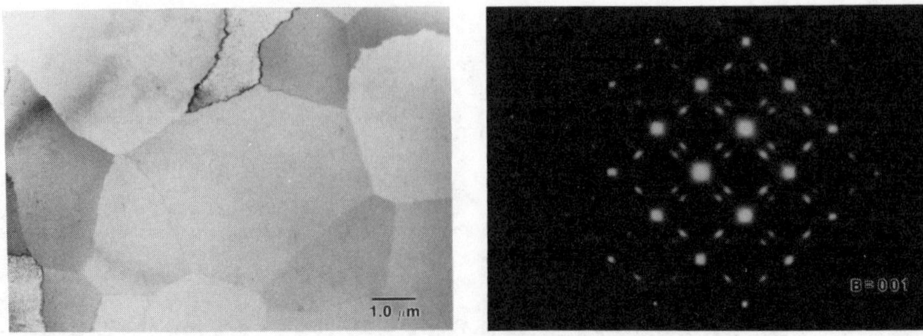

a b

Figure 4 - (a) Transmission electron micrograph from a melt spun Ti-8.5wt.%Si (low oxygen) sample, and (b) corresponding selected area diffraction pattern.

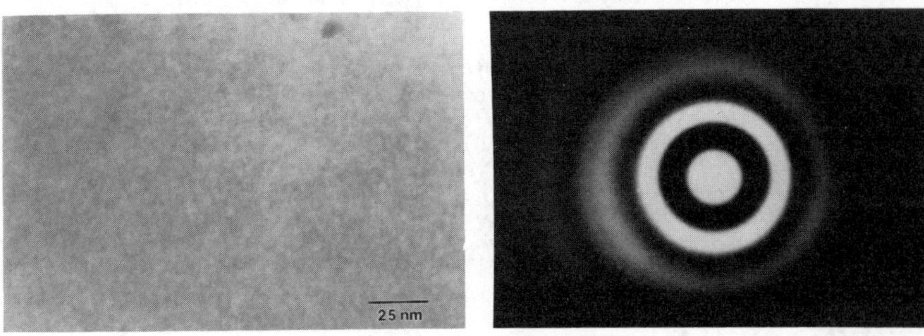

a b

Figure 5 - (a) Transmission electron micrograph from a melt spun Ti-10wt.%Si (low oxygen) sample, and (b) corresponding selected area diffraction pattern.

largest particles were approximately 10 nm in diameter (Fig. 6) making it difficult to accurately determine their crystal structure. However, Moire fringe analysis described in detail elsewhere (5), showed the particles to be hcp with lattice parameters of a_o = 6.19 Å and c_o = 4.30 Å.

Ti-Cu-Er Alloys

A typical transmission electron micrograph taken from a thin foil sample of the laser surface melted Ti-6wt.%Cu-1.5wt.%Er alloy is shown in Fig. 7 and can be seen to consist entirely of α'-Ti. Qualitatively, the microstructures of the 8 wt.%Cu and 10 wt.%Cu alloys appeared similar. Upon closer examination, fine particles (≈ 10 nm in diameter), thought to be erbia (Er_2O_3), are visible within the martensite plates in all three alloys (see Fig. 8). After heat treating a laser surface melted sample of the hypoeutectoid alloy (Ti-6Cu-1.5Er) for 1 hr at 500°C, there is evidence of heterogeneous precipitation at sub-grain boundaries and the homogeneous precipitation of thin, coherent plates within the α'-Ti matrix, Fig. 9. Williams et al (6), in a study of phase transformations in a Ti-6wt.%Cu alloy,

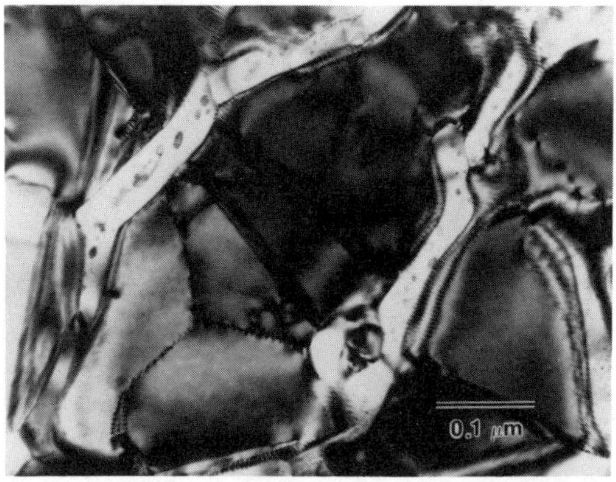

Figure 6 - Transmission electron micrograph from a melt spun Ti-6wt.%Si (high oxygen) sample.

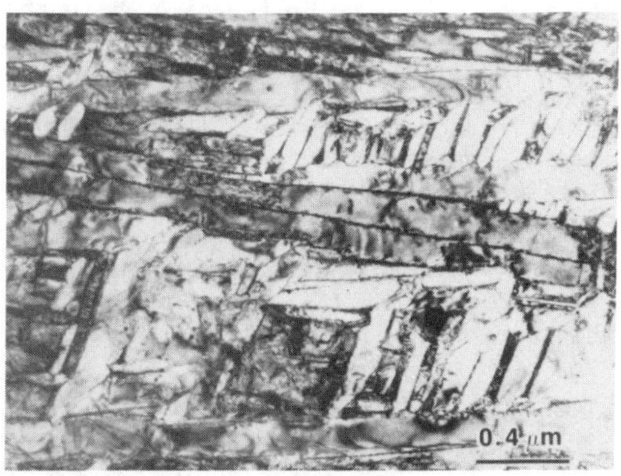

Figure 7 - Transmission electron micrograph from a laser treated Ti-4wt.%Cu-1.5wt.%Er sample.

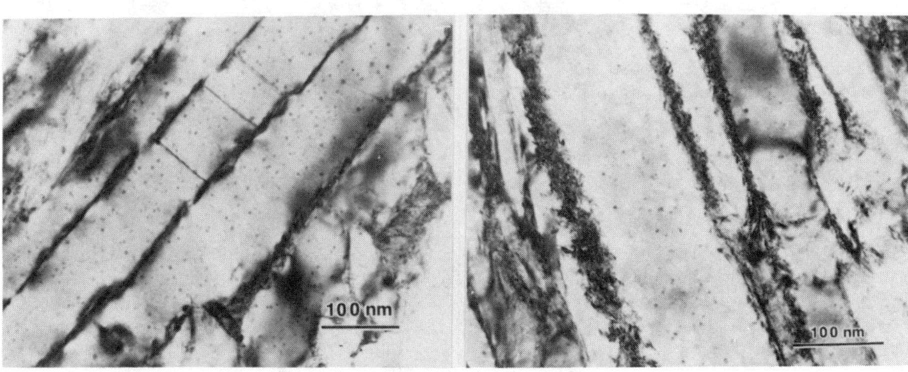

Figure 8 - Transmission electron micrographs taken from laser treated samples of (a) Ti-4Cu-1.5Er (wt.%) and (b) Ti-8Cu-1.5Er (wt.%).

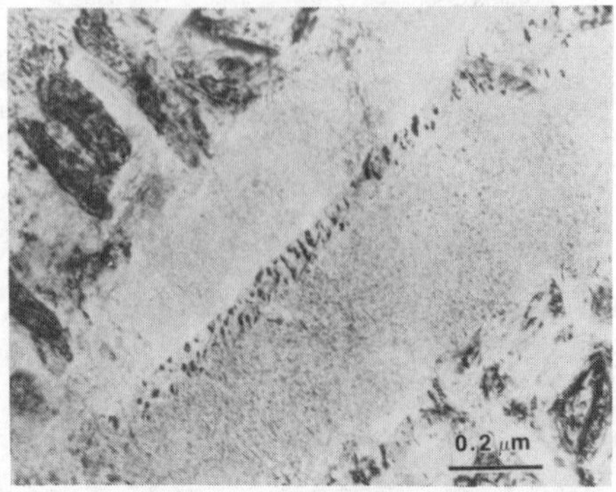

Figure 9 - Transmission electron micrograph taken from a Ti-6Cu-1.5Er (wt.%) specimen heated treated for 1 hr at 500°C after laser treating.

identified similar precipitate types as being the intermetallic Ti_2Cu; these precipitates are therefore assumed to be the same Ti_2Cu phase.

In the hypereutectoid alloys, the microstructure consists of regions of α'-Ti, which again contain fine Ti_2Cu precipitates (Fig. 10), and regions of a lamellar eutectoid of α-Ti and Ti_2Cu plates. Precipitation of the nearly-spherical Ti_2Cu phase is also evident within the eutectoid regions, as shown in the micrograph of Fig. 11, which was taken from a thin foil sample of the Ti-8wt.%Cu-1.5wt.%Er alloy. Throughout the microstructure fine particles, believed to be erbia, were also present.

After heat treatment at 700°C for 10 hrs following a "pre-treatment" of 1 hr at 500°C each alloy exhibits a fine dispersion of erbia particles (≈ 20 nm in diameter) within a recrystallized α-Ti matrix, Fig. 12.

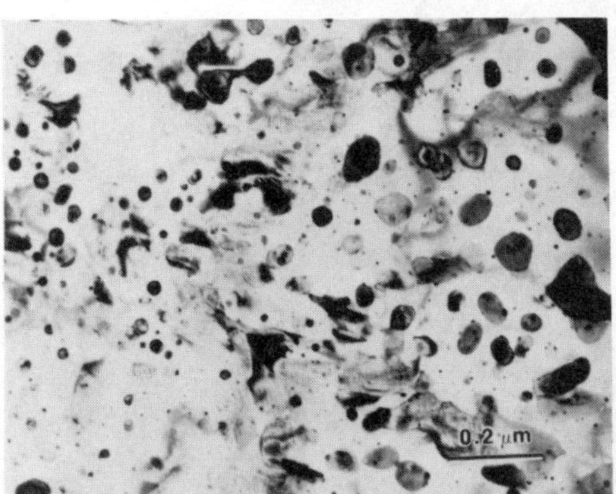

Figure 10 - Transmission electron micrograph taken from a Ti-8Cu-1.5Er (wt.%) specimen heat treated for 1 hr. at 500°C after laser treating.

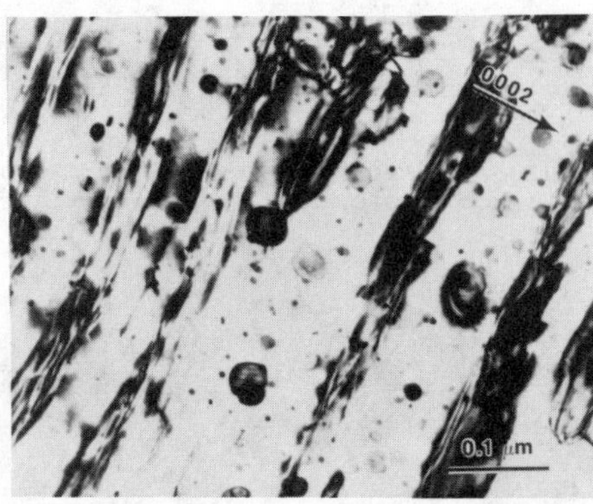

Figure 11 -Transmission electron micrograph taken from a Ti-8Cu-1.5 Er (wt.%) specimen heat treated for 1 hr at 500°C after laser treating, showing an eutectoid region.

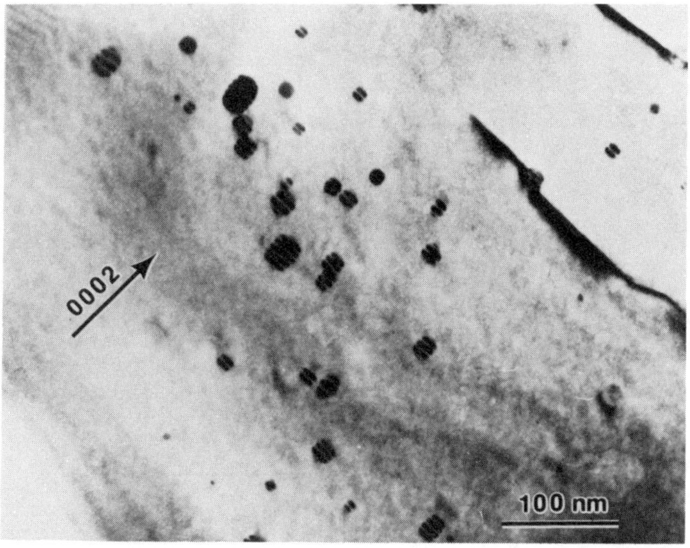

Figure 12 - Transmission electron micrograph taken from a Ti-6Cu-1.5Er (wt.%) specimen heat treated for 10 hrs at 700°C after a "pre-treatment" of 1 hr at 500°C following laser surface melting.

Discussion

Ti-Si Alloys

Melt spinning produces rapidly solidified microstructures which are quite different from those produced using laser surface melting. By considering metastable features of the Ti-Si phase diagram it can be deduced that these microstructural variations arise due to basic differences in the degree of undercooling achieved prior to nucleation for the two techniques.

Microstructural observations allow two basic conclusions to be drawn concerning laser surface melting of these alloys. Firstly, the presence of martensitic α', and the fine scale of the microstructures imply rapid cooling. Secondly, and in contrast, the degree of undercooling achieved prior to nucleation appears to have been rather small. This conclusion is based on the fact that in all cases the structures observed are consistent with those predicted by the equilibrium phase diagram. Also, analysis of the α' dendritic regions using EDX reveals very little Si to be in solution. Solute trapping has therefore not taken place to any significant degree; extended solid solubilities were not produced in any of the LSM alloys studied. This also points to a relatively small degree of undercooling prior to solidification.

The high temperature, Ti-rich region of the Ti-Si phase diagram is shown in Fig. 13, with several metastable lines added. The extended liquidus curve T_L is represented by a dotted line, and a T_0 curve and a glass transition region T_g (after ref. 7) have been added. In the case of the LSM specimens, it appears as though the degree of undercooling is insufficient to cool below the T_0 or T_g curves before the onset of nucleation and growth. The microstructures observed could therefore be predicted by the equilibrium phase

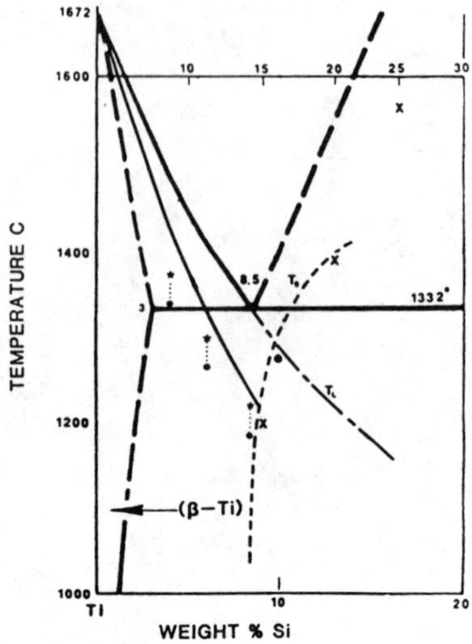

Figure 13 - High temperature, Ti-rich portion of the Ti-Si phase diagram showing the metastable lines T_L, T_0 and T_g. See text.

diagram, with the high cooling rate serving basically to refine the as-cast structure. Melt spinning, however, appears to yield a greater degree of undercooling. If the liquid can cool below the T_0 curve prior to nucleation then, from free energy considerations, conditions are favorable for the nucleation and growth of single phase β-Ti. This β may or may not transform to α-Ti at lower temperatures depending on alloy composition. The MS specimens of Ti-4wt.%Si, Ti-6wt.%Si, and Ti-8.5wt.%Si apparently do cool below the T_0 curve during solidification and single phase β does indeed nucleate (denoted by (•) in Fig. 13). The increase in temperature due to recalescence (as noted by *) is insufficient to overcome the large undercooling present, resulting in a structure of columnar β grains. The high Si content enables this β to be retained in the Ti-6wt.%Si and Ti-8.5wt.%Si alloys, with ω-Ti forming during the quench, whilst the Ti-4wt.%Si alloy transforms to martensitic α'.

The Ti-10wt.%Si alloy easily falls within the glass transition region noted by Suryanarayana (7) who produced amorphous structures for alloys ranging from 15at.%Si (9.4wt.%) to 20 at.%Si (12.5wt.%) (see Fig. 13).

The relatively small degree of undercooling achieved in these alloys using the LSM process is of interest since high cooling rates are usually associated with this technique. The reason for this may be explained as follows. The undercooling at the solid/liquid interface in LSM is of a kinetic origin. Thus, as heat is extracted through the substrate, an undercooling is established at this interface; the interface velocity will increase in response to this undercooling. If the rate of heat extraction is very rapid, the solid/liquid interface will be unable to achieve sufficiently rapid velocities, and so a relatively high degree of kinetic undercooling will result. For example, in the case of Al alloys, the coefficient of thermal conductivity is high, and so high degrees of kinetic undercooling may be expected. Indeed, Zindel et al (10) have shown essentially identical microstructures obtained for both LSM and MS, implying similar degrees of undercooling. However, in the case of Ti alloys, the coefficient of thermal conductivity is low, and so small degrees of kinetic undercooling will result.

Oxygen has been noted to affect the α-β transformation temperature in several alloys in such a way as to stabilize the α-phase (8). The microstructures seen for the oxygen-contaminated MS ribbon may be explained on this basis. As the molten sample is contaminated with oxygen during melt-spinning the alloy composition changes from a simple Ti-Si binary to a more complex Ti-Si-O system. The oxygen serves to stabilize the α-phase, causing the α-β transformation to shift to higher temperatures. When the charge is now ejected onto the copper wheel the undercooling achieved prior to nucleation is sufficiently large that the material begins to nucleate and grow the α-phase directly along with β. Si has a low solid solubility in the α-phase and so is rejected into the intercellular regions.

The β-phase is not retained on a large scale for two reasons. Firstly, if the concentration of oxygen is high β may never nucleate. Secondly, suppose an α grain and a β grain nucleate and grow adjacent to each other, then the α grain will quickly reject Si into the intercellular region between the two nuclei until a point is reached where the intermetallic Ti_5Si_3 begins to form. Once this occurs there is now a large driving force for the Si in the β-grain to diffuse to the intermetallic particle. This leaves an insufficient amount of Si in the β-grain for it to be retained to room temperature.

The third phase observed in the oxygen-rich samples could not be indexed as any known oxide of titanium silicide although its presence in the Ti_5Si_3 regions and similar crystal structure to the silicide suggests it may be oxygen rich Ti_5Si_3.

Ti-Cu-Er Alloys

Previous studies have shown no precipitation to occur during the rapid solidification of Ti-Er alloys (e.g. ref. 2). However, in this investigation fine erbia particles were present in Ti-Cu-Er alloys after LSM. Thus erbia is able to nucleate and grow before the $\beta \rightarrow \alpha$ transition occurs. This is thought to be the result of β-stabilization by Cu, which extends the temperature range over which β is present during cooling, and enhanced Er diffusion. It has recently been shown (9) that Er diffusion in β-Ti is much greater than in the α-phase. Thus, the precipitation of erbia is possible upon cooling.

After heat treating for 1 hr at 500°C erbia particles were evident in all the alloys. However, there appears to have been a decrease in the particle number density (i.e. number of particles per unit volume) with little increase in the mean particle size. This suggests that dissolution of the initial erbia particles has occurred upon heat treatment as particle coarsening cannot account entirely for the observed reduction in the volume fraction of erbia. This is thought to be due to the strained (metastable) nature of the α'-Ti matrix. After further heat treatment for 10 hrs at 700°C, the erbia then reprecipitates within the recrystallized α-Ti matrix.

Acknowledgements

This research was supported in part by the US DoE, Division of Materials Sciences, under contract DE AC02 76ER01198. The electron microscopy was performed using the facilities of the Center for the Microanalysis of Materials, operated by the Materials Research Laboratory of the University of Illinois. Thanks are due to Dr. Stephen Court for his helpful suggestions.

References

1. S.M.L. Sastry, T.C. Peng and L.P. Beckerman, Metall. Trans. A 15A (1984) 1465-1473.

2. D.G. Konitzer, J.T. Stanley, M.H. Loretto and H.L. Fraser, Acta Metall., in press.

3. R.G. Rowe, T.F. Broderick, E.F. Koch and F.H. Froes, ASM Conf. on "Rapidly Solidified Materials", San Diego, CA (1986).

4. R.G. Rowe and E.F. Koch, ibid.

5. L.S. Chumbley, Ph.D. Thesis, University of Illinois (1986).

6. J.C. Williams, D.H. Polonis and R. Taggart, "The Science, Technology and Applications of Titanium", Pergamon Press (1970) 733.

7. C. Suryanarayana, A. Inoue and T. Matsumoto, J. Mater. Sci. 15 (1980) 1193.

8. M.K. McQuillan, Met. Rev. 8 29 (1963) 41.

9. D.G. Konitzer, J.P.A. Lofvander, S,A, Court, R. Kirchheim and H.L. Fraser, submitted to Acta Metall.

10. J.W. Zindel, D.C. Van Aken, R.D. Field, P. Kurath and H.L. Fraser in "Mechanical Behavior of Rapidly Solidified Materials", TMS Publication, (1985) 189.

AGING RESPONSE OF RAPIDLY SOLIDIFIED

TITANIUM-TUNGSTEN ALLOYS

S. Krishnamurthy[a], D. Eylon[b], I. Weiss[c], A. G. Jackson[d], and F. H. Froes[e]

[a]Metcut-Materials Research Group, P.O. Box 33511
Wright-Patterson Air Force Base, OH 45433

[b]Metcut-Materials Research Group
now with the University of Dayton
Graduate Materials Engineering, Dayton, OH 45469

[c]Wright State University
School of Engineering
Dayton, OH 45435

[d]Systems Research Laboratories, Inc.
2800 Indian Ripple Road
Dayton, OH 45440

[e]Air Force Wright Aeronautical Laboratories
Materials Laboratory, AFWAL/MLLS
Wright-Patterson Air Force Base, OH 45433

Abstract

The effects of rapid solidification and aging of Ti alloys with monotectoid forming W and further additions of eutectoid forming elements such as Ni and Si were investigated. The alloys (in wt%) Ti-36W, Ti-19W-2.6Ni, and Ti-17W-2Ni-0.3Si were rapidly solidified (RS) into ribbons by the chill block melt spinning (CBMS) technique. The RS ribbons were heat treated at two sub-eutectoid temperatures (550°C and 650°C) for varying times (1 minute to 170 hr) to evaluate aging response. Aging of Ti-36W at 550°C and 650°C resulted in hardness peaks of 650Hv and 600Hv, respectively, after 1 hr. The effect of further alloying with Ni, an active eutectoid former, was evident in the aging of Ti-19W-2.6Ni. It showed a broadening of the 550°C age hardening peak although the hardness maximum still occurred at 1 hr. The combined effect of active and sluggish eutectoid forming additions was studied in the Ti-17W-2Ni-0.3Si alloy. A hardness maximum of 600Hv was observed after 10 minutes of aging at 550°C. The latter two alloys did show a gradual hardness decrease when aged at 650°C with no apparent peak. The hardness peaks in aged Ti-36W alloy were associated with the precipitation of omega phase from the parent beta phase. The hardness drop is due to rapid coarsening and dissolution of omega phase leading to equilibrium phases of HCP alpha Ti and BCC beta W. Additions of Ni as well as Si appear to suppress the formation of omega in the as-produced ribbons. The peak broadening in the ternary alloy containing Ni could be attributed to the formation of a "pre-omega" phase. However, the exact mechanism, as well as the peak shift in the quaternary alloy with Ni and Si additions, were not ascertained.

Introduction

Rapid solidification (RS) processing of titanium alloys has been recently investigated [1-3]. RS has been applied to many compositions including beta-eutectoid forming alloys that are prone to segregation when produced by ingot metallurgy (IM) methods [4, 5]. These studies have demonstrated several advantages of the RS route over the conventional IM approach such as improved homogeneity, microstructural refinement and formation of novel structures. The present work on RS binary and higher order alloys based on the Ti-W system is an extension of this alloy study program.

In the past, some interest was shown in the development of high strength titanium alloys containing W additions. Early work by Cross [6] on Ti-W binary alloys containing up to 20wt%W (prepared by powder metallurgy method) indicated an increase in tensile strength with increasing W level. The Ti-20W alloy in furnace-cooled condition exhibited a tensile strength of $105kg/mm^2$ and 4-5% elongation. More interesting properties were reported by Craighead et al., [7] in binary titanium alloys with small ternary additions of W. The addition of 1W to Ti-5Mn raised the tensile strength from $110kg/mm^2$ to $135kg/mm^2$ with a minimal elongation loss. However, the microstructural details are unavailable from these programs.

The phase diagram of arc-melted Ti-W was first investigated by Maykuth et al. [8]. A miscibility gap between the beta Ti and the beta W solid solutions was reported which covered a wide composition range and persisted up to a peritectic reaction L + (W) → (beta Ti) at 1880°C. Further it was found that the addition of W caused a gradual decrease in the beta transus temperature of titanium from 882°C to about 715°C where the beta phase containing ~28W undergoes a eutectoid decomposition: beta Ti (28W) → alpha Ti (0.8W) + W (4Ti). A re-evaluation of this phase diagram at higher temperatures was conducted by Rudy and Windisch [9] in which powder metallurgy alloys were used and metallographic and x-ray analyses were conducted. A complete range of solid solutions between beta Ti and beta W above a critical point at about 1250°C with no peritectic reaction was reported. The peritectic observed by Maykuth et al. [8] was attributed to insufficient homogenization of arc-melted alloys. The phase diagram assessed after this re-examination [9] is shown in Figure 1 [10]. The reaction (beta Ti) → (alpha Ti) + W should, now, be termed a "monotectoid" instead of a "eutectoid" decomposition.

It has been shown that the beta phase cannot be retained by quenching of Ti-W alloys containing less than ~20W [8, 11-13]. Quenching from the beta phase field may lead to three different metastable phases depending upon alloy composition. For alloys containing up to 8W, quenching leads to the formation of HCP martensite α' that is normally observed in binary titanium alloys. When alloys containing 8-20W are quenched, α'' martensite having an orthorhombic structure is formed. Quenching of alloys with 20-30W results in a beta + ω structure [12, 13]. These omega precipitates are coherent with the beta matrix and have an ellipsoidal morphology associated with a low misfit of the two structures [14].

Previous work on aging of Ti-W alloys with >20W at 700°C and lower temperatures [8] showed that the precipitation reactions in these alloys are extremely sluggish. Using results of light microscopy, it was suggested that the aging time for the completion of precipitation is almost 300 hr for a Ti-20W alloy. Longer times were noted at higher W levels [8]. A more detailed study of the sequence of precipitation from α'' martensite was made in a Ti-14W alloy by using TEM [15]. Tempering of this martensite at 550°C for 16 hr resulted in the precipitation of thin alpha platelets throughout the structure. With further tempering, a cellular reaction was initiated at the prior beta grain boundaries. The product of the cellular reaction was a lamellar alpha Ti + beta W structure which consumed the α + α'' mixture. This precipitation reached 95% completion after 36 hr.

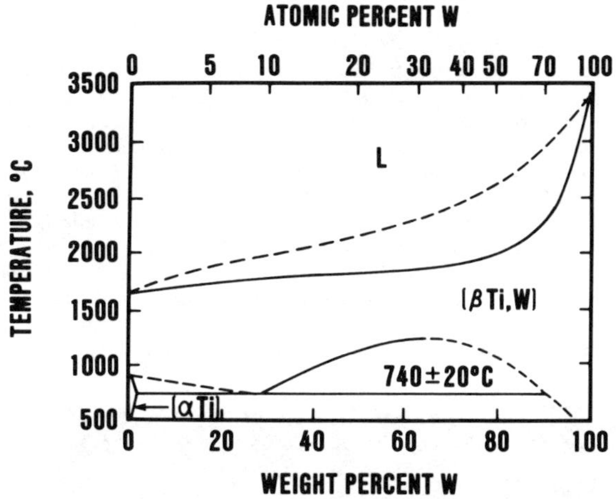

Figure 1 - The titanium-tungsten phase diagram [10].

Based on these considerations, it is clear that data on precipitation phenomena in Ti-W alloys with a retained beta or beta + ω microstructure are lacking. Furthermore, there is no published information about age hardening in these alloys. The effect of other beta-stabilizing additions on precipitation kinetics of Ti-W alloys is also unknown. The present work was undertaken to address these questions and to explore the benefits of applying RS processing to this group of alloys.

The alloys selected were (in wt%): Ti-36W, Ti-19W-2.6Ni, and Ti-17W-2Ni-0.3Si. The addition of the beta-stabilizers Ni and Si was made primarily to determine their effect on precipitation kinetics and aging response. Ni and Si are active and sluggish eutectoid formers, respectively, and were added to result in a range of aging rates.

Experimental Procedure

A 150 gm cigar-shaped button (25 mm in diameter) of each alloy composition was tungsten arc-melted from the elemental components. The buttons were remelted and solidified at least five times to ensure homogeneity and were swaged at 1040°C through successive dies to a final diameter of 10 mm. The swaged bars were remelted and rapidly solidified by the chill block melt spinning (CBMS) process [16] into ribbons which were about 2 mm wide and 50 µm thick. The RS ribbons were aged in vacuum at 550°C and 650°C for times ranging up to 170 hr.

The microstructures of as-produced and heat treated ribbons were examined by light microscopy, SEM, and STEM techniques. Aging kinetics were followed by microhardness measurements on mounted ribbon cross-sections, using a 136° apex diamond pyramid indentor at a test load of 50 gm. The hardness readings were taken around the middle plane of the ribbon thickness to minimize errors due to surface effects and structural inhomogeneities near the surfaces. At least 20 indentations were made on each specimen and average hardness values calculated.

Results and Discussion

The longitudinal cross-sections of as-produced alloy ribbons are shown in Figure 2. The Ti-36W alloy ribbon (Figure 2a) consists of a homogeneous region extending from the quenched wheel-side (bottom of picture) up to about the middle plane where fine columnar dendrites are observed. Beyond this region, the microstructure consists of equiaxed beta W dendrites (light areas marked by arrows) surrounded by beta Ti matrix leaner in W. The formation of these equiaxed dendrites indicates that the solidification rate of this tungsten-rich alloy ribbon was not rapid enough to completely suppress the dendritic segregation. By comparison, the microstructure of Ti-19W-2.6Ni (Figure 2b) and Ti-17W-2Ni-0.3Si (Figure 2c) ribbons appear to be more homogeneous. The ternary alloy showed columnar grains, each consisting of several columnar dendrites. However, the quaternary alloy showed only columnar dendrites with no grain boundaries.

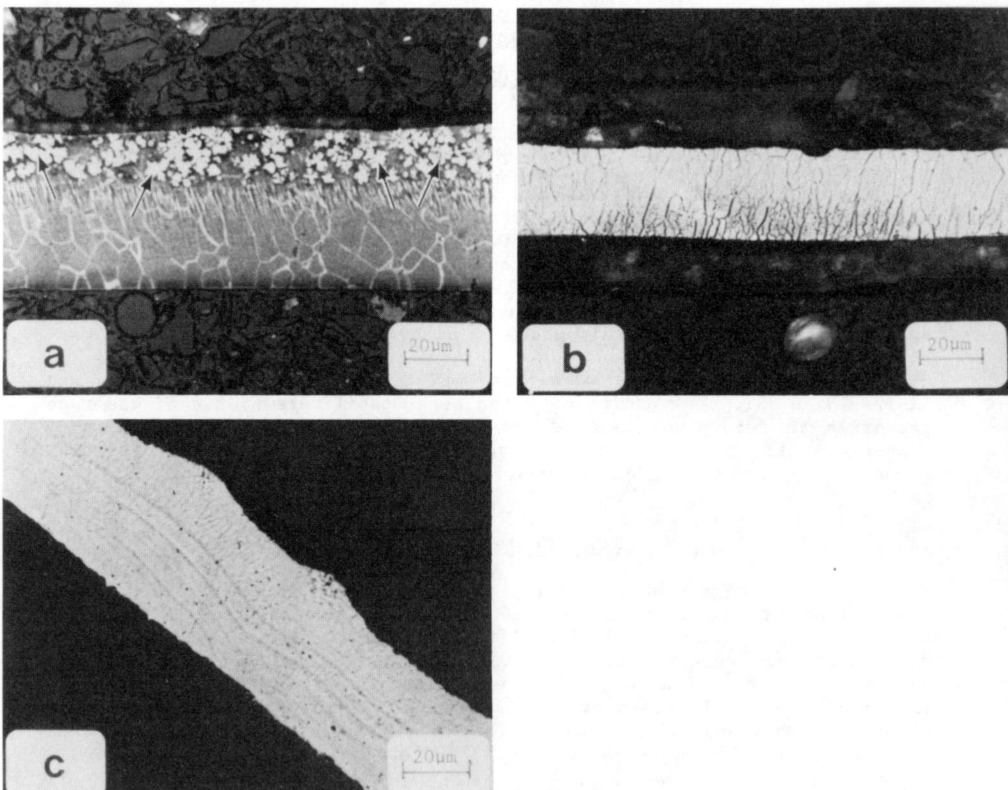

Figure 2 - Optical micrographs of as-produced RS ribbons: (a) Ti-36W, (b) Ti-19W-2.6Ni, and (c) Ti-17W-2Ni-0.3Si.

The aging response of the three RS alloys at 550°C and 650°C is plotted in Figures 3 through 5. Each alloy exhibits an age hardening peak at 550°C, whereas only the binary alloy shows a hardness peak for the 650°C aging. The age hardening effect is most noticeable in the binary alloy Ti-36W where hardness maxima occur after 1 hr of aging. In the case of Ti-19W-2.6Ni, aging at 550°C produces a wide hardness peak between aging times of 1 minute and 1 hr. Upon aging at 650°C, the hardness of this material gradually decreases. The hardness maximum in Ti-17W-2Ni-0.3Si occurs after 10 minutes at 550°C. Aging at 650°C led to a moderate drop in hardness.

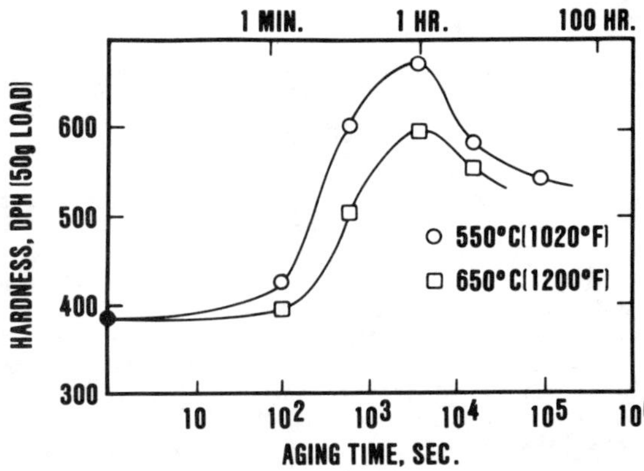

Figure 3 - The variation of microhardness with aging time for Ti-36W RS ribbons aged at 550°C and 650°C.

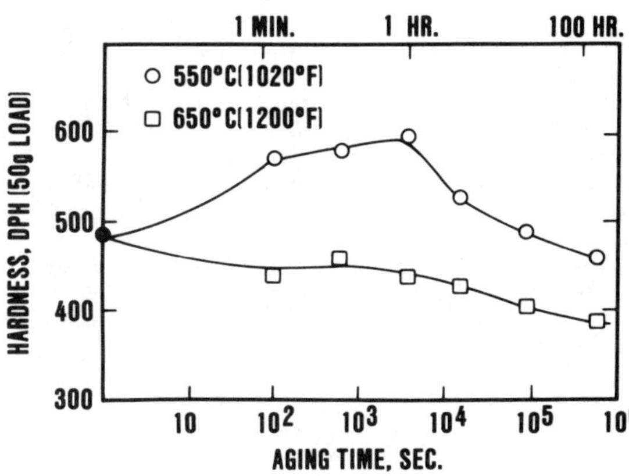

Figure 4 - The variation of microhardness with aging time for Ti-19W-2.6Ni RS ribbons aged at 550°C and 650°C.

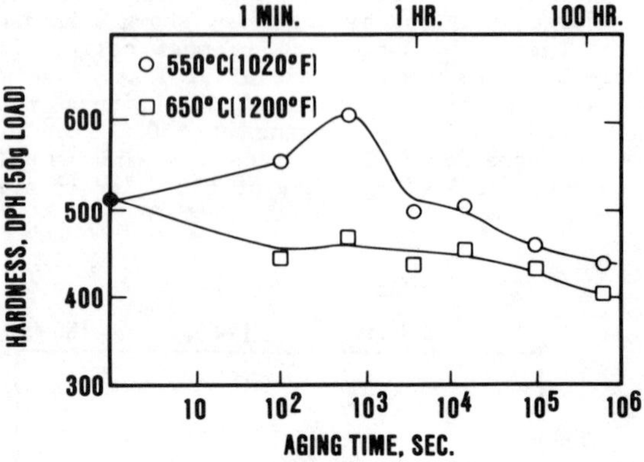

Figure 5 - The variation of microhardness with aging time for Ti-17W-2Ni-0.3Si RS ribbons aged at 550°C and 650°C.

Optical metallographic characterization of the aged ribbons indicates a coarser precipitation product at the prior-beta grain boundaries accompanied by an unresolved fine-scale precipitation within the beta grain structure (Figure 6). Proper etching of the Ti-36W alloy was very difficult as shown in Figure 6a. The coarser precipitation at grain boundaries in the other two alloys is evident from Figures 6b and 6c. Determination of the phases present in these rapidly solidified alloys required an analysis of TEM electron diffraction patterns which will be reported elsewhere [17]. TEM work on Ti-36W alloy ribbons in the as-solidified condition and 550°C/10 minutes aged condition has shown fine precipitates in both cases. These were identified as ω-phase through diffraction pattern analysis from various grains. The morphology of the omega phase is ellipsoidal, major axis along <111> beta, as expected, since the mismatch in lattice parameters between beta W and beta Ti is small, less than 5%. Examination of the ternary and quaternary alloys has revealed that both Ni and Si tend to suppress the formation of omega phase in the as-solidified ribbons. Aging of the Ti-19W-2.6Ni alloy ribbons at 550°C for ∿1 minute produced an unidentified phase which upon continued aging transforms to either omega or alpha Ti, as evidenced by the hardness peak broadening. A similar "pre-omega" age hardening effect has been previously observed in a Beta III alloy [18]. Beyond the hardness peak, microstructural coarsening is considered to be the cause for the drop in hardness. The "pre-omega" hardening effect is absent when aging is carried out at 650°C. Aging of the quaternary alloy containing both Ni and Si shows a marked peak at 550°C.

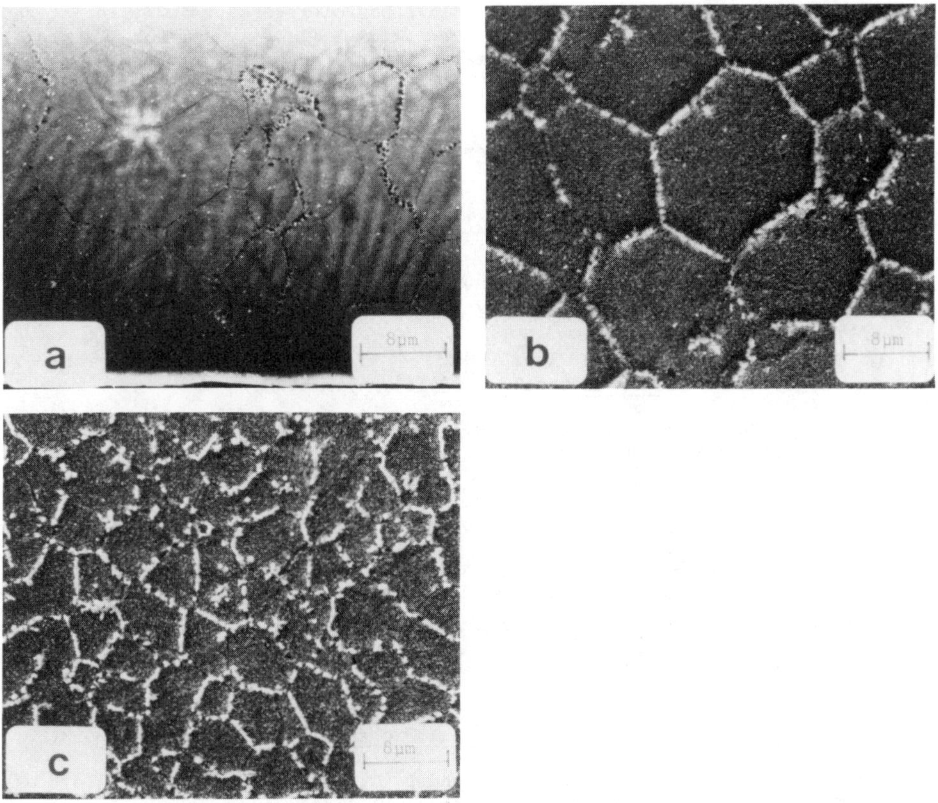

Figure 6 - Scanning electron micrographs of RS ribbons aged at 550°C for 24 hr: (a) Ti-36W, (b) Ti-19W-2.6Ni, and (c) Ti-17W-2Ni-0.3Si.

Summary

From the results for the binary alloy Ti-36W [17], it is clear that ω was present immediately following solidification. Aging produced a higher volume fraction of ω resulting in a peak in hardness. Continued aging induced a dissolution of the ω and production of alpha Ti and beta W which led to softening along with the microstructural coarsening. Addition of Ni to alloy appears to suppress the initial formation of ω. Aging at 550°C produced a transformation to an unidentified phase which upon further aging yielded either ω or alpha Ti, as evidenced by the peak broadening of the hardness curve. Longer aging yielded results similar to those seen in the binary alloy. Si addition shows a strong effect on shifting the 550°C aging hardness peak to shorter times.

Acknowledgments

S. Krishnamurthy, A. G. Jackson, and D. Eylon gratefully acknowledge support provided by AFWAL Contract Nos. F33615-82-C-5001, F33615-83-C-5073, and F33615-82-C-5078. The authors would like to express their appreciation to Ms. Karen A. Sitzman for typing the text, R. E. Omlor for TEM foil preparation, S. Ernst for hardness measurements, and R. D. Brodecki for SEM work.

References

1. F. H. Froes and J. R. Pickens, "Powder Metallurgy of Light Metals for Demanding Applications," Journal of Metals, 36(1)(1984), 14-28.

2. S. M. L. Sastry, T. C. Peng, P. J. Meschter, and J. E. O'Neal, "Rapid Solidification Processing of Titanium Alloys," Journal of Metals, 35(9)(1983), 21-28.

3. S. H. Whang, "Rapidly Solidified Ti Alloys Containing Novel Additions," Journal of Metals, 36(4)(1984), 34-40.

4. S. Krishnamurthy, A. G. Jackson, D. Eylon, R. R. Boyer, and F. H. Froes, "Rapidly Solidified Microstructures and Precipitation in a Ti-5.5w/o Ni Alloy," in Rapidly Quenched Metals, ed. S. Steeb and H. Warlimont (Amsterdam: Elsevier Science Publishers B.V., 1985), 945-948.

5. S. Krishnamurthy, I. Weiss, P. E. Blosser, D. Eylon, and F. H. Froes, "Aging Response of a Rapidly Solidified Beta-Eutectoid Ti-9 wt% Co Alloy," in Strength of Metals and Alloys, ed. H. J. McQueen, J-P. Bailon, J. I. Dickson, J. J. Jonas, and M. G. Akben, (Oxford: Pergamon Press, 1985), 1627-1632.

6. H. C. Cross, "Mechanical Properties of Wrought Titanium Alloys Made by Arc Melting or by Sintering," Metal Progress, 55(3)(1949), 356-358.

7. C. M. Craighead, O. W. Simmons, and L. W. Eastwood, "Ternary Alloys of Titanium," Journal of Metals, (March 1950), 514-538.

8. D. J. Maykuth, H. R. Ogden, and R. I. Jaffee, "Titanium-Tungsten and Titanium-Tantalum Systems," Journal of Metals, (February 1953), 231-237.

9. E. Rudy and St. Windisch, in Trans. AIME, 242 (1968), 953-954.

10. J. L. Murray, "The Ti-W (Titanium-Tungsten) System," in Bulletin of Alloy Phase Diagrams, 2(2)(1981), 192-196.

11. P. Duwez, "The Martensite Transformation Temperature in Titanium Binary Alloys," in Trans. ASM, 45 (1953), 934-940.

12. Iu. A. Bagariatskii, G. I. Nosova, and T. V. Tagunova, J. Inorg. Chem. USSR, 3 (1958), 330-341.

13. Iu. A. Bagariatskii, G. I. Nosova, and T. V. Tagunova, "Factors in the Formation of Metastable Phases in Titanium-Base Alloys," Sov. Phys. Dokl., 3 (1958), 1014-1018.

14. H. Ikawa, S. Shin, M. Miyagi, and M. Morikawa, "Some Fundamental Studies on the Phase Transformation from Beta Phase to Alpha Phase in Titanium Alloys," in Titanium Science and Technology, vol. 3, ed. R. I. Jaffee and H. M. Burte (New York: Plenum Press, 1973), 1545-1556.

15. J. C. Williams and B. S. Hickman, "Tempering Behavior of Orthorhombic Martensite in Titanium Alloys," Met. Trans. 1 (1970), 2648-2650.

16. S. H. Whang and B. C. Giessen, "An Arc Furnace Melt Spinner for the RSR Processing of Refractory and Reactive Alloys," in Rapid Solidification Processing: Principles and Technologies, III, ed. R. Mehrabian (Gaithersburg, MD: National Bureau of Standards, 1982), 439-442.

17. S. Krishnamurthy, A. G. Jackson, I. Weiss, and F. H. Froes, "Aging Response of Rapidly Solidified Titanium-Tungsten Alloys with Nickel and Silicon Additions," in Rapidly Solidified Materials, ed. P. W. Lee and R. S. Carbonara (Metals Park, OH: American Society for Metals, 1985), 121-127.

18. F. H. Froes, C. F. Yolton, J. M. Capenos, M. G. H. Wells, and J. C. Williams, "The Relationship Between Microstructure and Age Hardening Response in the Metastable Beta Titanium Alloy Ti-11.5Mo-6Zr-4.5Sn (Beta III)," Met. Trans. A, 11A (1980), 21-31.

FUNDAMENTAL STUDIES OF THE MICROSTRUCTURES OF

RAPIDLY SOLIDIFIED Ti_3Al+Zr AND Ti_3Al+Nb

K. R. Teal[a], A. G. Jackson[b], D. Eylon[c], and F. H. Froes[a]

[a]Air Force Wright Aeronautical Laboratories
Materials Laboratory, AFWAL/MLLS
Wright-Patterson AFB, OH 45433

[b]Systems Research Laboratories, Inc.
2800 Indian Ripple Road, Dayton, OH 45440
work conducted at AFWAL, Materials Laboratory

[c]Metcut-Materials Research Group
now with the University of Dayton
Graduate Materials Engineering
Dayton, OH 45469

Abstract

The microstructures of Ti_3Al+1Zr (α_2+Zr) and Ti_3Al+Nb (α_2+Nb) produced by ingot metallurgy (IM) and by rapid solidification (RS) are compared. Grain refinement was observed in both RS Ti_3Al+Zr and RS Ti_3Al+Nb alloys, and heat treatment of RS Ti_3Al+Nb ribbons produced refined α_2 microstructures compared with those observed in the IM material. Ti_3Al+Nb RS ribbons yielded good room temperature tensile strength and elongation, and were successfully compacted by extrusion at 925°C (1700°F), resulting in a fine α_2 structure.

Introduction

Research toward developing ingot metallurgy (IM) titanium aluminide aerospace alloys has been pursued for the last twelve years. The impetus came from the fact that stoichiometric Ti_3Al is an ordered intermetallic compound with high temperature strength, stiffness and oxidation resistance [1]. The higher aluminum content makes the aluminides somewhat less dense than titanium alloys and appreciably less dense than the nickel-base superalloys which they are intended to replace, and at the same time increases the oxidation protection [1]. However, the disadvantage of IM titanium aluminides is their very low room temperature ductility [1]. The objective of the present work was to study the potential of using rapid solidification (RS) technology for titanium aluminide base alloys, since RS offers many advantages above that of ingot metallurgy, among these being refinement of the microstructure, extension in solid solubility and promotion of disordered structures [2] which could lead to improvements in ductility. The two alloys selected were Ti_3Al+Zr close to the stoichiometric composition, and a Ti_3Al+Nb alloy representative of this class of alloys. A microstructural comparison between IM and RS products was conducted in both the Ti_3Al+Zr and Ti_3Al+Nb alloys. This was a necessary step to identify initially what RS can offer and in what manner it can be used to improve the room temperature ductility of these ordered alloys.

Experimental

Material and Characterization

The Ti_3Al+Zr alloy was beta extruded at 1200°C (2200°F) to bar form. The Ti_3Al+Nb material was a β-processed cast ingot, which was extruded and swaged below the beta transus temperature at 990°C (1800°F) to rod. Rods of both alloys were converted into RS ribbon in vacuum by the pendant drop melt extraction (PDME) process [3]. Heat treatments of the RS Ti_3Al+Nb ribbons were carried out in evacuated quartz capsules. For consolidation purposes, Ti_3Al+Nb ribbons were cut into approximately 10 mm length segments, chopped to form 100 grams of 0.8 mm mesh size flakes in a commercial fiber chopper, and placed into a Ti-6Al-6V-2Sn can with a CP titanium insert, sealed, and extruded 10:1 below the beta transus temperature at 925°C (1700°F) to produce a fully dense compact. Metallographic samples were prepared by conventional techniques and were etched with Kroll's reagent. Optical photomicrographs were taken to show the general microstructure. A more detailed microstructural analysis was conducted by scanning and transmission electron microscopy using a JEOL 100CX and a 2000FX microscopes.

Ribbon Tensile Testing

Room temperature tensile testing was conducted on four as-produced RS Ti_3Al+Nb ribbon segments using a ribbon tensile tester equipped with a 50 lb load cell at a crosshead speed of 0.125 mm/minute. The ribbon tensile samples were cut to approximately 40 mm in length and were mounted on aluminum projector slide frames prior to testing to allow proper fixturing in the ribbon tester grips. The width and thickness of the samples are listed in Table I.

Table I. Tensile Test Results of Unpolished PDME Ti_3Al+Nb Ribbons

(Crosshead Speed 0.005 in./min)

Sample	Thickness, in.	Width, in.	0.2% YS, ksi	0.2% YS, MPa	UTS, ksi	UTS, MPa	Elongation, %
1	.0038	.0675	123.2	849.5	138.2	952.9	3.7
2	.0038	.0690	133.5	920.5	147.2	1014.9	4.8
3	.0035	.0684	135.8	936.3	151.7	1045.9	6.3
4	.0032	.0487	139.9	964.6	148.4	1023.2	3.2
Avg.	.0036	.0634	133.1	917.7	146.4	1009.2	4.5

Results

Ti_3Al+Zr Microstructure

Figure 1 shows the microstructures of the as-received Ti_3Al+Zr material in both IM and RS ribbon forms. The IM method produced large (75 μm) α_2 grains; whereas, α_2 grain refinement to about 10 μm grain diameter occurred in the as-quenched RS ribbon.

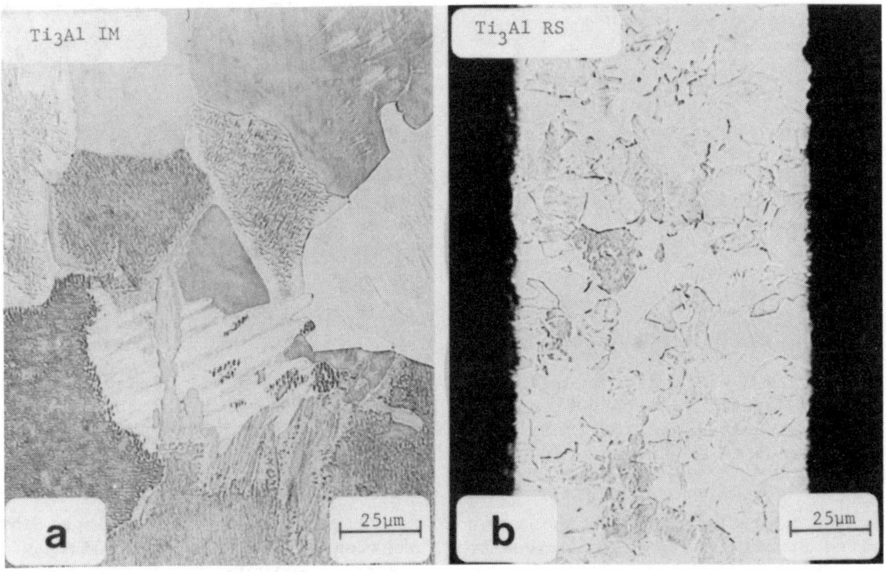

Figure 1 - Optical photomicrographs of (original magnification 600x): (a) IM Ti_3Al+Zr and (b) RS Ti_3Al+Zr (longitudinal section).

Ti_3Al+Nb Microstructure

Optically the β-processed IM Ti_3Al+Nb alloy displayed large lenticular α_2 lathes, as compared to the fine columnar structure which is beta (Figure 2) with small amounts of α_2, the latter which can be identified using TEM techniques [4] in the as-quenched RS ribbon. An equiaxed α_2 microstructure (Figure 2c) was produced from the IM material by extruding and swaging below the beta transus temperature at 990°C (1800°F) followed by a 990°C/2 hr/FC (furnace cool) heat treatment. However, annealing the as-quenched RS material at 900°C (1650°F) resulted in a much finer equiaxed α_2 structure in comparision with the IM alloy (Figure 2d).

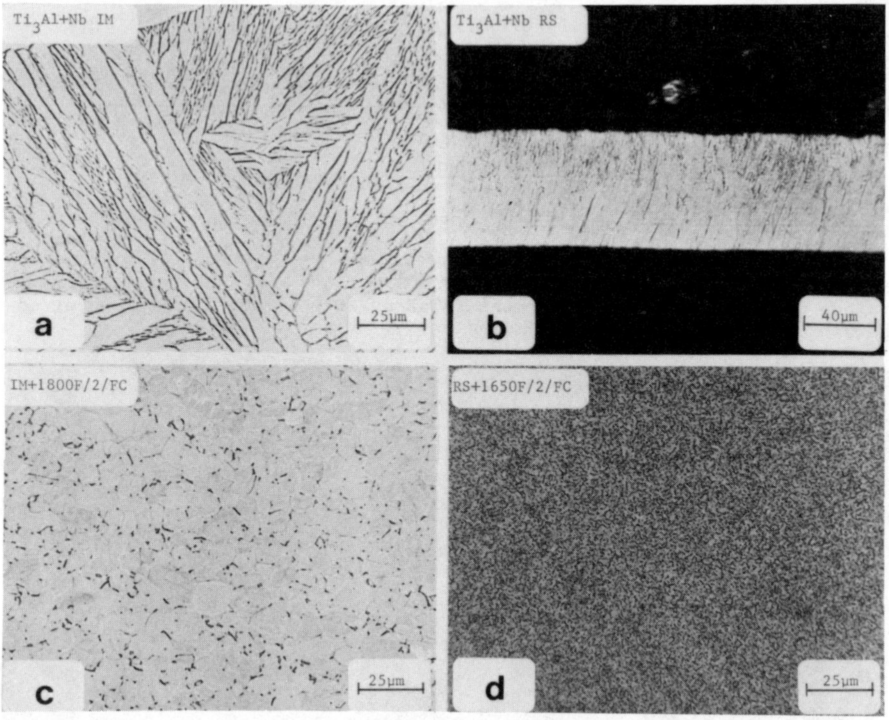

Figure 2 - Microstructures of Ti_3Al+Nb (original magnifications are noted): (a) Beta processed IM (600x); (b) As-produced RS ribbon, longitudinal section (250x); (c) 990°C (1800°F) extruded, as-heat treated IM (600x); and (d) 900°C (1650°F) annealed RS ribbon (600x).

The microstructure of the as-quenched RS ribbons can be developed into desired structures through various heat treatments resulting in a fairly wide range of microstructural control as demonstrated in Figure 3. The as-quenched RS Ti_3Al+Nb ribbon displayed a very fine α_2 structure (2 μm) after aging at low 700°C (1300°F) and intermediate 900°C (1650°F) temperatures as shown in Figures 3a and 3b. After heating at a temperature closer to the β-transus, 1170°C (2140°F), coarsening of the grains occurred, resulting in a duplex microstructure of equiaxed and lenticular α_2 (Figure 3c). This microstructure is similar to the α+β processed structure obtained in α+β titanium alloys.

Figure 3 - Optical photomicrographs (original magnification 600x) of: (a) RS Ti_3Al+Nb heat treated at 700°C (1300°F)/ 126 hr/AC; (b) RS Ti_3Al+Nb heat treated at 900°C (1650°F)/2 hr/ FC; and (c) RS Ti_3Al+Nb heat treated at 1170°C (2140°F)/0.2 hr/FC.

Antiphase Domain Size

Since the antiphase domain (APD) size is diffusion controlled and because of the short time spent at elevated tempeatures in the solid state during RS, the RS process was expected to have a strong influence on the APD size.

Examination of the Ti_3Al_2+Zr IM material and as-quenched RS ribbons by TEM showed that RS produced finer antiphase domains than quenched ingot material (Figure 4). The RS process was not expected to suppress the ordering reaction since the diffusion distances required are a few nanometers or less.

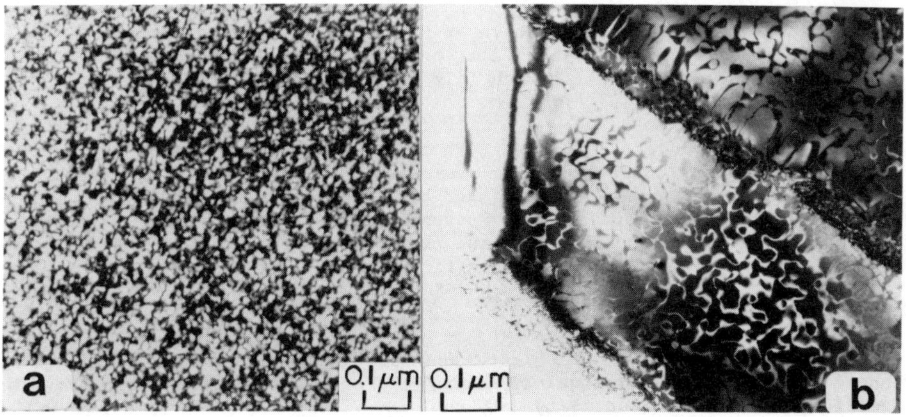

Figure 4 - TEM micrographs (bright-field) of: (a) Antiphase domain structure in RS Ti_3Al_2+Zr and (b) Antiphase domain structure in ingot Ti_3Al_2+Zr heat treated at 990°C (1800°F)/2 hr/FC.

Figure 5 is a diffraction pattern and higher order Laue zone (holz) ring from the Ti_3+Zr RS material for the $[0\bar{1}11]$ zone. Analysis of the pattern and ring shows that the ring is associated with the first order Laue zone (folz) and that the folz is shifted along a $(01\bar{1}2)$ vector by about $(0.5 \pm 0.2)g(01\bar{1}2)$, in good agreement with the theoretically predicted shift of 0.41g(0112), using a = 0.570 nm and c = 0.465nm [5].

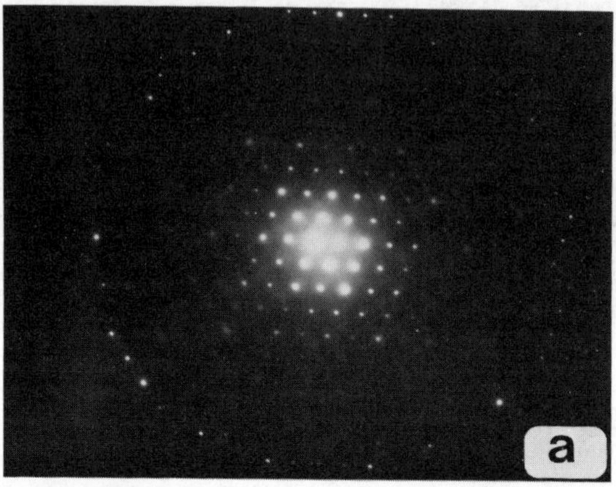

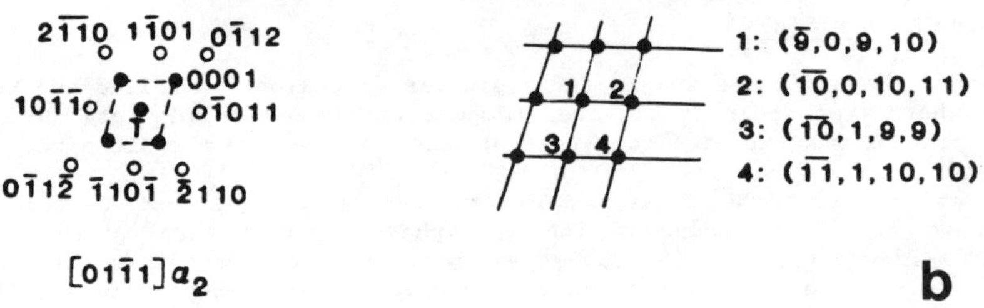

Figure 5 - Diffraction pattern of RS Ti_3Al_2+Zr showing (a) zero order laue zone (zolz) and first order laue zone (folz) ring and (b) schematic of (a) with folz shift indicated.

TEM examination of the RS Ti_3Al_2+Nb alloy showed the presence of equiaxed α_2 in a BCC β matrix. The optical observation of a columnar beta structure (Figure 2) as compared with the equiaxed beta structure in the TEM foil [4] as resolved by noting that the foil was prepared from the transverse orientation. Hence the column structures seen in the cross section are viewed end-on in the TEM foil. The α_2 present is on a much finer scale.

Ribbon Tensile Testing

Table I shows the tensile test results for the as-produced RS Ti_3Al+Nb ribbons. The elongation values obtained are encouraging for titanium aluminides since the ductility of the IM alloy at room temperature is extremely low. These elongations also represent minimum values, since in ribbon material elongation is limited due to geometrical reasons.

Extruded RS Ti_3Al+Nb

Figure 6a shows the RS Ti_3Al+Nb core and can of the extruded bar indicating the small diameter obtained from the extruded RS ribbons. A higher magnification (Figure 6b) reveals the fine α_2 structure of the ribbon compacted by extrusion, and the lack of porosity, indicating a fully consolidated product.

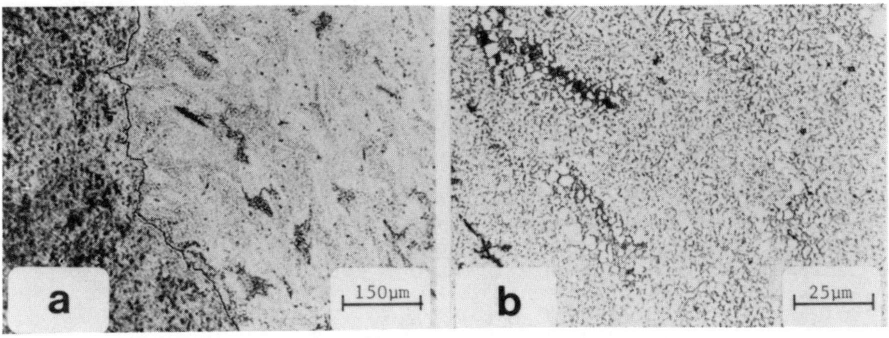

Figure 6 - Optical photomicrographs (original magnifications noted) of: (a) Compacted chopped RS Ti_3Al+Nb ribbon by extrusion (100x) and (b) Same as (a) except at 600x.

Discussion and Conclusions

Evaluation of the RS Ti_3Al+Zr ribbon indicates a significant α_2 size refinement in the RS material compared to the IM alloy. The observation of differences in APD size in the RS and IM Ti_3Al+Zr material (Figure 4) can be rationalized by noting first that the upper temperature for ordering of α_2 is 1150°C (2100°F) [6]. The quench drops the temperature of the material rapidly below the transformation temperature, causing nucleation of the α_2. As the temperature continues to drop, the APD's grow until diffusion is no longer significant, at about 700°C (1290°F) or less. Since, the RS material is only in the growth temperature range for a short length of time, a fine domain size results compared with IM alloys which has been subjected to thermal treatments or slower quench rates.

In the Ti_3Al+Nb alloy, a fine columnar structure of grain containing β and α_2 was present in the as-produced RS ribbon; whereas, the cast material of the same composition exhibited large lenticular α_2 grains. By annealing the RS ribbons at a low (700°C [1300°F]) or at an intermediate temperature (900°C [1650°F]) fine equiaxed α_2 microstructure resulted (Figures 3a and 3b), which is much finer than the equiaxed structure obtained in the IM material by hot-working followed by a 990°C (1800°F) heat treatment (Figure 2c).

Hence, various heat treatments with the RS Ti_3Al+Nb ribbon produced refined α_2 microstructures which are comparable to structures observed in conventional α+β titanium alloys (Figure 3c) and demonstrates that microstructural control is obtainable when dealing with RS products. Room temperature tensile testing of the as-produced Ti_3Al+Nb RS ribbons showed that the alloy possesses yield strengths comparable to those found in Ti-6Al-4V and reasonable ductility values (4-6%), which would be expected to be higher in thicker material. RS Ti_3Al+Nb ribbons were successfully compacted by extrusion at 925°C (1700°F) to produce a fully dense consolidated product exhibiting a fine α_2 microstructure.

Acknowledgments

This work was partially supported by AFWAL Materials Laboratory Contract Nos. F33615-83-C-5073 and F33615-82-C-5078. The authors thank R. Sweeney and I. A. Martorell for preparation of the button melts and the extrusion work, J. G. Paine for the conversion of the alloys to ribbon, W. A. Houston for metallographic assistance, D. K. Patrick for the tensile testing, J. S. Paine for typing the manuscript, and to Dr. H. A. Lipsitt for helpful discussions.

References

1. H. A. Lipsitt, "Titanium Aluminides - An Overview," in High-Temperature Ordered Intermetallic Alloys, vol. 39, ed. C. C. Koch, C. T. Liu, and N. S. Stoloff (Pittsburgh, PA: Materials Research Society, 1985), 351-364.

2. F. H. Froes and R. G. Rowe, "Rapidly Solidified Titanium," in Rapidly Solidified Alloys and Their Mechanical and Magnetic Properties, ed. B. C. Giessen, D. Polk, and A. I. Taub (New York: North Holland, 1986), in print.

3. S. J. Savage and F. H. Froes, "Production of Rapidly Solidified Metals and Alloys," Journal of Metals, 36(4)(1984), 20-33.

4. A. G. Jackson, K. R. Teal, D. Eylon, F. H. Froes and S. J. Savage, "Microstructure of Rapidly Solidified Ti_3Al Alloys," in Rapidly Solidified Alloys and Their Mechanical and Magnetic Properties, ed. B. C. Giessen, D. Polk, and A. I. Taub (New York: North Holland, 1986), in print.

5. A. G. Jackson (submitted to J. of Microscopy).

6. P. L. Martin, H. A. Lipsitt, N. T. Nuhfer, and J. C. Williams, "The Effects of Alloying on the Microstructure and Properties of Ti_3Al and TiAl," in Titanium '80, Science and Technology, vol. 2, ed. H. Kimura and O. Izumi (Warrendale, PA: The Metallurgical Society, 1980), 1245-1254.

COMPARISON OF MELT SPUN AND CONSOLIDATED Ti_3Al-Nb ALLOYS

WITH AND WITHOUT A DISPERSOID

R. G. Rowe, J. A. Sutliff, and E. F. Koch

General Electric Corporate Research and Development
PO Box 8
Schenectady, NY 12301

Abstract

Titanium aluminide alloys containing 0, 5 and 7.5 a/o Nb and with and without rare earth additions were rapidly solidified by melt spinning. All alloys with rare earth additions exhibited a fine dispersed oxide or sulfide phase after aging. The pattern of dispersoid distribution suggested segregationless solidification in at least part of the columnar growth region, and cellular and dendritic segregation patterns in the upper portion of the ribbon. Ribbon was consolidated by HIP and extrusion at temperatures below the beta transus temperatures of the alloys. Consolidated alloys exhibited preferential dispersoid depletion and coarsening at grain boundaries which suggested transient Ostwald Ripening near a migrating grain boundary. High Nb alloys had the greatest resistance to grain boundary coarsening. Room temperature ductility of dispersoid-bearing alloys was less than for the same alloy without a dispersoid.

Introduction

The ordered compound Ti_3Al has good strength and creep resistance at high temperatures, but no ductility below 600°C (1-3). Refinement of slip length has been proposed as a means of modifying the low temperature deformability of Ti_3Al (4-7). The introduction of a dispersed phase may accomplish this. We have evaluated the effect of rapid solidification and the introduction of rare earth dispersoids on the microstructure and properties of Ti_3Al-Nb alloys. Alloys based upon Ti_3Al with 0, 5, 6, 7.5 and 10.5 at.% Nb were rapidly solidified both as base alloys and with the addition of dispersoids. Two different dispersoids were introduced, Er_2O_3, and Ce_2S_3 (8,9).

Experimental Procedure

The alloy compositions tabulated in Table 1 were rapidly solidified by melt spinning (10). Some of the alloys had aluminum losses, and there was some pickup of tungsten from the non-consumable arc melting electrode. Aluminum and tungsten levels were, hence, somewhat variable. This may have led to increased scatter in mechanical behavior, but averaged over the entire test matrix, did not influence the trends which we observed.

Table I. Alloy Compositions (Atom %)

Alloy	Ti	Al	Nb	Er	O	Ce	S	W
Ti_3Al	Bal.	25.0			0.05			0.16
Ti_3Al-Er	Bal.	24.8		0.6	0.2			0.08
Ti_3Al-Ce-S	Bal.	23.6			0.6	0.5	0.2	0.05
Ti_3Al-5Nb	Bal.	24.6	4.7					0.16
Ti_3Al-5Nb-Er	Bal.	20.0	4.7	0.5	0.4			0.19
Ti_3Al-5Nb-Ce-S	Bal.	23.5	4.9		0.6	0.6	0.2	0.04
Ti_3Al-7.5Nb	Bal.	23.3	7.7		0.3			0.13
Ti_3Al-7.5Nb-Er	Bal.	24.8	7.5	0.6	0.4			0.10
Ti_3Al-7.5Nb-Ce-S	Bal.	21.75	7.6		0.5	0.7	0.15	0.09
Ti_3Al-10.7Nb-Er	Bal.	25.3	10.7	0.5	0.3			

Melt spun ribbon was consolidated by hot isostatic pressing (HIP) at from 850°C to 950°C and extruded at 925°C. Alloys without Nb were extruded at a temperature of 1108°C.

Experimental Results

As-Melt-Spun and Aged Ribbon. Selected area electron diffraction (SAD) was used to analyze the crystal structure of as-melt-spun ribbon. Alloys based upon Ti_3Al had an as-melt-spun structure of ordered α_2 (hcp DO_{19}). As-melt-spun alloys based upon Ti_3Al-Nb had an ordered $\beta+\beta_2$ structure (bcc with B2 ordering)(11). Diffraction patterns of Ti_3Al-Nb alloys also had features which have been identified as indicative of the precipitation of omega phase (12). Aging at 750°C transformed the $\beta+\beta_2$ structure to ordered α_2, demonstrating the metastability of the as-melt-spun structure. There was no evidence of beta phase in aged ribbon, regardless of niobium content.

All as-melt-spun ribbons contained a dispersion of 10 to 50 nm particles which were identified by energy dispersive spectroscopy (EDS) and selected area electron diffraction, to be Er_2O_3, Ce_2S_3, or $Ce_4O_4S_3$, depending upon alloy additions (9,10,13). Heat treatment of the ribbon for one hour at 750°C resulted in further dispersoid precipitation. The aged microstructure is shown for Ti_3Al-2Er and Ti_3Al-10Nb-2Er, in Figures 1 (a) and (b), respectively. Dispersoid particles which precipitated during aging were approximately one third to one fifth the diameter of those found in as-melt-spun ribbon, and had a higher particle density. The alloys containing niobium, which quenched to a $\beta+\beta_2$ microstructure, had the finest aged dispersion.

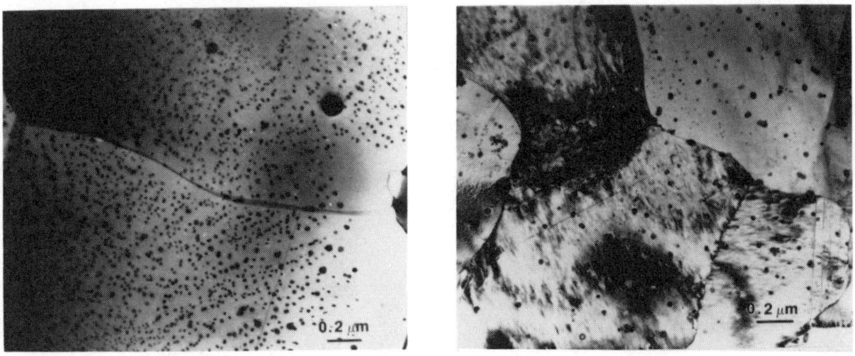

(a) Ti_3Al-Er, (b) Ti_3Al-10Nb-Er

Figures 1. TEM micrographs of dispersoid distributions in melt spun ribbon aged at 750°C/1hr.

Consolidated Ribbon. The microstructure of compacts which were consolidated by HIP at 850°C was similar to that of annealed ribbon of the same composition. Dispersoid particles in grain interiors exhibited little coarsening at this temperature. Particles were coarsened on or near the grain boundaries during the anneal, however (14,15). A zone of dispersoid particle depletion adjacent to the grain boundaries occurred in all samples, particularly in Ti_3Al-Er. The degree of dispersoid coarsening at grain boundaries decreased with increasing niobium content. The effect of niobium on grain boundary coarsening was more pronounced at higher temperatures. Figures 2 (a) and (b) are TEM micrographs of Er_2O_3 dispersoid-bearing alloys with 6 and 10.7 at.% Nb respectively, following HIP at 950°C.

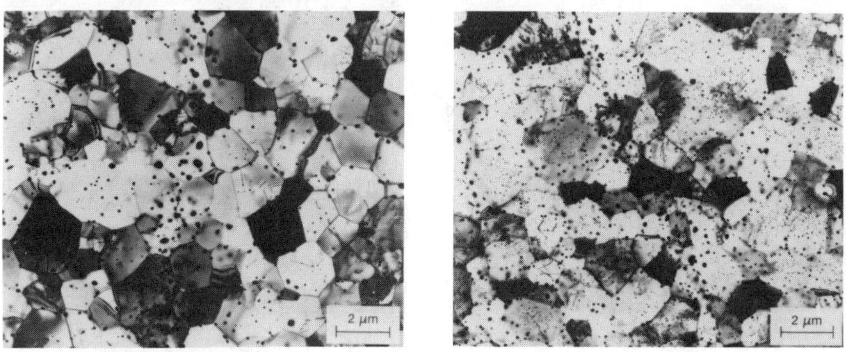

(a) Ti_3Al-6Nb-Er (b) Ti_3Al-10.8Nb-Er.

Figures 2. TEM micrographs of dispersoid distributions in 950°C HIP'ed alloys. A band of coarse particles from a prior ribbon boundary of the HIP compact is evident in both micrographs.

Some grain boundaries were observed to have dispersoid coarsening on one side and not the other. Figure 3(a) shows Ti_3Al-2Er after HIP at 850°C. This suggested that coarsening was greatest when the dispersoid particles were intercepted by a moving grain boundary. Large particles at the grain boundary were separated by a denuded zone from fine Er_2O_3 in the grain interior. The large particles appeared to have been dragged along with the grain boundary. Some particles, at A, were left behind by the grain boundary. Another particle, at B, appeared to be dropping behind as the boundary moved away, Figure 3(b).

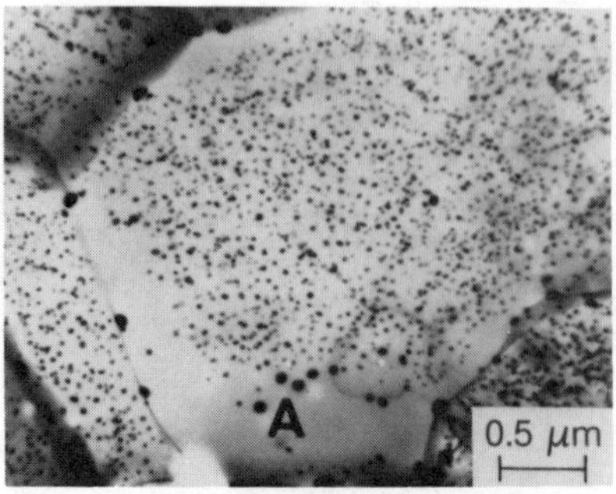

(a)

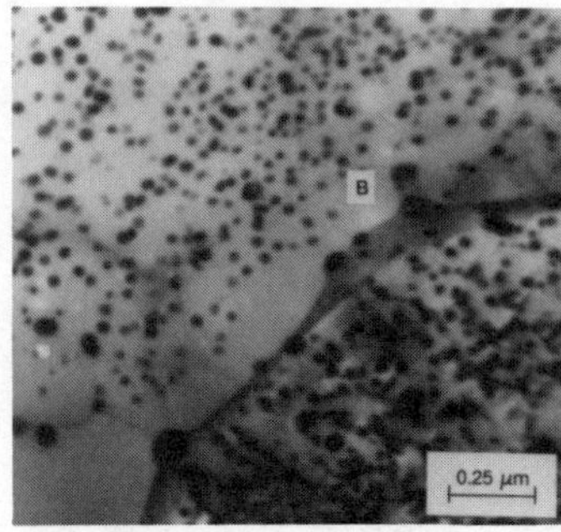

(b)

Figure 3. TEM micrographs of Ti_3Al-Er as-HIP'ed at 850°C.

The microstructure of Ti_3Al-Ce-S which was HIP'ed at 900°C is shown in Figure 4 (a). This alloy also had a bimodal dispersoid size distribution. Particles 20 to 40 nm were mixed with much finer particles 2.5 to 10 nm. In some cases, the fine particles were aligned in rows which appeared to correspond to solidification cell boundaries. The width of the cells was on the order of 0.1 to 0.15 micrometers. Figure 4 (b) shows the microstructure of Ti_3Al-7.5Nb-Ce-S as-HIP'ed at the same temperature. In this case, a homogeneous distribution of Ce_2S_3 particles suggested a region of segregationless solidification. The average dispersoid particle diameter was 5 nm. Grain boundary dispersoid coarsening was also evident in this alloy, but not to the same extent as for alloys with Er_2O_3 dispersoids.

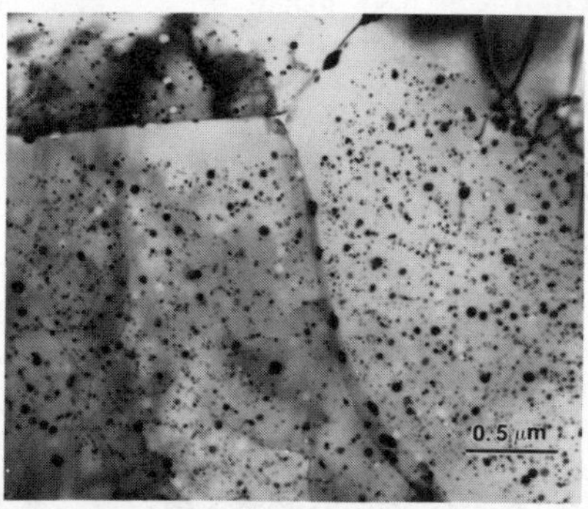

(a) Ti_3Al-Ce-S

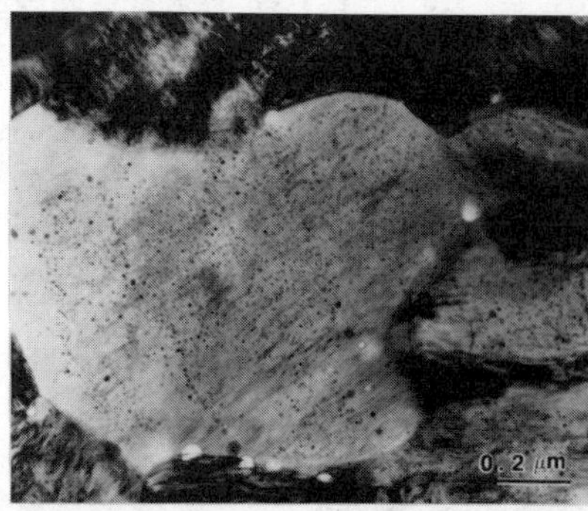

(b) Ti_3Al-7.5Nb-Ce-S

Figures 4. TEM Micrographs of As-900°C-HIP compacts.

Dispersoid Coarsening During Extrusion. Extrusion of HIP'ed alloys at 925°C caused severe coarsening of Er_2O_3 to an average diameter approaching 0.5 μm. The microstructure of Ti_3Al-5Nb-Er is shown in Figures 5 after (a) HIP at 900°C and (b) HIP at 900°C followed by extrusion (6:1) at 925°C. This same extreme dispersoid coarsening occurred in all alloys extruded at this temperature. Extrusion at 870°C caused dispersoid coarsening also, resulting in a dispersoid diameter much larger than observed after HIP at 900°C (13).

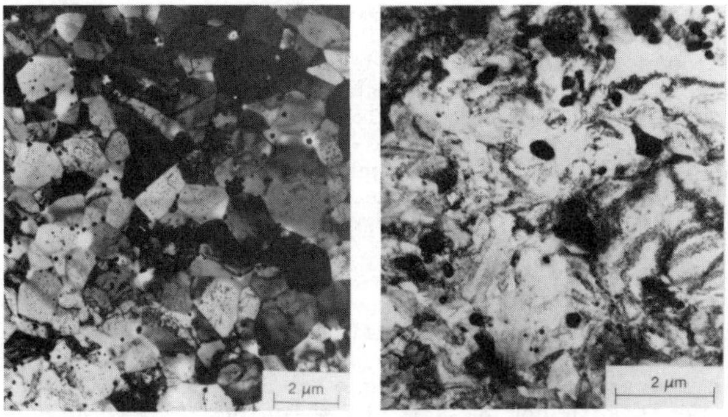

(a) HIP at 900°C

(b) HIP at 900°C followed by extrusion at 925°C.

Figure 5. The dispersoid distribution in melt spun Ti_3Al-5Nb-Er after HIP or HIP plus extrusion.

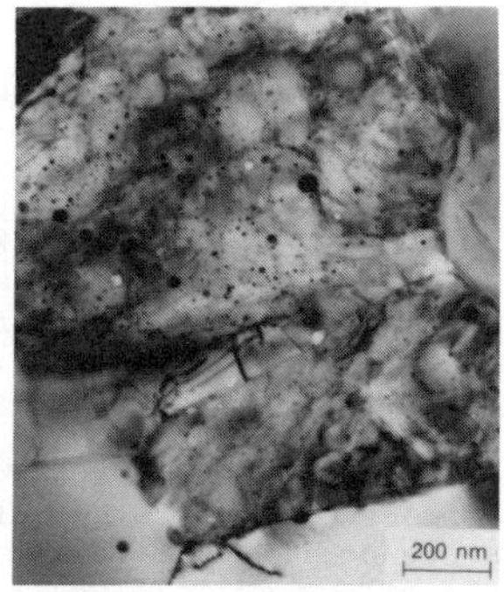

Figure 6. TEM micrograph showing the dispersoid distribution in melt spun Ti_3Al-5Nb-Ce-S after HIP at 900°C followed by extrusion at 925°C.

Extrusion under identical conditions produced little coarsening of the Ce-S based dispersoid. Figure 6 is a TEM micrograph of as-extruded Ti_3Al-5Nb-Ce-S. The dispersoid was extremely fine and closely spaced. There was little dispersoid coarsening in any area, but regions of dispersoid depletion were mixed with regions of very high dispersoid density. The dispersoid diameter ranged from 5-25 nm.

Room Temperature Ductility. The fracture toughness of consolidated samples with the coarse extruded dispersoid, such as in Figure 5(b), was lower than the same alloy without a dispersoid (10). This indicated that a very coarse dispersoid does not improve the low temperature deformability of Ti_3Al or Ti_3Al-Nb alloys. Because of this, it was necessary to test aged melt spun ribbons of the alloys in order to evaluate the effect of a fine dispersion on room temperature ductility. Ribbon bend tests (10) were conducted on as-melt-spun and aged ribbon of the Ti_3Al-Nb alloy series for niobium levels of 0 to 7.5 at.%. Although ribbon bend tests were far from quantitative because of specimen thickness and surface roughness variations, they provided a comparative test of ductility. All ribbons were tested with the wheel contact side of the ribbon facing toward the outside of the bend in order to maintain a relatively constant outer diameter surface condition. The results of bend tests on as-melt-spun ribbon are tabulated in Table 2(a) which gives the calculated outer fiber strain, ribbon outside bend diameter divided by its thickness (16), at the first detectable crack. The ductility results are an average of 25 tests for each value. It can be seen that as-melt-spun ribbon of some of the alloys was very ductile, capable of being bent back upon itself without cracking. The ductility of as-melt-spun alloys increased with the level of Nb. Furthermore, both Er_2O_3 and Ce_2S_3 dispersoid-bearing alloys had lower as-melt-spun ductility than alloys without a dispersoid.

Table 2. Percent outer fiber strain for melt spun Ti_3Al-Nb ribbon bent to failure (13).

	(a) As-Melt-Spun			(b) Aged Ribbon		
	0 Nb	5 Nb	7.5 Nb	0 Nb	5 Nb	7.5 Nb
Base Alloy	1.2	Ductile	Ductile	1.0	1.3	3.5
Er_2O_3 Dispersoid	1.0	4.3	8.3	1.0	1.5	1.8
Ce-S Dispersoid	1.1	4.1	2.0	1.3	1.2	1.8

As mentioned earlier, the as-melt-spun crystal structure of Ti_3Al-Nb alloys was $\beta+\beta_2$. Aging at 750°C produced fully ordered α_2, and precipitation of an ultrafine dispersoid. The bend test results on aged ribbon are shown in Table 2(b). The ductility of the aged alloys was consistently lower than for as-melt-spun alloys except for Ti_3Al-base alloys which had an ordered α_2 structure in both cases. Again, the bend ductilities of alloys with Er_2O_3 and Ce-S dispersoids were comparable, and generally lower than the same alloys without a dispersoid.

Discussion of Results

The dispersoid distributions observed in as-melt-spun and annealed ribbons can be understood in the context of the rapid solidification and solid state cooling precipitation events. It is reasonable to assume that all of the alloys solidified to a metastable bcc beta solid solution. After solidification, the structure of the matrix phase and the rate of solid state cooling determined the Er_2O_3 dispersoid size distribution. In the alloy without niobium the as-melt-spun structure was ordered α_2 (DO_{19}), while in the niobium-bearing alloys it was $\beta+\beta_2$. This difference in transformation path led to generally different dispersoid distributions in the Ti_3Al and Ti_3Al-Nb alloys. The Ti_3Al-Nb alloys exhibited a finer distribution of secondarily precipitated particles. This is reflected in the difference between 750°C aged Ti_3Al-Er and Ti_3Al-10Nb-Er, Figures 1 (a) and (b), respectively.

Finer secondary precipitation in the Ti_3Al-Nb alloys, is consistent with a longer time period in the beta phase, and secondary precipitation from α, or α_2 at a lower temperature. During solid state cool-down, Ti_3Al alloys containing a dispersoid transformed to the hexagonal phase at a high temperature (alpha transus temperature 1150°C, (17) and M_s above 1060°C (18)), while the highest temperature that Ti_3Al-Nb alloys were exposed to after transformation to the alpha phase was 750°C, during the aging heat treatment. Hence, the finer secondary precipitation of Er_2O_3 in these alloys.

<u>Grain Boundary Dispersoid Coarsening</u>. Microstructural features, such as in Figure 3, indicated that particle coarsening and the formation of particle depleted zones were influenced by grain boundary motion. As-HIP'ed or as-annealed samples of all alloys showed examples of grain boundary depletion. In some cases, depletion was the same on both sides of a grain boundary, suggesting a diffusive transport region on both sides of the boundary. Many cases of asymmetrical dispersoid depletion were also observed. In these instances, two phenomena appeared to have occurred. First, from the evidence of Figure 3(a), dispersoid particles were captured and dragged along with the migrating grain boundary. Second, while particles were captured by the grain boundary, two-dimensional Ostwald ripening occurred by grain boundary diffusion, dissolving the fine particles in favor of larger particles which can be seen in Figures 3, 4(a) and 6. The existence of coarsening at grain boundaries while little occurred in the bulk suggested discontinuous coarsening, where grain boundary diffusion dominates the coarsening process (19,20). Behind a moving grain boundary, fine dispersoid particles were depleted almost completely, leaving a denuded zone.

Large, spherical particles behind the grain boundary, at point A in Figure 3 appear to have either coarsened at the grain boundary and left behind, or coarsened by short circuit dislocation diffusion. A single dislocation can be seen to connect these particles. The same dislocation terminates at one of the large particles which is migrating with the grain boundary. Dislocation short circuit diffusion has been shown to cause preferential coarsening of particles where bulk diffusion is orders of magnitude slower (21).

The dragging of particles along with a moving grain boundary has been observed in other alloy-dispersoid systems, but only for the case of small particles capable of rapid internal mass transport. Particle mobility was only observed for low viscosity glassy phases in which internal diffusion was rapid. It was observed that crystalline dispersoids such as Al_2O_3 did not migrate in copper at 1075°C, which was attributed to the low self diffusivity of Al_2O_3 at this temperature (24).

The low solubility of Al and O in copper in the presence of Al_2O_3 prevented any significant diffusional transport of the particles by bulk diffusion. The reactivity of titanium and titanium aluminide reduces the solubility product of the oxide formation reaction (9,22) relative to that of less reactive metals, and permits greater mass transport in the matrix phase. Ostwald ripening of Er_2O_3 and Ce_2S_3 dispersoids in Ti and Ti_3Al alloys is evidence of this. This difference apparently permits migration of the crystalline oxide in Ti_3Al along with the moving grain boundary (20).

The driving force for particle dragging by a moving grain boundary is the maximum surface tension force, $F_p = \pi R \gamma$ where R is the particle radius, and γ is the surface energy of the grain boundary (19). The large particles which are dragged along with the grain boundary will continue to grow by two-dimensional Ostwald ripening or discontinuous coarsening (19,23) until their mobility is no longer high enough to maintain the same velocity as that of the grain boundary. The velocity, v, of a particle migrating under the action of a driving force, F_p is given by $v = F_p M$ where M is the particle mobility. The functional relationship between particle mobility and particle diameter has not been determined, but it is expected to follow a negative power relationship, as for self diffusion of the matrix phase (21 Particles above a critical diameter should drop behind the boundary. It can be seen that since the driving force on the particles is a function of the grain boundary energy, then the critical particle size should depend upon both the grain boundary surface energy as well as on mass transport factors. Hence, the size of large particles behind a grain boundary is influenced by surface energy as well as by elemental diffusivities.

Niobium Effect. Ti_3Al-Nb alloys with Nb contents of 7.5 to 10.5 at.% had much less grain boundary dispersoid depletion than alloys without Nb (13). The microstructure of two Ti_3Al alloys containing 6 and 10.5 at.% Nb are shown after HIP at 950°C (1740°F) in Figures 2 (a) and (b), respectively. The alloy containing 10.5 at.% Nb had virtually no dispersoid depletion, while the alloy with 6 at.% Nb had a severely coarsened dispersoid in all but the centers of the largest grains. The effect of niobium on grain boundary coarsening was observed to be the same from 850 to 950°C (1740°F) in a series of alloys with different Nb contents. The difference did not appear to relate to the formation of beta phase at grain boundaries since the alloy with 10.5 at.% Nb should have had beta phase present at a lower temperature than the alloy with only 6 at.% Nb. The inhibiting effect of high Nb levels upon grain boundary depletion may be related to a number of properties which are specific to the grain boundary (20). Both diffusivities and surface energies directly influence the rate of particle coarsening and the driving force and mobility of particles.

Dispersoid Coarsening During Extrusion. Extrusion at from 870°C to 950°C caused extensive Er_2O_3 dispersoid coarsening. Extrusion at 925°C eliminated almost all fine particles and caused coarsening to dispersoid sizes as large as 0.5 μm diameter. Adiabatic heating during extrusion could result in transient formation of beta phase and cause accelerated Ostwald ripening such as has been observed in beta heat treated alloys (14). Since the post-extrusion microstructures were all characteristic of a recrystallized single phase rather than the acicular structure which results from transformation of beta to alpha titanium, transient formation of beta phase would have to have been followed by recrystallization or grain growth during the post-extrusion cool down. Because of this, the precise cause of the coarsening during extrusion remains undefined.

Room Temperature Ductility. The ductility of as-melt-spun alloys appeared to be dominated by matrix structure effects. Alloys which had an ordered $\beta+\beta_2$ structure had significantly higher room temperature ductility than the Ti_3Al alloys with an ordered α_2 structure. Transformation of Ti_3Al-Nb alloys to an ordered DO_{19} structure reduced the room temperature ductility of all alloys. Thus, the α_2 structure appeared to have inherently less ductility than the ordered $\beta+\beta_2$.

The ductility of alloys without a dispersoid increased with increasing niobium content as has been previously reported for ingot metallurgy alloys of similar composition (2). This trend of increasing ductility with increasing niobium level was also observed for the dispersoid-bearing alloys, although the dispersoid reduced the ductility of the niobium-bearing alloys in both as-melt-spun and aged conditions. Because of this, it appeared that the ductility of the matrix remained the major factor contributing to ductility in both dispersoid-bearing and dispersoid-free alloys.

Conclusions

- Heat treatment or consolidation of melt spun Ti_3Al-Nb ribbon at 750°C to 900°C resulted in a high density of Er_2O_3 and Ce_2S_3 particles with a particle size between 5 and 20 nm.
- Dispersoid depletion and coarsening appeared to be dominated by grain boundary effects, rather than by bulk Ostwald ripening.
- Grain boundary depletion of fine dispersoids was observed, in some cases, to be related to grain boundary migration, and particle dragging along with the grain boundary.
- Alloys with the highest niobium content exhibited the least grain boundary coarsening.
- Extrusion, caused much greater coarsening of Er_2O_3 than Ce_2S_3 or cerium oxysulfide dispersoids.
- As-melt-spun Ti_3Al-Nb alloys which had an ordered $\beta+\beta_2$ structure, were significantly more ductile than after aging and transformation to the ordered α_2 structure.
- The room temperature ductility of dispersoid-bearing Ti_3Al-Nb alloys was less than that of the corresponding alloy without a dispersoid, and generally related to the ductility of the matrix phase.

Acknowledgements

The authors would like to acknowledge the contributions of Mr. Robert O. Auer to this study. He assisted in the development of the titanium melt spinning system, and the melt spinning of all of the alloys of this study. In addition, he conducted the ribbon bend tests.

Part of this work was supported by the General Electric Company Aircraft Engine Group under the Independent Research and Development Program.

References

1. S.M.L. Sastry and H.A. Lipsitt, "Ordering Transformations and Mechanical Properties of Ti_3Al and Ti_3Al-Nb Alloys", Met Trans A, 8A, pp. 1543-1552, (1977).
2. J.C. Williams and M.J. Blackburn, "The Structure, Mechanical Properties and Deformation Behavior of Ti-Al and Ti-Al-X Alloys", in Ordered Alloys, Ed. B.H. Kear, C.T. Sims, N.S. Stoloff, and J.H. Westbrook, Proc. 3rd Bolton Landing Conf., Sept. 1969, Claitors, 1970.

3. S.M.L. Sastry and H.A. Lipsitt, "Plastic Deformation Of TiAl and Ti_3Al", in Titanium '80, Science and Technology, Vol. 2, Proc. Int. Conf. on Titanium, Kyoto, Japan, May 1980, pp. 1231-1243, Pub. TMS/AIME, Warrendale, PA, 1980.

4. S.M.L. Sastry and H.A. Lipsitt, "Cyclic Deformation of Ti_3Al", Acta. Met. 25, pp 1279-1288 (1977).

5. P.L. Martin, H.A. Lipsitt, N.T. Nuhfer, and J.C. Williams, "The Effects of Alloying on the Microstructure and Properties of Ti_3Al And TiAl", in Titanium '80, Science and Technology, Proc. 4th Int Conf on Titanium, Kyoto, Japan, Ed: K. Kimura and O. Izumi, Met Soc. AIME, Warrendale, Pa., 1980, Vol. 2, pp. 1245-1254.

6. J.C. Williams, "Phase Transformations in Ti Alloys - A Review of Recent Developments", in Titanium and Titanium Alloys - Scientific and Technological Aspects, Proc. Third Int. Conf., Moscow, USSR, 1976, Ed. J.C. Williams and E.F. Belov, Plenum Press, N.Y., 1976, p. 1477.

7. M.G. Mendiratta, S.M.L. Sastry and J.V. Smith, "Effect of Grain Size Upon Flow and Fracture in a Precipitation-Strengthened Ti-8 wt% Al-0.25 wt% Si Alloy", J. Matls Sci, 11, pp. 1835-1842 (1976).

8. S.M.L. Sastry, P.J. Meschter, and J.E. O'Neal, Met Trans A 15A, pp 1451-1463 (1984).

9. R.G. Rowe and E.F. Koch, "Rapidly Solidified Titanium Alloys Containing Cerium Sulfide and Oxysulfide Dispersions", in Rapidly Solidified Materials, Edited by P.W. Lee and R.S. Carbonara, Proc. 1986 Int. Conf., San Diego, CA, Feb., 1986, Amer. Soc. for Metals, Metals Park, Ohio, 1986 pp. 115-120.

10. J.A. Sutliff and R.G. Rowe, "Rare Earth Oxide Dispersoid Stability and Microstructural Effects in Rapidly Solidified Ti_3Al and Ti_3Al-Nb", in Rapidly Solidified Alloys and Their Mechanical and Magnetic Properties, Ed. B.C. Giessen, D.R. Polk, and A.I. Taub, Proc. Mat. Res. Soc. Symp., Boston, MA, Dec., 1985, To be published.

11. T. Hamajima, G Leutjering and S. Weissmann, "Microstructure and Phase Relationships for Ti-Mo-Al Alloys", Met. Trans., 3, pp. 2805-2810 (1972).

12. R. Strychor and J. C. Williams, "Phase Transformations in Ti-Al-Nb Alloys", in Solid-Solid Phase Transformations, Proc. Int. Conf., Pittsburgh, PA, Aug., 1981, Ed. H.I, Aaronson, D.E. Laughlin, R.F. Sekerka, and C.M. Wayman, TMS-AIME, Warrendale, PA, 1982, pp. 249-253.

13. R.G. Rowe, J.A. Sutliff, and E.F. Koch, Dispersoid Modification of Ti_3Al-Nb Alloys in Rapidly Solidified Alloys and Their Mechanical and Magnetic Properties, Ed. B.C. Giessen, D.R. Polk, and A.I. Taub, Proc. Mat. Res. Soc. Symp., Boston, MA, Dec., 1985 To be published.

14. M.F.X. Gigliotti, R.G. Rowe, G.E. Wasielewski, G.K. Scarr, and J.C. Williams, "Tensile and Creep Properties of Rapidly Solidified Titanium Alloys Containing Complex Matrices and Fine Dispersoids', in Rapidly Solidified Alloys and Their Mechanical and Magnetic Properties, Ed. B.C. Giessen, D.R. Polk, and A.I. Taub, Proc. Mat. Res. Soc. Symp., Boston, MA, Dec., 1985 To be published.

15. R.G. Rowe, T.F. Broderick, E.F. Koch, and F.H. Froes, "Microstructural Study of Rapidly Solidified Titanium Containing Erbium and Boron", in Rapidly Solidified Materials, Edited by P. W. Lee and R. S. Carbonara, Proc. 1986 Int. Conf. San Diego, CA, Feb., 1986, Amer. Soc. for Metals, Metals Park, Ohio, 1986 pp. 107-114.

16. G.E. Dieter, Mechanical Metallurgy, Mc Graw-Hill, New York, 1961, p. 559

17. R.D. Shull, A.J. McAlister and R.C. Reno, "Phase Equilibria in The Titanium-Aluminum System", in: Titanium Science and Technology, Ed: G. Lutjering, U. Zwicker, and W. Bunk, Proc. Fifth Int. Conf. on Titanium, Munich, FRG, Sept, 1984, DGM, Oberursel, FRG, 1985

18. E.W. Collings, The Physical Metallurgy of Titanium Alloys, Amer. Soc. for Metals, Metals Park, Ohio, 1984, p.94.

19. J.D. Livingston and J.W. Cahn, "Discontinuous Coarsening of Aligned Eutectoids", Acta Met., 22, pp. 495-503 (1974).

20. R.G. Rowe, J.A. Sutliff, and E.F. Koch, "Coarsening of Dispersed Particles in Rapidly Solidified Ti_3Al-Nb by Grain Boundary Migration", to be published.

21. H. Kreye, "Einfluss von Versetzungen auf die Umlosung von Teilchen", Z. Metallkde., 61, pp 108-112 (1970).

22. G. Boureau and P. Gerdanian, "Thermodynamic Study of Interstitial Solid Solutions of Oxygen in Titanium at 1000 C", Acta Met., 24, pp 717-723 (1976)

23. A.J. Ardell, "On The Coarsening of Grain Boundary Precipitates", Acta Met., 20, pp. 601-609, (1972).

24. M.F. Ashby and R.M.A. Centamore, "The Dragging of Small Particles by Migrating Grain Boundaries in Copper", Acta. Met., 16, pp. 1081-1092 (1968).

DISLOCATIONS IN Nb-CONTAINING Ti_3Al

S.A. Court, S.A. Skewes, M.H. Loretto* and H.L. Fraser

Department of Materials Science, University of Illinois, Urbana, IL 61801
*Department of Metallurgy and Materials, University of Birmingham
Birmingham B15 2TT, United Kingdom

Abstract

A study has been made of the types of dislocations observed in the alloy Ti-25Al-5Nb (at.%) which has been deformed in compression at room temperature to ~ 3.5%. Dislocations with $\underline{b} = 1/3 \langle 11\bar{2}0\rangle$ and $\underline{b} = 1/6 \langle 11\bar{2}0\rangle$ lying on $\{1\bar{1}00\}$ have been observed. Also, dislocations with $\underline{b} = 1/3 \langle 11\bar{2}0\rangle$ lying in networks on (0001) are also present. These latter dislocations yield double images when imaged with diffracting vectors of the type $20\bar{2}0$, $40\bar{4}0$ and $20\bar{2}1$; this is a consequence of the value of $|\mathbf{g} \cdot \mathbf{b}|$ and should not be taken to imply that the partial dislocations are widely separated.

Introduction

It is widely accepted that the slip systems in Ti_3Al are of the type $\langle 11\bar{2}0\rangle$ $\{1\bar{1}00\}$, $\langle 11\bar{2}0\rangle$ $\{0001\}$ together with some systems that have dislocations with c-components which provide the necessary five independent slip systems. As part of a study of Ti_3Al and Ti_3Al with Nb substituted for Ti we have been examining the contrast shown by dislocations on $\{1\bar{1}00\}$ and on (0001) and in this preliminary report we show images which suggest that dislocations of Burgers vector $\frac{1}{3}$ $\langle 11\bar{2}0\rangle$ and $\frac{1}{6}$ $\langle 11\bar{2}0\rangle$ may be gliding on $\{1\bar{1}00\}$ and contrary to earlier work the $\frac{1}{3}$ $\langle 11\bar{2}0\rangle$ dislocations on (0001) are not measurably dissociated.

Experimental Procedure

Samples of Ti-25Al-5Nb (at.%) have been deformed in compression at room temperature to ~ 3.5% and thinned discs have been prepared for TEM using standard techniques. Burgers vector analysis has been carried out using a Philips EM 430 and EM 420 operating at 200 and 120 kV, respectively.

Results

Fig. 1 shows a network of screw dislocations imaged with the three $\bar{2}200$ type reflections available when the electron beam is aligned nearly parallel to an [0001] direction, and $\underline{g}$ = $04\bar{4}0$. In agreement with the micrographs published by Yang (1), the in-contrast images appear to be double. In

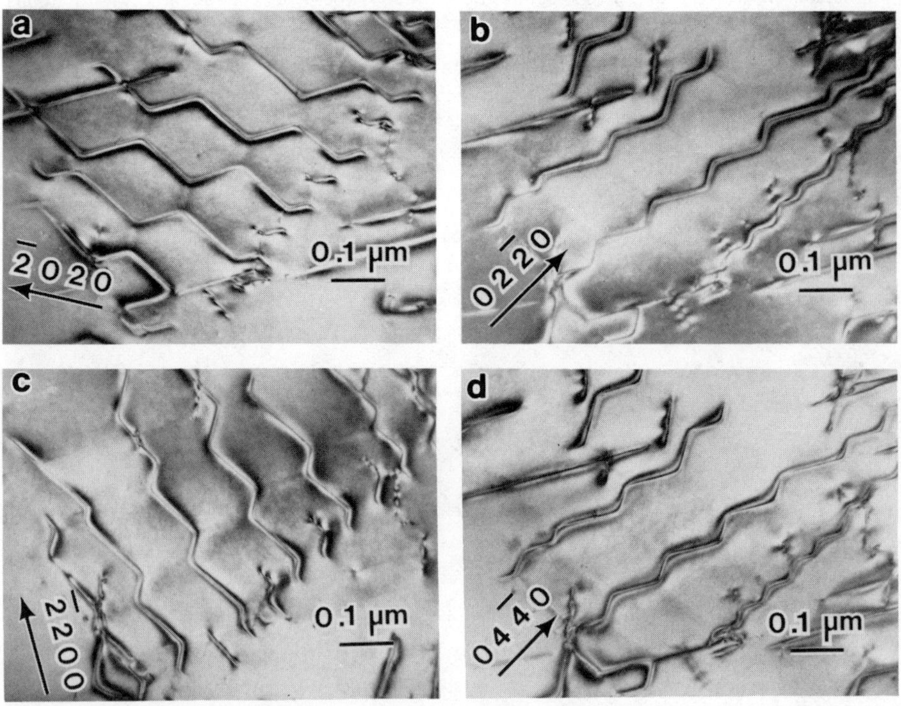

Figure 1 - Dislocation network imaged with (a) $\underline{g}$ = $\bar{2}020$ (b) $\underline{g}$ = $02\bar{2}0$ (c) $\underline{g}$ = $\bar{2}200$ and (d) $\underline{g}$ = $04\bar{4}0$. $\underline{B}$ near [0001].

contrast to his interpretation that the micrographs show splitting of the 1/3 <11$\bar{2}$0> dislocations into 1/6 <11$\bar{2}$0> partials, we have shown that the origin of the double images lies in the fact that for these imaging conditions $|\mathbf{g} \cdot \mathbf{b}| > 2$. Thus, Fig. 2 shows a network viewed with the same diffracting vector before and after tilting through ~ 45° from [0001]. The separation of the double images is not reduced after tilting, which suggests that two 1/6 <11$\bar{2}$0> partials on (0001) are not being imaged. Additionally, the two peaks show distinctly different contrast from each other and as the deviation parameter is increased so the difference between the two peaks becomes more obvious and on reversing $\underline{g}$ the characteristics of the two peaks reverse (Fig. 3). It should be noted that similar double peaks are observed with $\underline{g}$ = 02$\bar{2}$1 for a large range of deviation parameter (Fig. 4). It is thus concluded that the dislocations have $\underline{b}$ = 1/3 <11$\bar{2}$0> and since it has not been possible to observe individual images from the 1/6 <11$\bar{2}$0> partials it must be concluded that their separation is less than ~ 40 Å rather than the 150 Å implied by Yang (1).

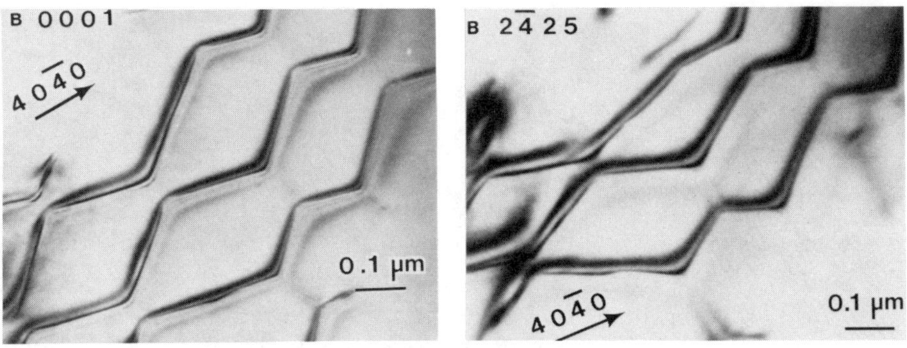

(a) (b)

Figure 2 - Dislocation network imaged with $\underline{g}$ = 40$\bar{4}$0, (a) $\underline{B}$ near [0001] and (b) $\underline{B}$ near [2$\bar{4}$25]. Note, there is no reduction in the separation of the double images after tilting through ~ 45°. (See text)

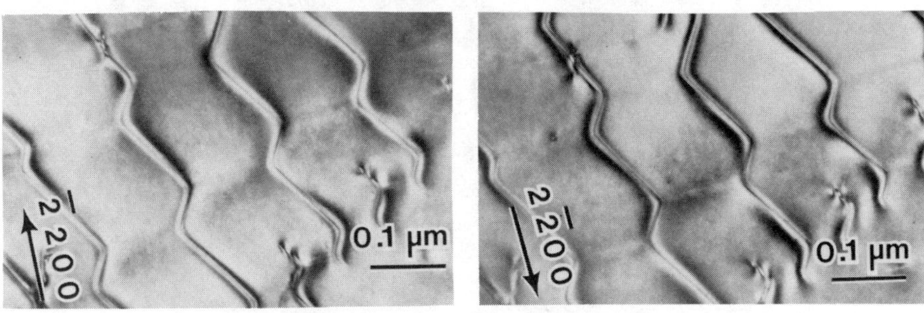

(a) (b)

Figure 3 - Dislocation network imaged with (a) $\underline{g}$ = $\bar{2}$200 and (b) $\underline{g}$ = 2$\bar{2}$00. Note the reversal of the peak contrast with ± $\underline{g}$. $\underline{B}$ near [0001].

If glide of dislocations with Burgers vectors parallel to <11$\bar{2}$0> occurs on {1$\bar{1}$00}, then, in principle, it is possible for two different Burgers vectors to become operative. Thus, dislocations of the type $\underline{b}$ = 1/3 <11$\bar{2}$0> may glide in the form of superdislocations made up of two partial dislocations of the type 1/6 <11$\bar{2}$0>. On one type of {1$\bar{1}$00} plane (2), these partials cause first nearest neighbor violations, and therefore disorder the material; hence the partials will glide in pairs. On the second type of $\{1\bar{1}00\}$ plane[2], the glide of 1/6 $\langle 11\bar{2}0\rangle$ causes only second nearest neighbor violations and so glide of single dislocations (with $\underline{b}$ = 1/6 $\langle 11\bar{2}0\rangle$) on these particular planes may contribute significantly to the plastic deformation of the compound.

To test this concept, images have been obtained, with $\bar{2}021$ diffraction vectors of dislocations of the $\langle 11\bar{2}0\rangle$ type. As shown above 1/3 $\langle 11\bar{2}0\rangle$ dislocations show double images with $02\bar{2}1$ diffraction vectors but single images would be expected with 1/6 $\langle 11\bar{2}0\rangle$ dislocations. Fig. 5 shows an image taken with $\underline{g}$ = $\bar{2}02\bar{1}$, and although some double images can be seen the vast majority of dislocations show single images even though the deviation from Bragg is very small. This implies that these dislocations, which trace analysis has shown to lie on $\{1\bar{1}00\}$, may have $\underline{b}$ = 1/6 $\langle 11\bar{2}0\rangle$. However, detailed diffraction contrast studies are needed to show the presence of the

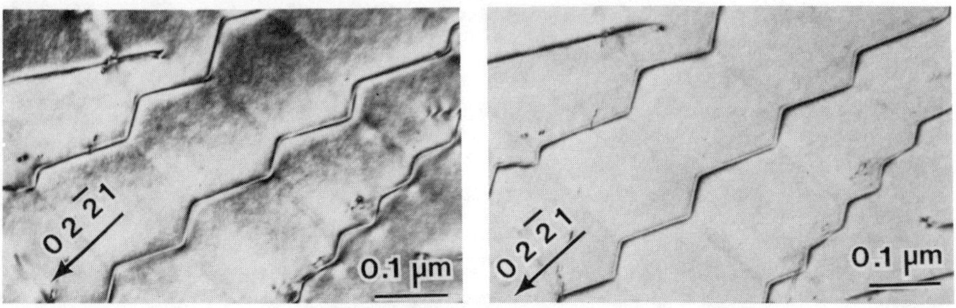

(a) (b)

Figure 4 - Dislocation network shown in Fig. 1 imaged with $\underline{g}$ = $02\bar{2}1$ for (a) small deviation parameter (s) and (b) large s. The dislocations show double peak contrast characteristics. $\underline{B}$ near [0001].

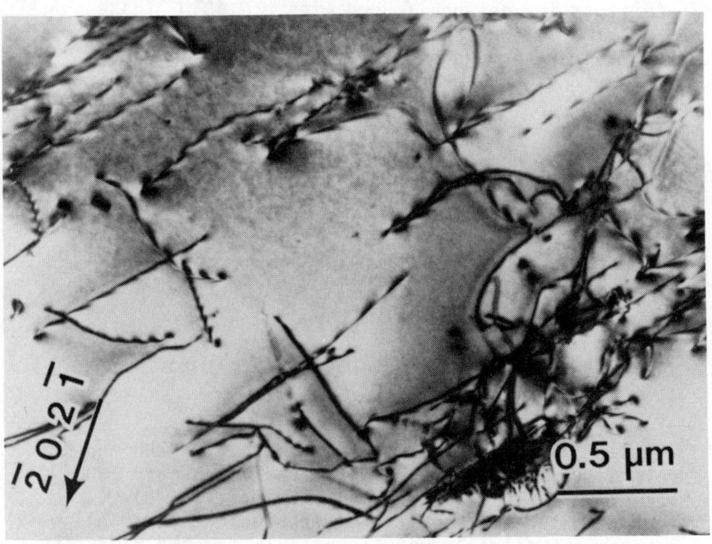

Figure 5 - Bright field micrograph taken with $\underline{g}$ = $\bar{2}021$, showing both double and single dislocation images. $\underline{B}$ near $[1\bar{2}10]$.

fault caused by the second nearest neighbor violations, this most conveniently being done by use of superlattice dark field imaging. This is the subject of further study.

Conclusions

It may be concluded from this preliminary study that dislocations having Burgers vectors given by 1/3 $<11\bar{2}0>$ and 1/6 $<11\bar{2}0>$ lying on $\{1\bar{1}00\}$ are present in deformed samples of Ti_3Al containing Nb. Furthermore, the images of dislocations with $\underline{b}$ = 1/3 $<11\bar{2}0>$ exhibit double images when $|\mathbf{g} \cdot \mathbf{b}| > 2$; there appears to be no evidence to suggest that they are widely dissociated. From the present work, it seems that the separation of the two 1/6 $<11\bar{2}0>$ partial dislocations is < 40 Å.

Acknowledgements

The electron microscopy was performed in the Center for the Microanalysis of Materials, a facility operated by the Materials Research Laboratory of the University of Illinois. Stimulating discussions with Dr. I.P. Jones are gratefully acknowledged.

References

1. W.J.S. Yang, J. Mater. Sci. Letts. 1 (1982), 199-202.

2. J.C. Williams and M.J. Blackburn, in "Ordered Alloys", B.H. Kear et al., eds. (Claitor's Publishing Division, Baton Rouge, LA, (1970) 425-445.

THE EFFECTS OF HEAT TREATMENT ON THE FATIGUE CRACK GROWTH BEHAVIOR OF Ti_3Al + Nb

M. A. STUCKE and H. A. LIPSITT

MATERIALS LABORATORY
AIR FORCE WRIGHT AERONAUTICAL LABORATORIES
WRIGHT-PATTERSON AFB, OH 45433

Abstract

The effect of different cooling rates from the beta and alpha-two + beta solution temperatures on the subsequent fatigue crack growth behavior of a titanium aluminide alloy is presented. Round compact tension specimens were machined from extruded powder and fatigue crack growth tests were conducted at 650°C. Data for the different cooling rates and solution treatments were compared to each other and to a nickel base superalloy. The results show that there is no effect of these cooling rates on the fatigue crack growth behavior. When comparing these results with the results of a nickel base superalloy, the fatigue crack growth rate is much higher although the slopes of the curves are very similar. The apparent threshold stress intensity range is found to be about 6.5 $MPa\sqrt{m}$.

Introduction

Titanium aluminide research has been a subject of interest in the Materials Laboratory for more than ten years (1). A number of papers have been published on the subject alloy, Ti-16wt%Al-10wt%Nb (2-5). The emphasis has been on the dislocation behavior, the influence of heat treatment variables on microstructure and ductility and on gathering mechanical property data for possible static component applications in turbine engines. The fatigue crack growth behavior and the effect of different cooling rates have not been studied in any detail. The different heat treatments may change the fatigue behavior of the titanium aluminide with the possible change in the microstructure. Therefore, the cooling rates were selected to obtain a change in the morphology of the microstructural elements and to determine if there is any effect on the fatigue crack growth behavior.

Experimental Procedure

Powder Ti-16wt% Al-10wt% Nb was produced by the rotating electrode process, canned, hot compacted and subsequently extruded at 1204°C. The extrusion ratio was selected at 6:1 for a 25.4 mm diameter bar. The bar was then sectioned for heat treatment trials. The bars were solutioned in either the beta phase field at 1204°C or below the beta transus in the alpha-2 plus beta phase field at 1121°C. Then, the bars were quenched into three different media: water, molten salt at 760°C, or air cooled. Some of the bars were then aged at 760°C for approximately 8 hours. The heat treatments are listed in Table I. Round sub-compact tension specimens were machined for the fatigue crack growth tests. All of the specimens were tested at 650°C with a load ratio, R, of 0.1 and a frequency of 1 Hz. A computer controlled servo-hydraulic MTS machine was used for the experiment with an extensometer to determine crack length from compliance measurements. Constant load tests were performed for all the specimens except for two threshold tests, which were decreasing load tests.

Table I Heat Treatments of Ti-16wt%Al-10wt% Nb

Heat Treatment	Solution Temperature °C (°F)	Cooling Media	Age Temp/Time
1. (Fig. 1a)	1121°C (2050°F)	Water	Age 760°C/8hrs/AC
2. (Fig. 2b-c)	1121°C (2050°F) 1204°C (2200°F)	Molten Salt 760°C (1400°F)/ 30 min	--
3. (Fig. 3d-e)	1121°C 1204°C	Air	Age 760°C/8hrs/AC

The fracture surfaces of the tested specimens were observed in the scanning electron microscope in the low growth, high growth and fast fracture regions. The microstructure was also observed by optical microscopy for each heat treatment.

Results and Discussion

The microstructures resulting from the different heat treatments are shown in Figure 1a-e. Water quenching is a faster cooling rate than the air and/or salt quench. The intermediate cooling rate is an air cool and the slowest is the salt quench. The beta transus for this alloy has been shown to be about 1182°C (6), so several solution heat treatments were done below the beta transus, at 1121°C, and several were done above the beta transus, at 1204°C. All of the resulting microstructures are single phase alpha-two. Figure 1a shows the microstructure of the sample water quenched from below the beta transus. The alpha-two has a lenticular morphology. Figures 1b and 1c are microstructures of the air cooled samples that were solutioned above and below the beta transus. Figure 1b shows a coarser microstructure than Figure 1a and 1c shows a very coarse Widmanstatten type structure and the outline of the large prior beta grains. The microstructures shown in Figures 1d and 1e resulted from solution at 1121°C and 1204°C, respectively, followed by a salt quench to 760°C. Figure 1d shows a coarse lenticular phase and Figure 1e shows a fine Widmanstatten type structure and large prior beta grains.

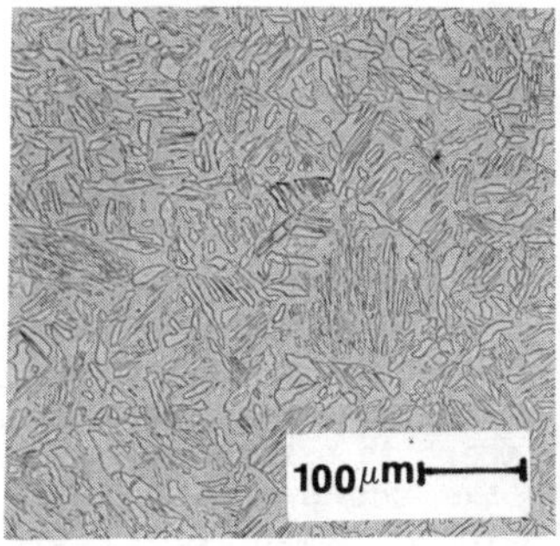

a)

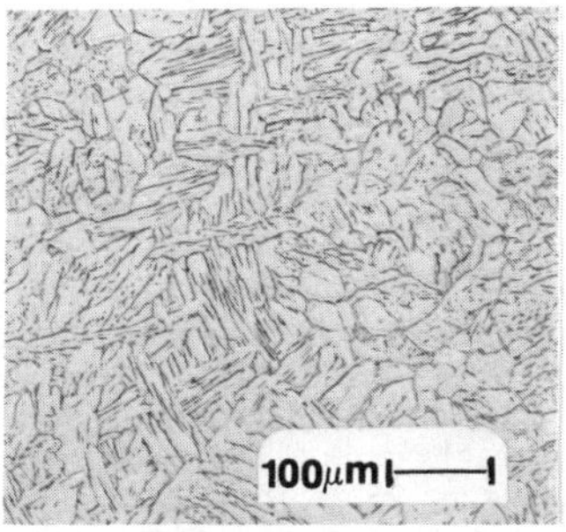

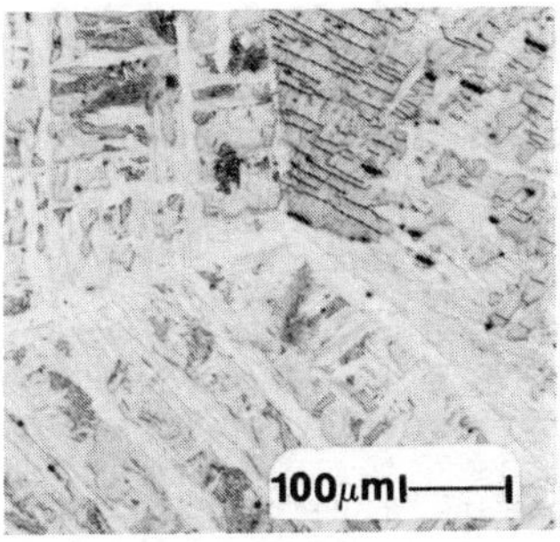

b) c)

Figure 1 - Micrographs of heat treated Ti_3Al+Nb specimens,
a) 1121°C/1 hr/WQ, 760°C/8 hrs/AC, b) 1121°C/1 hr/AC, 760°C/8 hrs/AC,
c) 1204°C/1 hr/AC, 760°C/8 hrs/AC, d) 1121°C/1 hr/SQ to 760°C/½ hr/AC,
e) 1204°C/1 hr/SQ to 760°C/½ hr/AC

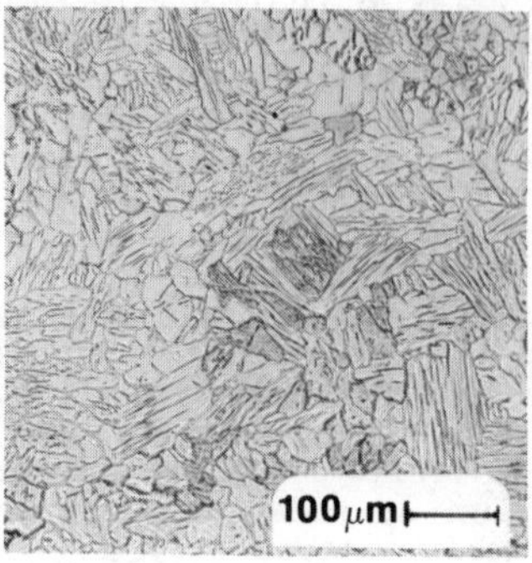

d)

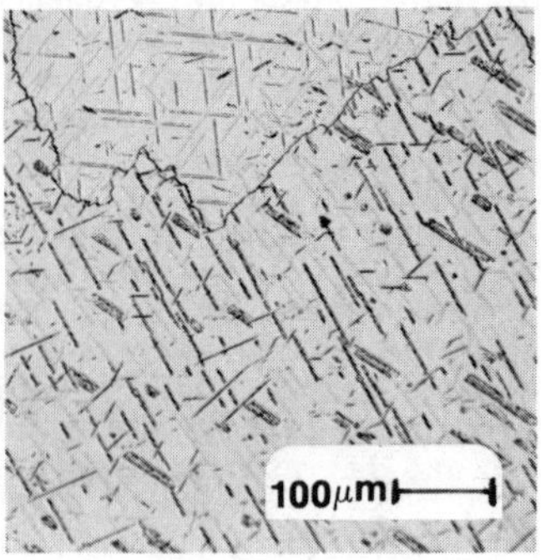

e)

Figure 1 d) - e) Continued

Figure 2 is a plot of crack growth rate versus stress intensity range. The data illustrate that there is essentially no difference in fatigue crack growth behavior between the cooling rates selected. Also, there is no effect of solution treatment temperature for this titaniun aluminide composition. The heat treatment can vary from solutioning above or below the beta transus and the cooling rate can vary from a relatively slow rate (air) to a fast rate (water) and not change the fatigue crack growth properties. Thus, the fatigue crack growth rate is insensitive to a wide range of microstructural changes. Figure 3 is the same type of plot, but for a decreasing load threshold test. The load was slowly decreased to reduce the stress intensity range until the threshold stress intensity, the point at which a crack growth rate reduce to 10^{-10} m/cycle, is reached. The true threshold should indicate that crack growth at 10^{-10} m/cycle occurs at some critical value of the stress intensity factor. However, it was observed that crack arrest occurred at 2×10^{-8} m/cycle at a delta K of 6.5 MPa$\sqrt{m}$. The crack stopped growing completely and the test could not be run to the true threshold. Similar observations of crack arrest in elevated temperature fatigue threshold tests have been made by Allison (7) on a titanium alloy and Zawada and Nicholas (8) on a nickel base superalloy. Figure 4 is a comparison of the fatigue crack growth behavior of titanium aluminide and a nickel base superalloy, Rene 95. This plot shows quite clearly, that at 650°C , the titanium aluminide has a fatigue crack growth rate an order of magnitude larger than that the nickel base material. For example, at a delta K of 20 MPa$\sqrt{m}$, Rene 95 has a crack growth rate of 4×10^{-7} m/cycle where Ti_3Al + Nb is 4×10^{-6} m/cycle. However, notice that the slopes of the crack growth rate curves are very similar. Also, the density of the titanium aluminide is about one-half of the density of the nickel material, so the direct comparison does not reveal all the information needed for a complete evaluation.

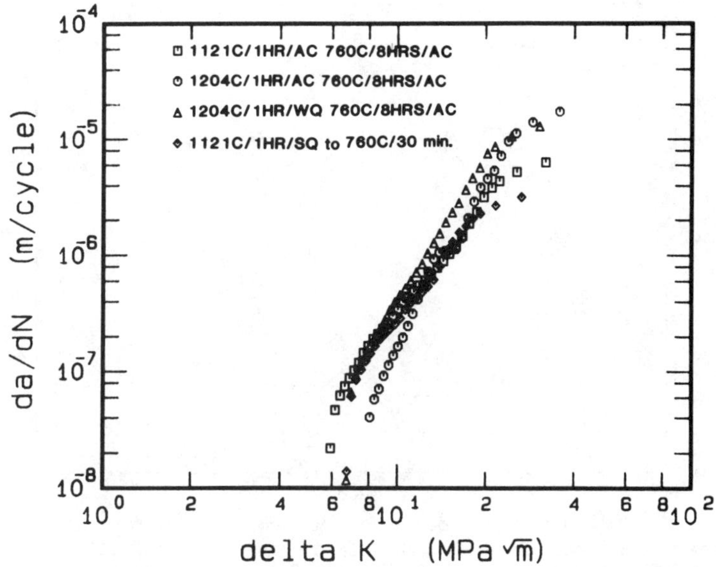

Figure 2 a) - b) - Fatigue Crack Growth Rate Data at 650°C for all the heat treatments.

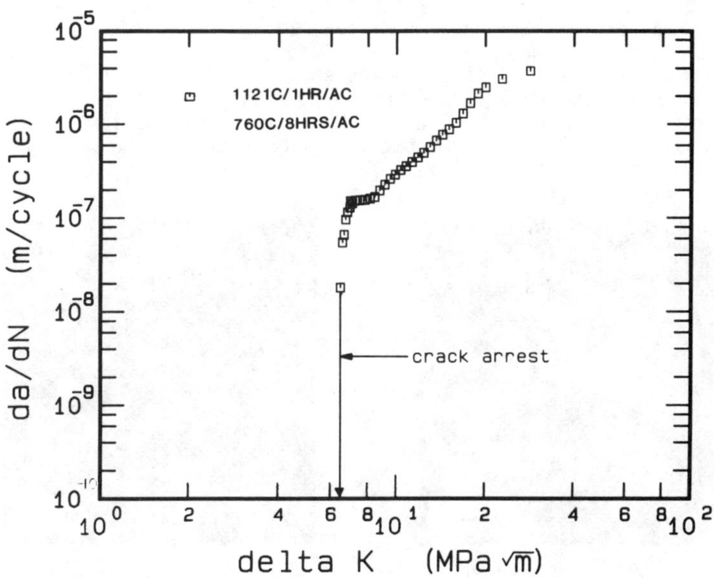

Figure 3 a) - b) - Fatigue Crack Growth Rate Data of Decreasing Load Threshold Test at 650°C.

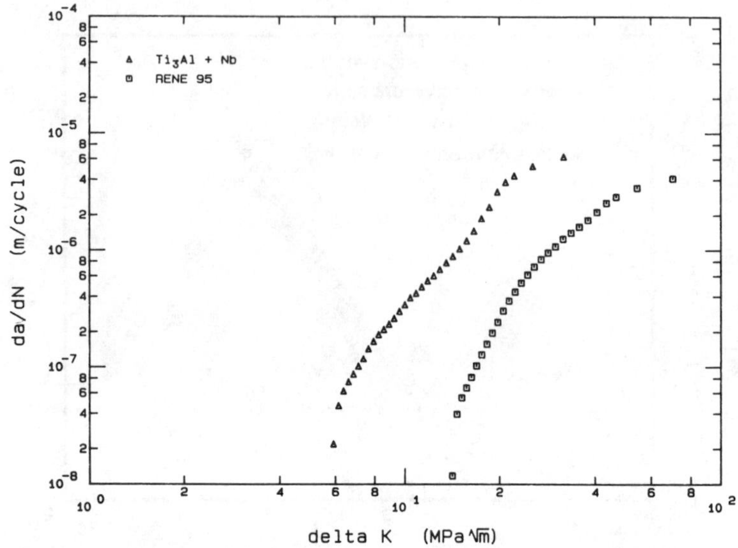

Figure 4 - Fatigue Crack Growth Data of Rene 95 and Ti_3Al+Nb Illustrating the Difference in Fatigue Behavior at 650°C.

Figures 5-7 show the fracture surfaces from several of the fatigue crack growth specimens. The observation of the fracture surfaces from Figure 5a-b reveals a flat, cleavage type appearance with some secondary cracking. Figures 6a-b show the same type of cleavage behavior in another specimen as well as some evidence of fatigue striations. The fracture surfaces from the threshold specimen are shown in Figure 7a-b and indicate a rough surface and secondary cracking. The differences in the fracture surfaces may be because the low crack growth rates allow time for an enhanced environmental attack to occur at the crack tip which changes the appearance in the threshold region.

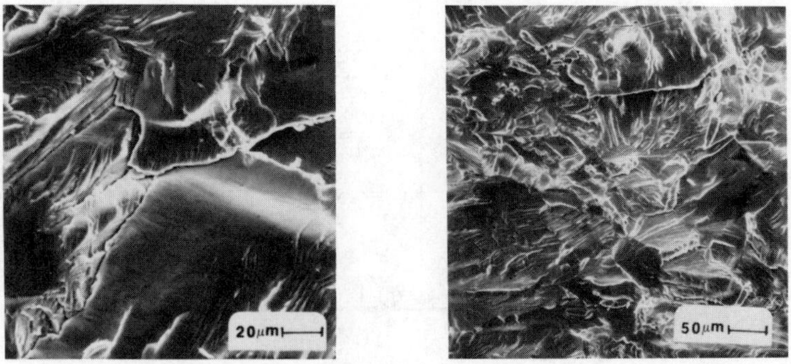

a) b)

Figure 5 a) - b) - Fracture Surfaces of Fatigue Crack Growth Specimen Tested at 650°C (Heat treatment: 1121°C/1 hr/AC, 760°C/8 hrs/AC)

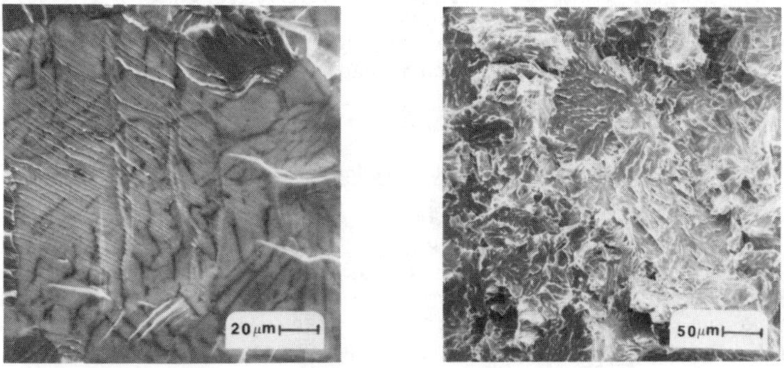

a) b)

Figure 6 a) - b) - Fracture Surfaces of Fatigue Crack Growth Specimen Tested at 650°C (Heat treatment: 1121°C/1 hr/SQ to 760°C/½ hr/AC).

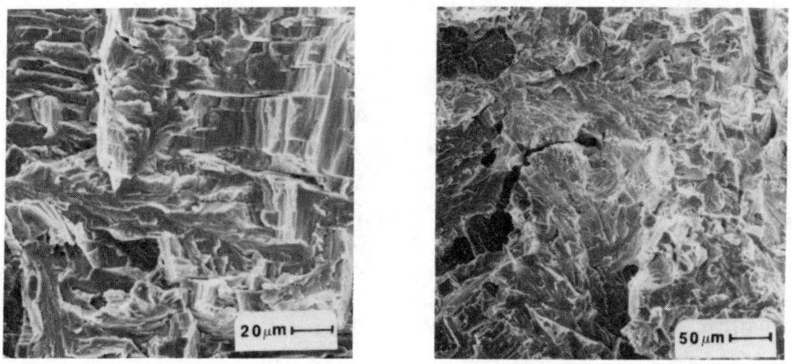

a) b)

Figure 7 a) - b) - Fracture Surfaces of Threshold Specimen Tested at 650°C (Heat treatment: 1121°C/1 hr/SQ to 760°C/½ hr/AC).

Conclusions

The fatigue crack growth behavior of Ti_3Al + Nb has been studied. The quench rate was varied to produce a considerable variation in the microstructure. Even with a large change in the microstructure from a Widmanstatten to a thick lenticular morphology, the fatigue crack growth behavior remained unchanged at a test temperature of 650°C. The apparent threshold condition has been determined to be 6.5 Mpa$\sqrt{m}$. The fatigue crack growth behavior in the titanium aluminide was compared with a nickel base superalloy and found to an order of magnitude higher. However, the nickel base material is twice the density of the titanium aluminide. The fracture surfaces were also analyzed and found to have a flat, cleavage type appearance in all specimens except the threshold test. The fracture surface of the threshold specimen was rougher in appearance which may be due to an interaction with the environment in the low growth regions. More work needs to be done in the lower temperature regions, so the fatigue crack growth behavior in titanium aluminides can be directly compared with that of elevated temperature titanium alloys.

Acknowledgement

This research was performed under Materials Laboratory in-house research project number 2302P101, Air Force Wright Aeronautical Laboratories, Wright-Patterson Air Force Base, OH 45433. The authors wish to thank Mr D. Johnson, of University of Dayton Research Institute, for conducting the fatigue crack growth tests, and Dr T. Nicholas of AFWAL/MLLN, Materials Laboratory, for reviewing the paper.

References

1. Lipsitt, H.A., "Titanium Aluminides - An Overview," Materials Research Society Symposium Proceedings, 39, (1985), 351-364.

2. Shafrik, R.E., "Dynamic Elastic Moduli of the Titanium Aluminides," Met. Trans., 8A(6) (1977), 1003-1006.

3. Sastry, S.M.L., and Lipsitt, H.A., "Ordering Transformations and Mechanical Properties of Ti_3Al and Ti_3Al - Nb Alloys," Met. Trans. A, 8A(10) (1977), 1543-1552.

4. Mendiratta, M.G., and Lipsitt, H.A., "Steady State Creep Behavior of Ti_3Al - base Intermetallics," Journal of Mat. Sci., 15 (1980), 2985-2990.

5. Yang, W.J.S., "Observations of Superdislocation Networks in Ti_3Al - Nb," Journal of Mat. Sci. Let., 1 (1982) 199-202.

6. Teal, K., private communication, AFWAL/MLLS, Air Force Wright Aeronautical Laboratories, WPAFB, Ohio, 1985.

7. Allison, J.E., and Williams, J.C., "Near-Threshold Fatigue Crack Growth Phenomena at Elevated Temperature in Titanium Alloys," Scripta Met., 19(6) (1985), 773-778.

8. Zawada, L.P., and Nicholas, T., "The Effect of Loading History on Closure Behavior in Rene' 95," (Paper presented at 18th National Fracture Mechanics Symposium of ASTM, Boulder, Colorado, 25 June 1985, 15).

POWDER METALLURGY OF TITANIUM ALUMINIDE COMPONENTS*

C. F. Yolton, T. Lizzi, V. K. Chandhok and J. H. Moll

Crucible Materials Corporation
Crucible Research Center
P. O. Box 88
Pittsburgh, Pa. 15230

Abstract

Titanium alloys based on intermetallic compounds in the titanium-aluminum binary system have attractive properties which make them excellent candidates for applications in gas turbine engines. These properties include good high temperature strength and ductility, good oxidation resistance and a low density. Titanium aluminide alloys can potentially be substituted for nickel-based alloys currently used at intermediate temperatures resulting in a substantial weight savings with no loss in performance. A niobium modified alpha two titanium aluminide alloy is one of the alloys under consideration. While this alloy has very attractive elevated temperature properties, it is a difficult alloy to hot work and fabricate into components. Powder metallurgy production of near-net shape components is one means of circumventing fabrication difficulties with this alloy. In this paper, several near-net shape components produced by the ceramic mold process using Plasma Rotating Electrode Process (PREP) powder are described. Room and elevated temperature mechanical properties of the powder metallurgy product are presented.

*Presented at the 1986 Annual Powder Metallurgy Conference and published with the permission of MPIF and APMI.

Introduction

Titanium aluminide alloys have been receiving increased attention for potential applications in advanced gas turbine engines (1). The titanium aluminide alloys under development fall into two general classes. One is based on the compound Ti_3Al and the other is based on the TiAl compound. Primary advantages of the titanium aluminide alloys are low density, a high strength-to-weight ratio at elevated temperatures, a higher modulus than conventional titanium alloys, good elevated temperature creep-rupture strength and good oxidation resistance. Principle disadvantages of the titanium aluminides are low ductility at room temperature, hot working is more difficult than conventional titanium alloys and the alloy cost may be high because of expensive alloying elements such as niobium or expensive processing methods. Design studies have indicated that substantial weight savings can be obtained by utilization of titanium aluminides in engine hardware (2). Weight savings of 43 percent have been projected for replacement of a superalloy with a Ti_3Al base aluminide in a turbine engine stator support ring (1).

One of the alloys under development is a niobium modified Ti_3Al (alpha two) composition. This alloy has 1-2 percent ductility at room temperature and has useful creep strength and oxidation resistance up to 705°C (1300°F). While this alloy can be processed using conventional titanium fabrication methods, it is more difficult to process than standard titanium alloys (2).

A potential method of reducing material and fabrication costs for titanium alloys is to produce near-net shapes by powder metallurgy (P/M) techniques. One promising approach to producing P/M near-net shapes is to hot-isostatically-press (HIP) prealloyed powder using the ceramic mold process (3). Work with conventional alloys such as Ti-6Al-4V and Ti-6Al-6V-2Sn has shown that this process is capable of producing shapes with properties comparable to cast plus wrought product at a reduced cost (4,5).

The current work was undertaken to assess the potential of applying the P/M near-net shape process to the Ti_3Al base alloy. Objective of the program was twofold. The first objective was to establish HIP processing parameters for this alloy and the second was to characterize mechanical properties of the HIP plus heat treated P/M product.

Experimental

The prealloyed powder used for the production of shapes and mechanical property testing was produced by the plasma rotating electrode process (PREP) (6). Physical properties of the powder are given in Table I. The PREP powder was consolidated using the ceramic mold process. Details of this process are given elsewhere (3).

Table I. Physical Properties of PREP Ti_3Al Base Powder

Screen Analysis						
U. S. Standard Mesh	35	60	80	100	200	325
(μm)	(500)	(250)	(177)	(149)	(74)	(44)
Percent Undersize	100	96	56	27	1.4	nil
Tap Density	2.76 g/cc (62% of theoretical density)					
Flow Rate	30 sec (ASTM B213)					

Mechanical property test blanks were cut from the as-HIP shapes and then heat treated as individual blanks. The standard solution and aging heat treatment used was 1205°C (2200°F) 0.5 hour + 760°C (1400°F) 1 hour although in some cases a temperature of 1175°C (2150°F) was used for solution treatment. Nominal beta transus temperature for this alloy is 1175°C (2150°F). Heat treatments were done in both vacuum and air. Cooling rate from the solution treatment temperature was varied to determine the effect on mechanical properties.

Tensile tests were conducted using round specimens with a 6.3mm (0.25 inch) diameter x 25mm (1 inch) long gage length in accordance with ASTM-E8. Creep and stress rupture tests were conducted using specimens with a 4.44mm (0.175 inch) diameter x 18mm (0.7 inch) gage length. All tests were conducted in air. The tensile tests were conducted at room temperature, 425°C (800°F), 540°C (1000°F), 650°C (1200°F) and 760°C (1400°F). Elevated temperature tensile specimens were soaked for thirty minutes at the test temperature before starting the test. Stress rupture tests were conducted at 540°C (1000°F), 650°C (1200°F) and 760°C (1400°F). Creep testing was done at 760°C (1400°F).

Results and Discussion

Microstructure of the as-produced PREP powder is shown in Figure 1. The powder exhibits a fine dendritic type structure indicative of a rapidly solidified product. This dendritic structure is generally observed in titanium alloys with a high beta stabilizer content. Alloys with a low beta stabilizer content typically have a martensitic structure.

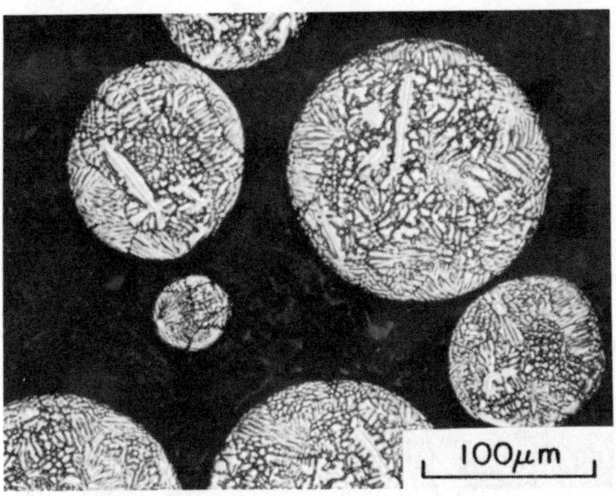

Figure 1. Microstructure of PREP Ti_3Al base powder. 250X

The PREP powder was HIPed at a pressure of 103MPa (15ksi) and temperatures of 955°C (1750°F), 1010°C (1850°F) and 1065°C (1950°F) for 6 to 8 hrs. Microstructures after HIPing at the three temperatures are shown in Figure 2. HIPing at 955°C (1750°F) resulted in very little densification. Increasing the temperature to 1010°C (1850°F) resulted in substantially greater densification but some porosity remained. HIPing at 1065°C (1950°F) produced full density in this alloy as indicated in Figure 2c. Microstructure after HIPing at 1065°C (1950°F) is equiaxed with a grain size of about 20μm.

Two niobium modified Ti_3Al near-net shapes HIPed at 1065°C (1950°F) using the ceramic mold process are shown in Figures 3 and 4.

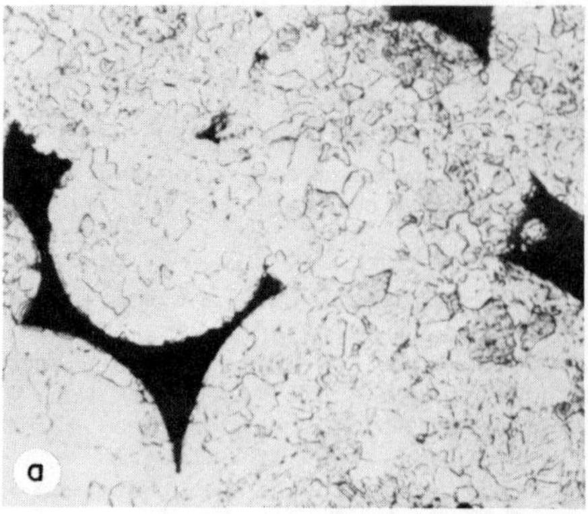

HIP 955°C (1750°F)

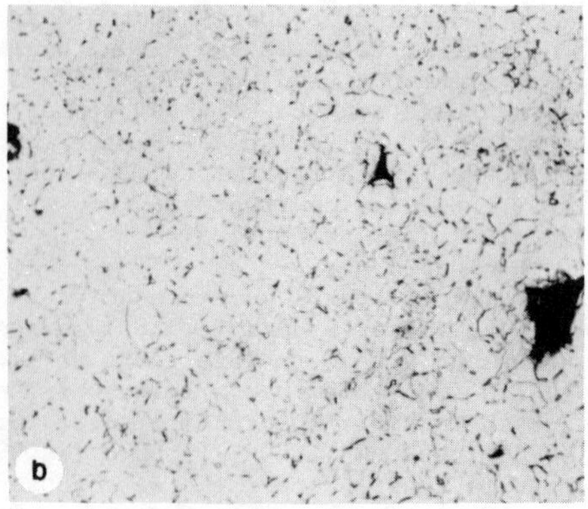

HIP 1010°C (1850°F)

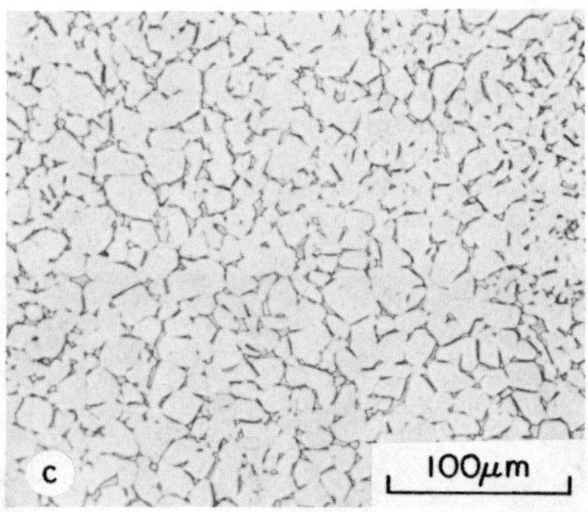

HIP 1065°C (1950°F)

Figure 2. Microstructure of PREP niobium modified Ti_3Al powder HIPed at (a) 955°C (1750°F), (b) 1010°C (1850°F), and (c) 1065°C (1950°F) and 103MPa (15ksi) for 6 to 8 hours. 250X

Both shapes are 100 percent dense as determined by microstructural examination of sections taken from the hub areas. The shape in Figure 4 is a radial compressor rotor for a small gas turbine engine. This rotor weighs 4kg (9 pounds) and is 178mm (7 inches) in diameter. The blades are 1.3mm (0.050 inch) thick at the tips. The shapes in Figure 4 are centrifugal compressor rotors for a small gas turbine engine. Weight of each rotor in the as-HIP condition is 4.3kg (9.5 pounds) and they are 216mm (8.5 inches) in diameter at the base. Minimum blade thickness in these rotors is 0.5mm (0.020 inch).

The mechanical property tests on the powder were taken from the base and hub area of the radial compressor rotor. Table II summarizes the room and elevated temperature tests on this material. Although tensile properties for this alloy in wrought form are not available in the open literature for comparison, the tensile strengths and ductilities of the P/M material have been found to compare favorably with the properties of cast plus wrought product (7).

Figure 3. As-HIP niobium modified Ti_3Al radial compressor rotor.

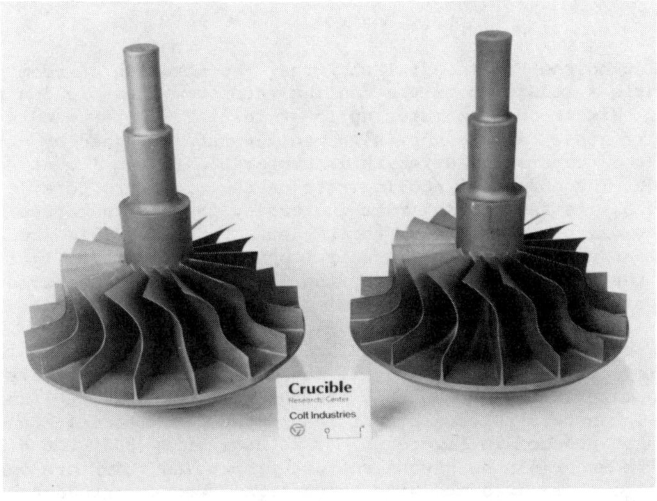

Figure 4. As-HIP niobium modified Ti_3Al centrifugal compressor rotors.

Table II. Room and Elevated Temperature Tensile Properties of P/M Niobium Modified Ti_3Al HIPed at 1065°C (1950°F)

Heat Treat-ment	Cooling* Rate (C/min)	Test Temperature °C (°F)	Tensile Strength MPa	(ksi)	Yield Strength MPa	(ksi)	Elongation (%)	Reduction of Area (%)
A	50	RT	606	(88)	510	(74)	2	2
A	120	RT	592	(86)	530	(77)	<1	2
B	>280	RT	772	(112)	661	(96)	1	3
C	60	425 (800)	572	(83)	344	(50)	11	12
C	60	540 (1000)	517	(75)	324	(47)	12	15
C	60	650 (1200)	468	(68)	282	(41)	13	20
C	60	760 (1400)	400	(58)	248	(36)	18	22
B	>280	760 (1400)	558	(81)	303	(44)	14	20
D	>17000	760 (1400)	386	(56)	241	(35)	14	15

Heat Treatment:

A - 1175°C (2150°F) 0.5Hr Ar Cool + 760°C (1400°F) 1 Hr Ar Cool + 595°C (1100°F) 8 Hr Vacuum Cool.

B - 1205°C (2200°F) 0.5Hr Fan Air Cool + 760°C (1400°F) 1 Hr Air Cool.

C - 1205°C (2200°F) 0.5Hr Ar Cool + 760°C (1400°F) 1 Hr Vacuum Cool.

D - 1205°C (2200°F) 0.5Hr Oil Quench + 760°C (1400°F) 1 Hr Vacuum Cool.

*Cooling rates are from solution treatment temperature to 595°C (1100°F).

The results on the P/M material show that the strength at room and elevated temperature is sensitive to the cooling rate from the solution treatment temperature. Higher cooling rates up to at least 280°C per minute result in higher strengths. At the very high cooling rate obtained by oil quenching, the elevated temperature strength is comparable to the lowest cooling rate evaluated. The effect of cooling rate on the microstructure is illustrated in Figure 5. As the cooling rate increases, the Widmanstatten platelets become finer which gives rise to the higher strengths. This increase in strength has been attributed to slip length refinement in Ti_3Al type alloys (8). At the highest cooling rate, the Widmanstatten transformation appears to be suppressed and the strengthening effect is not obtained.

The results of stress rupture and creep tests on the niobium modified Ti_3Al alloy are shown in Tables III and IV. All of these tests were conducted on material which was heat treated at 1205°C (2200°F) and argon cooled. The stress rupture properties of the P/M material are compared with wrought product in Figure 6. It is noteworthy that the HIP plus heat treated P/M material has stress rupture properties which are essentially identical to those attained with conventional wrought product.

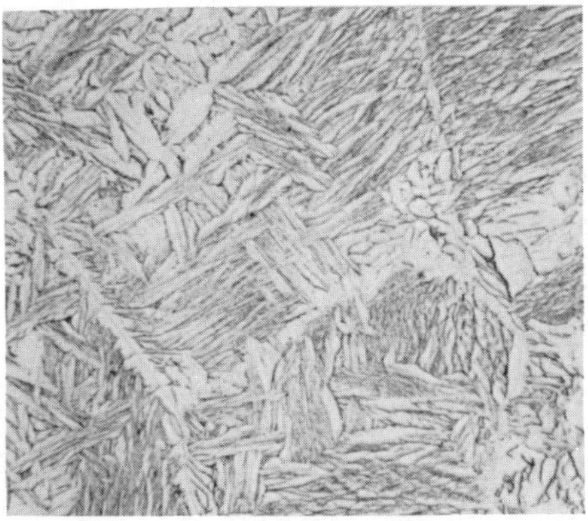

Air Cool (50°C/min)

Fan Air Cool (>280°C/min)

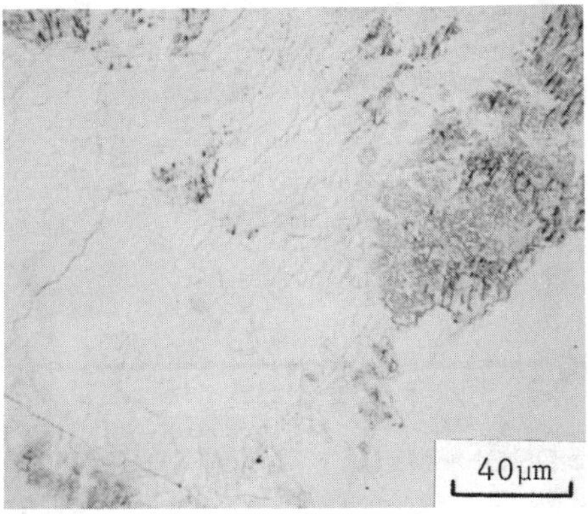

Oil Quench (>17000°C/min)

Figure 5. Effect of cooling rate from the beta field on microstructure of P/M niobium modified Ti_3Al. 400X

Table III. Stress Rupture Properties of P/M Niobium Modified Ti_3Al HIPed at 1065°C (1950°F)

Test Temperature °C	(°F)	Stress MPa	(ksi)	Time Failure (Hr)	Elongation (%)	Reduction of Area (%)
540	(1000)	482	(70)	1.0	7.9	14.5
540	(1000)	482	(70)	2.8	-	-
650	(1200)	345	(50)	11.8	18.6	18.6
650	(1200)	345	(50)	28.7	7.9	17.4
760	(1400)	172	(25)	218.7	13.1	18.7
760	(1400)	172	(25)	170.9	13.1	15.7

Heat Treatment: 1205°C (2200°F) 0.5Hr Ar Çool +
760°C (1400°F) 1Hr Vacuum Cool.

Table IV. Creep Properties of P/M Niobium Modified Ti_3Al HIPed at 1065°C (1950°F)

Powder	Test Temp. °C	(°F)	Stress MPa	(ksi)	Plastic Deformation 23Hr (%)	50Hr (%)	165Hr (%)
PREP	760	(1400)	138	(20)	0.95	1.53	3.55
PREP	760	(1400)	138	(20)	1.00	1.57	3.95

Heat Treatment: 1205°C (2200°F) 0.5Hr Ar Cool +
760°C (1400°F) 1Hr Vacuum Cool.

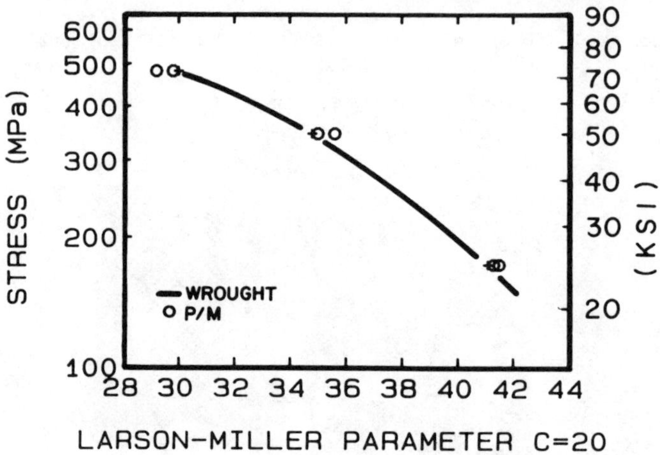

Figure 6. Stress rupture properties of niobium modified Ti_3Al.

In summary, the work discussed here shows that fully dense complex shaped parts can be produced using prealloyed Ti_3Al base powder. Also, the properties of this material are essentially equivalent to conventionally cast and wrought material. Thus, the P/M process offers a potentially lower cost method for producing Ti_3Al base components through reduced input material and less machining.

Conclusions

- Complex shapes can be produced with the niobium modified Ti_3Al alloy using the ceramic mold process.
- Full density can be achieved in the niobium modified Ti_3Al alloy using HIP conditions of 1205°C (1950°F) and 103MPa (15ksi) for 6 to 8 hours.
- Tensile and stress rupture properties of P/M niobium modified Ti_3Al compare favorably with cast plus wrought product.
- Room and elevated temperature strength of P/M niobium modified Ti_3Al increases with increased cooling rate from the solution treat temperature.

Acknowledgements

This work was supported under a joint program between Teledyne-CAE Turbine Engines and Crucible Research Center.

References

1. H. A. Lipsitt, "Titanium Aluminides - An Overview," High Temperature Ordered Intermetallic Alloys, C. C. Koch, C. T. Liu and N. S. Stoloff Eds., Mat. Res. Soc. Symp. Proc., Vol. 39, Materials Research Society, 1985.

2. J. Holowach and T. K. Redden, "Ti/Al Design/Cost Trade-Off Analysis," General Electric Co., AFAPL-TR-78-74, October 1978.

3. V. C. Petersen, V. K. Chandhok and C. A. Kelto, "Hot Isostatic Pressing of Large Titanium Shapes," Powder Metallurgy of Titanium Alloys, F. H. Froes and John E. Smugeresky, Eds., TMS-AIME Publication, Warrendale, PA, 1980.

4. A. A. Sheinker, "Prealloyed Titanium P/M Parts in the F-18 Aircraft," Presented at the 1984 AIME Spring Meeting, Los Angeles, CA, February 1984.

5. R. H. Witt and I. G. Weaver, "Titanium PM Components for Airframes," Titanium Net Shape Technologies, F. H. Froes and D. Eylon, Eds., TMS-AIME Publication, Warrendale, PA, 1984.

6. E. J. Kosinski, "The Mechanical Properties of Titanium PM Parts Produced From Superclean Powders," Progress in Powder Metallurgy 1982, 38, J. G. Bewley and S. W. McGee Eds., MPIF/APMI, Princeton, NJ, 1982.

7. T. Lizzi and C. F. Yolton, (unpublished research 1986).

8. P. L. Martin, H. A. Lipsitt, N. T. Nuhfer and J. C. Williams, "The Effects of Alloying on the Microstructure and Properties of Ti_3Al and TiAl," Titanium '80 Science and Technology, Proceedings of the 4th Int'l Conf. on Titanium, H. Kimura & O. Izumi Eds., TMS-AIME Publication, Warrendale, PA, 1980.

HIP COMPACTION OF TITANIUM ALLOY POWDERS AT

HIGH PRESSURE AND LOW TEMPERATURE (HPLT)

D. Eylon[a] and F. H. Froes[b]

[a]Metcut-Materials Research Group, now with the University of Dayton, Graduate Materials Engineering, Dayton, OH 45469

[b]Air Force Wright Aeronautical Laboratories, Materials Laboratory, AFWAL/MLLS, Wright-Patterson Air Force Base, OH 45433

Abstract

Compaction of rapidly solidified rotating electrode process (REP) Ti-6Al-4V (Ti-6-4) and plasma rotating electrode process (PREP) Ti-10V-2Fe-3Al (Ti-10-2-3) prealloyed spherical powders at low temperature and high pressure was studied. The Ti-6-4 powder was compacted at 650°C and 595°C and the Ti-10-2-3 was compacted at 595°C and 540°C, in both cases with a compaction pressure of 315MPa for 24 hr. Although all four compact conditions displayed a fully dense microstructure, interparticle debonding was observed in the lower temperature compacts after tensile testing, leading to very low elongation. The higher temperature compacts displayed a good combination of tensile strength and ductility the result of the very fine as-compacted microstructure. The fine microstructures are the result of the relatively low hot isostatic pressing (HIP'ing) temperatures used in this work which retained the fine structural scale of the as-produced powder particles. Recrystallization anneal resulted in the Ti-6-4 a fully equiaxed alpha structure with much improved tensile elongation which is predominantly the result of the improved metallurgical bond between the prior powder particles during this treatment. This work demonstrated that at relatively low temperatures under high pressure it is possible to obtain very fine microstructures with excellent combination of tensile strength and elongation.

Introduction

Rapid solidification (RS) requires production of material with, at least, one small dimension [1], to achieve high rates of solidification. As-produced RS materials are now available through various techniques, in powder, flake, ribbon or fiber form [2].

For most technical applications, the as-quenched RS products need to be consolidated into larger and more usable forms. However, many consolidation techniques and practices used in conventional powder metallurgy (PM) [3, 4] or metal matrix composite (MMC) production [5, 6], require relatively high temperatures. This can lead to morphological changes like grain growth or precipitation of coarse second phase particles and subsequent loss of properties and microstructural features obtained by the RS processing.

To prevent or restrict these changes, compaction at relatively low temperatures is required. Various techniques have been tried in recent years like lower temperature extrusion [7, 8], gun compaction [9, 10], explosive compaction [11], and press consolidation [12, 13]. In all these methods, the compaction at low temperature was achieved by increasing the process pressure, energy input, and exposure time or by reducing the strain rate. The objective of the present work was to explore low temperature hot isostatic pressing (HIP) for compacting rapidly solidified titanium alloy powders. HIP'ing was done at 315MPa (45ksi) pressure for 24 hr instead of the conventional 105MPa (15ksi) for 2-6 hr used in titanium alloy powder metallurgy.

Since RS titanium alloy particulates are not available yet in quantities large enough to produce relatively large size compacts, rotating electrode process (REP) Ti-6Al-4V (Ti-6-4) and plasma rotating electrode process (PREP) Ti-10V-2Fe-3Al (Ti-10-2-3) prealloyed spherical powders were used in this study. Recent work has estimated the cooling rates of Ti-6-4 REP spherical powder particles to be in the range of 10^4 to 10^6°C/sec [14]. Therefore, the REP and PREP alloy powders are essentially, RS materials. The fine microstructural features of both alloy powder particles are shown in Figures 1a and 1b. The alloys selected for this study are important for titanium technology in general and titanium powder metallurgy in particular. The Ti-6-4 is the most commonly used titanium alloy [15], while the Ti-10-2-3 is an emerging high strength metastable beta alloy [16]. A substantial body of work was developed in recent years on Ti-6-4 [4, 17] and Ti-10-2-3 [8, 18-20] powder metallurgy and baseline data on conventionally HIP'd compacts are available.

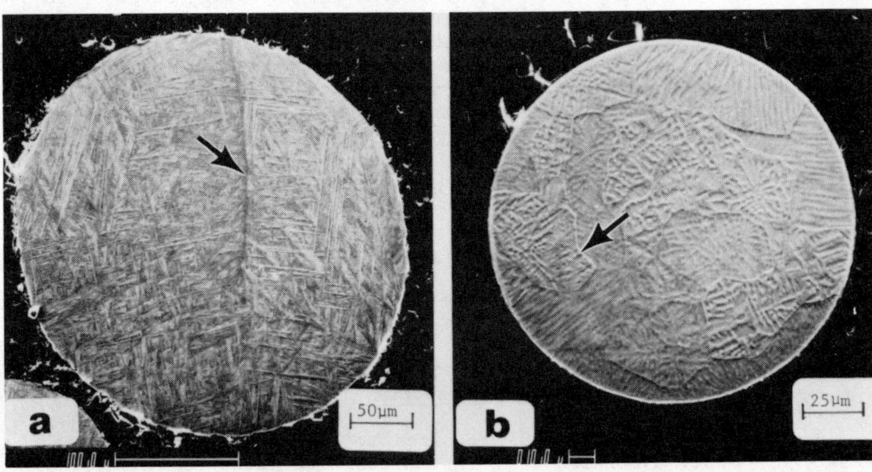

Figure 1 - SEM images showing microstructure of (a) REP Ti-6Al-4V powder particles and (b) PREP Ti-10V-2Fe-3Al powder particles. Arrows mark beta grain boundaries (original magnifications 250x and 500x, respectively).

In an earlier work [21], Ti-6-4 and Beta III (Ti-11.5Mo-6Zr-4.5Sn -- a lean metastable beta alloy) powders were HIP compacted to full density at 675°C (1250°F)/235MPa (34ksi)/24 hr. However, the metallurgical bond between the compacted particle was not sufficient leading to loss of tensile elongation. A higher pressure was utilized in the present work and the compaction temperature range was 540-650°C (1000-1200°F). At the higher compaction temperatures of this study, powders were compacted in the as-produced condition (Figure 2a). At the lower compaction temperatures, powders were deformed by the strain energizing process (SEP) [22] (Figure 2b) to promote better metallurgical bonding [21] due to the particle higher aspect ratio. Previous works showed that SEP'd powders can be consolidated at lower temperature [23] and result in recrystallized microstructures [24].

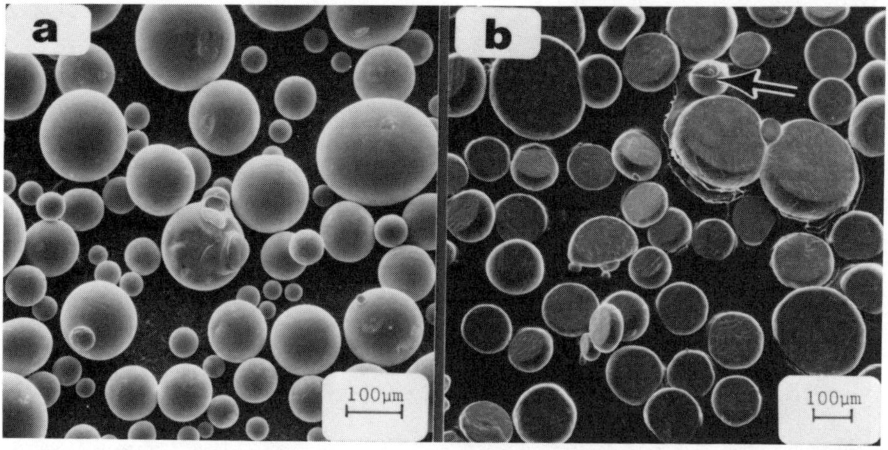

Figure 2 - REP Ti-6Al-4V -35 mesh (-420µm) powder particles (a) as-produced, (b) after 60% double pass SEP deformation. Note the undeformed small particle marked with an arrow (original magnifications 100x and 70x, respectively).

This study was focused on establishing the lowest practical HIP compaction temperature for the two alloy powders, evaluating the as-compacted microstructure, and the subsequent tensile and fatigue properties. The effect of post compaction heat treatments on microstructure and properties was also investigated.

Experimental Procedures and Results

Material and Compaction

REP [25] Ti-6-4 and PREP [26] Ti-10-2-3 -35 mesh (-420µm) powders were selected. Half of each powder was compacted in the as-produced condition (Figure 2a) and half was SEP'd [22-24] (Figure 2b) to promote better metallurgical bond at the lower compaction temperatures. Past work showed that due to some process limitations, not all powder particles get SEP deformation and small regions of non-crystallized structure exist in the compacts [24, 27]. To overcome this problem, a double pass high reduction (60%) SEP [28] was used. Nevertheless, a few very small particles (see arrow in Figure 2b) still remained undeformed. After the SEP'ing it was also discovered that some contamination was introduced to the powders from the SEP equipment.

The compaction was performed in a 315MPa (45ksi) HIP autoclave with work space of 140mm (5.6 in.) diameter x 280mm (11.2 in.) length. Powders were filled into welded mild steel cans (Figure 3). The final compact dimension after can removal was 50mm (2.0 in.) diameter x 80mm (3.2 in.) long.

Figure 3 - A steel can filled with powder after evacuation and HIP'ing.

The HIP'ing and powder conditions of the four HIP compacts are listed in Table I.

Table I. Powder Treatment and HIP'ing Conditions Used in This Work

Compact Designation	Alloy	Type of Powder	Powder Treatment	HIP Condition Temperature, °C (°F)	Pressure, MPa (ksi)	Time, hr
HPLT 1	Ti-6-4	REP	-	650 (1200)	315 (45)	24
HPLT 2	Ti-6-4	REP	SEP	595 (1100)	315 (45)	24
HPLT 3	Ti-10-2-3	PREP	-	595 (1100)	315 (45)	24
HPLT 4	Ti-10-2-3	PREP	SEP	540 (1000)	315 (45)	24

The chemical composition of the four compacted billets is listed in Table II.

Table II. Chemical Composition in Weight Percent (Balance Ti)

Compact Designation/ alloy/HIP temperature	Al	V	Fe	C	N, ppm	O, ppm	H, ppm
HPLT 1/Ti-6-4/650°C	5.93	4.02	0.17	0.038	158	2050	157
HPLT 2/Ti-6-4/595°C	6.18	4.04	0.18	0.006	150	2088	71
HPLT 3/Ti-10-2-3/595°C	3.05	10.02	2.42	0.005	117	1234	75
HPLT 4/Ti-10-2-3/540°C	2.95	9.67	2.62	0.024	141	1643	150

The photomicrographs of the four as-compacted conditions (Table I) are shown in Figures 4a-4d.

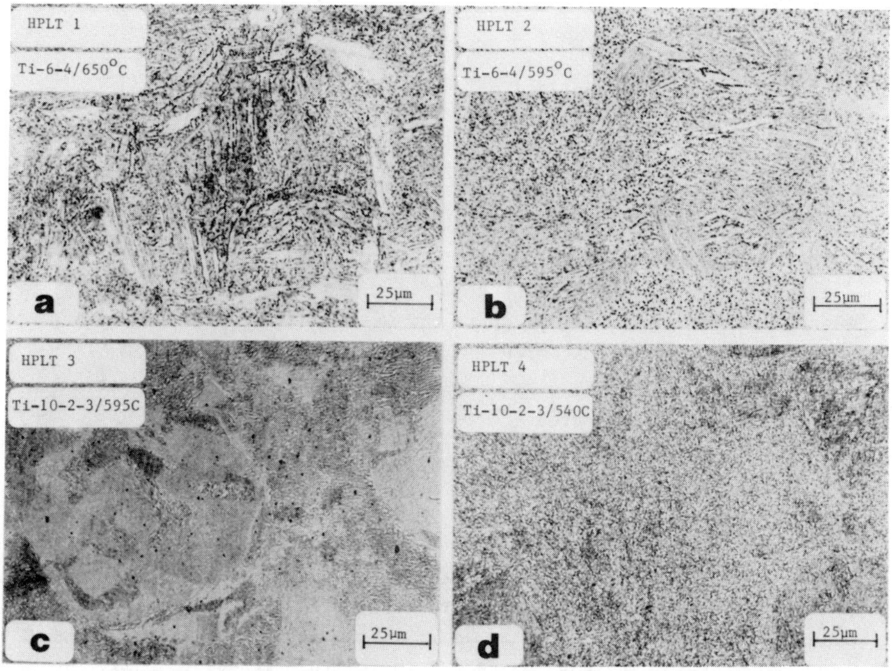

Figure 4 - Optical photomicrographs (original magnification 600x) of the following as-HIP'd conditions: (a) Ti-6-4, 650°C (HPLT 1), (b) Ti-6-4, 595°C (HPLT 2), (c) Ti-10-2-3, 595°C (HPLT 3), and (d) Ti-10-2-3, 540°C (HPLT 4).

Heat Treatment

The low HIP temperature resulted in a compacts with a very fine microstructure (Figures 4a-4d). Subsequently, heat treatment study was performed on all compacts to try and achieve (a) microstructure modification for mechanical property improvements and (b) improved metallurgical bond between the particles.

The heat treatments used and the subsequent evaluations are listed in Table III.

Table III. Compact Heat Treatment and Evaluation

Compact Designation, (alloy, HIP temperature)	Condition, °C (°F)/hr	Evaluation: Metallographic	Evaluation: RT Tensile
HPLT 1			
	As-compacted	+	+[a]
	760 (1400)/24/AC	+	
(Ti-6-4	815 (1500)/24/AC	+	+
650°C)	900 (1650)/2/SC to 1400/FC	+	
	955 (1750)/2/SC to 1400/FC	+	
HPLT 2			
	As-compacted	+	+
	705 (1300)/2/FC	+	+
(Ti-6-4	705 (1300)/24/FC	+	+
595°C)	815 (1500)/2/FC	+	+
	815 (1500)/24/FC	+	+
	955 (1750)/2/FC	+	+
HPLT 3			
	As-compacted	+	+
(Ti-10-2-3	760 (1400)/1/WQ + 510 (950)/8/AC	+	+
595°C)	760 (1400)/3/AC + 370 (700)/4/AC	+	+
	790 (1450)/3/AC + 370 (700)/4/AC	+	+
HPLT 4			
	As-compacted	+	+
(Ti-10-2-3	750 (1385)/1/WQ + 550 (1025)/8/AC	+	+
540°C)	760 (1400)/1/WQ + 510 (950)/8/AC	+	+

[a] Room temperature HCF test was conducted for this condition only.
AC = air cool; SC = slow programmed furnace cooling at 90°F/hr; FC = furnace cool; WQ = water quench.

The photomicrographs of Ti-6-4 and Ti-10-2-3 selected heat treatment conditions are shown in Figures 5a-5f and Figures 6a-6d, respectively.

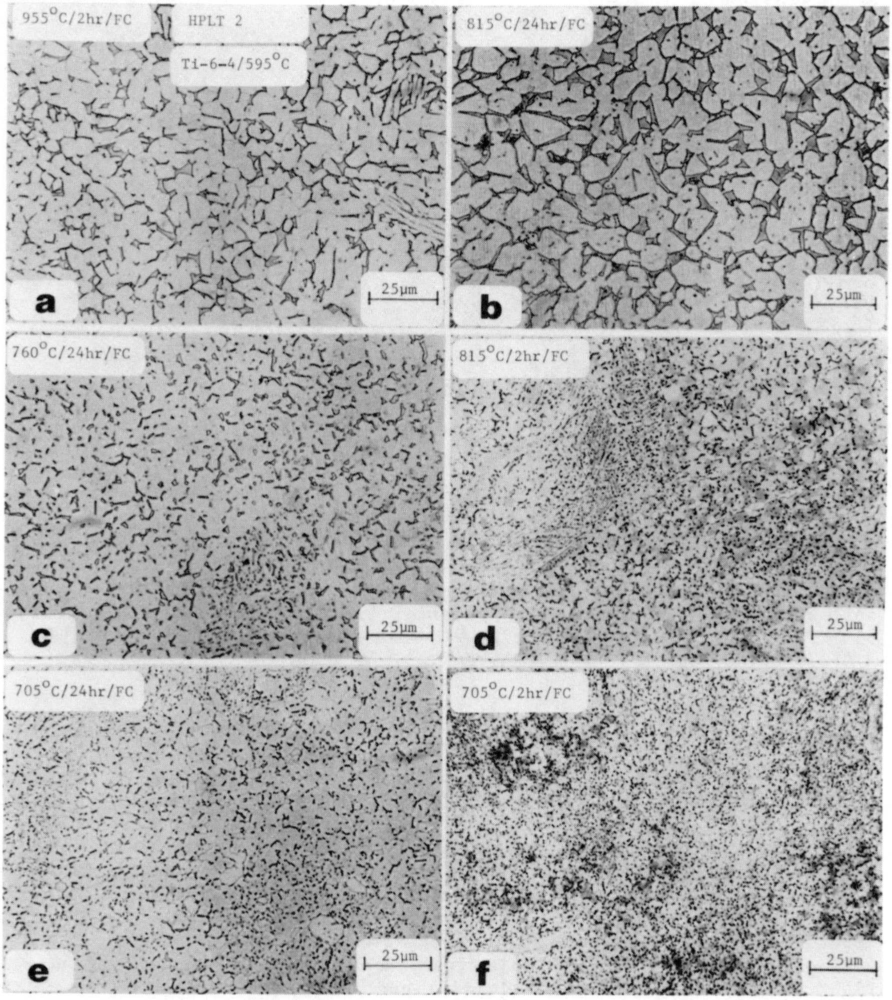

Figure 5 - Optical photomicrographs (original magnification 600x) of the following Ti-6-4 595°C HIP'd material treatments: (a) 955°C/2 hr/FC, (b) 815°C/24 hr/FC, (c) 760°C/24 hr/FC, (d) 815°C/2 hr/FC, (e) 705°C/24 hr/FC, and (f) 705°C/2 hr/FC.

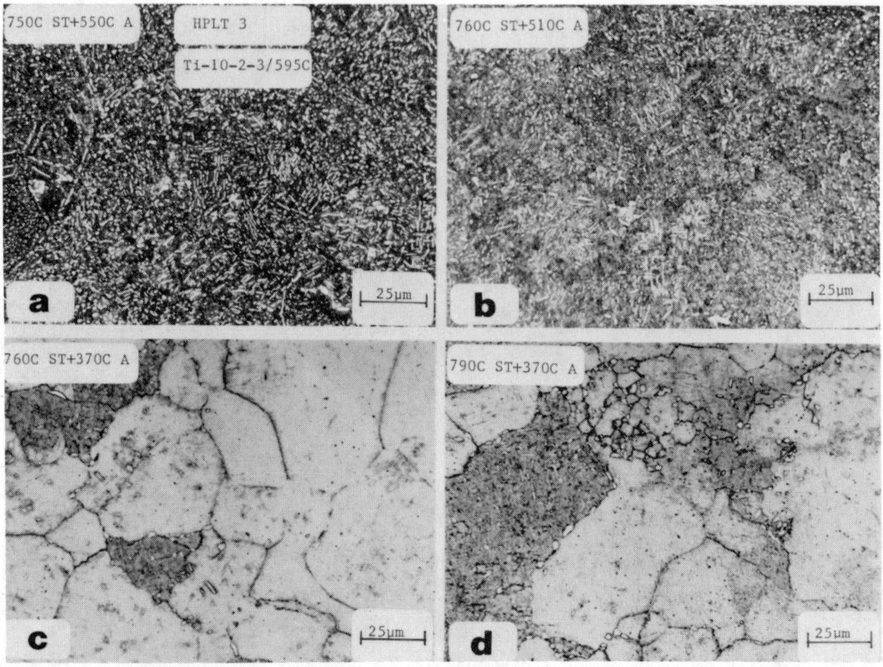

Figure 6 - Optical photomicrographs (original magnification 600x) of the following Ti-10-2-3 595°C HIP'd material treatments: (a) 750°C/1 hr/WQ + 550°C/8 hr/AC, (b) 760°C/1 hr/WQ + 510°C/8 hr/AC, (c) 760°C/3 hr/AC + 370°C/4 hr/AC, and (d) 790°C/3 hr/AC + 370°C/4 hr/AC.

Mechanical Testing

Room temperature tensile tests were performed on the as-HIP'd and heat treated conditions listed in Table III. Due to the small dimensions of the material available, both tensile and fatigue tests were conducted on subsize smooth bar specimens 2.5mm (0.1 in.) gage diameter x 17.5mm (0.7 in.) gage length. Tensile test strain rate was kept at 0.005mm/mm/min through the 0.2% yield point followed by 1.25mm (0.05 in.)/min cross head speed to failure. The tensile test results are listed in Table IV. The SEM fractographs of selected tensile samples are shown in Figure 7 and a section through a HPLT 4 sample fracture surface is shown in Figure 8. High cycle room temperature axial fatigue tests were conducted on an MTS servohydraulic machine at R (minimum stress/maximum stress) = 0.1; stress concentration factor K_t = 1; frequency = 5Hz, and trangular waveform. The fatigue results thus far, were done only for HPLT 1 as-compacted condition (Figure 8). Three additional conditions are in the process of testing.

Table IV. Tensile Results of the Various Microstructural Conditions

Compact Designation, (alloy, HIP temperature)	Heat Treatment, °C/hr	% YS, MPa (ksi)	UTS, MPa (ksi)	EL %	RA %	No. of Tests
HPLT 1						
	As-compacted	1082 (157)	1130 (164)	8	19	2
(Ti-6-4 650°C)	815/24/AC	937 (136)	1013 (147)	22	38	2
HPLT 2						
	As-compacted	N/A	1027 (149)	0.2	0	2
	705/2/FC	N/A	1034 (150)	0	1	3
(Ti-6-4	705/24/FC	1054 (153)	1068 (155)	1	5	3
595°C)	815/2/FC	1102 (160)	1123 (163)	1	4	3
	815/24/FC	992 (144)	1102 (160)	7	17	3
	955/2/FC	965 (140)	1027 (149)	8	26	3
HPLT 3						
	As-compacted	951 (138)	992 (144)	14	49	2
(Ti-10-2-3	760/1/WQ+510/8/AC	1226 (178)	1295 (188)	3	6	2
595°C)	760/3/AC+370/4/AC	N/A	1447 (210)	1	4	2
	790/3/AC+370/4/AC	1461 (212)	1564 (227)	1	1	2
HPLT 4						
	As-compacted	999 (145)	1006 (146)	1	3	2
(Ti-10-2-3	750/1/WQ+550/8/AC	1144 (166)	1164 (169)	1	2	3
540°C)	760/1/WQ+510/8/AC	N/A	1093 (159)	0	0	3

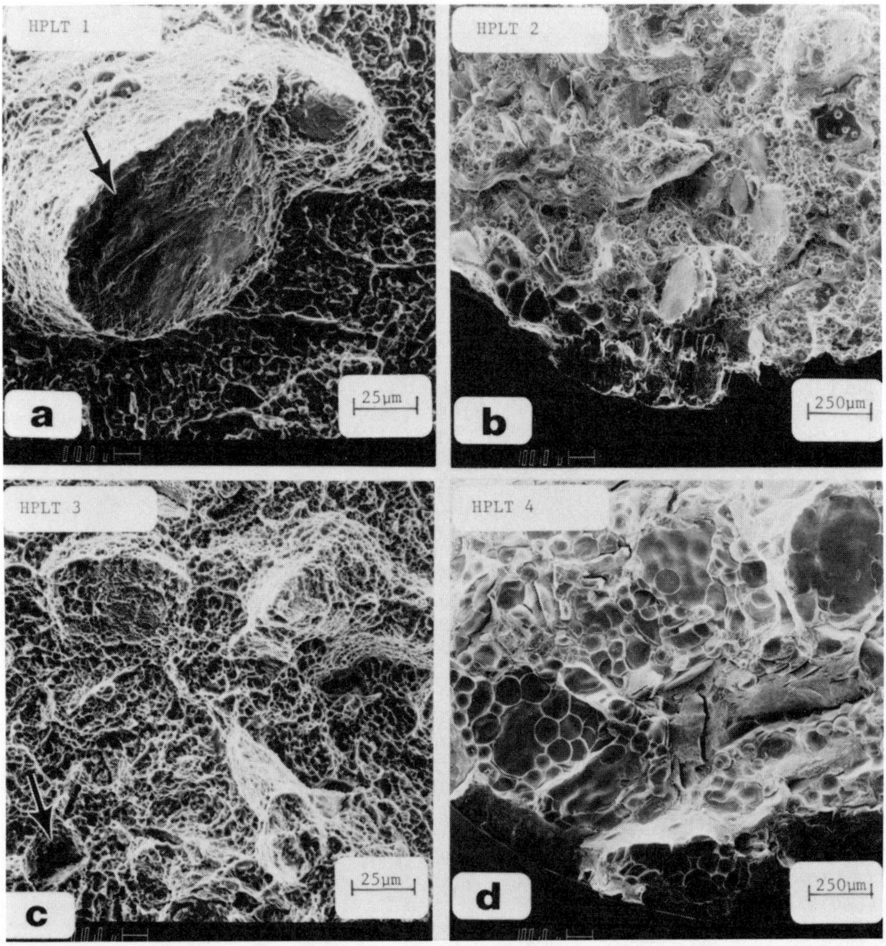

Figure 7 - SEM fractographs of tensile fracture of the following conditions (arrows mark isolated particle debonding): (a) HPLT 1, Ti-6-4, 650°C HIP, (b) HPLT 2, Ti-6-4, 595°C HIP, (c) HPLT 3, Ti-10-2-3, 595°C HIP, and (d) HPLT 4, Ti-10-2-3, 540°C HIP (original magnifications 500x, 50x, 500x, and 50x, respectively).

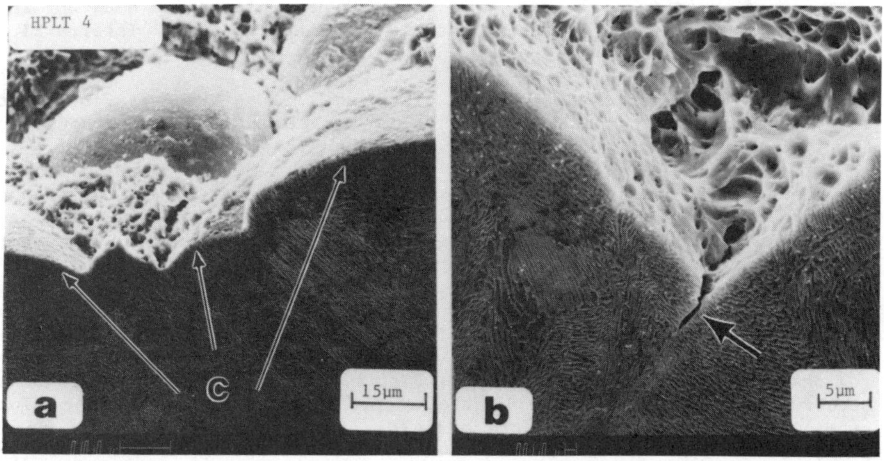

Figure 8 - A polished and etched section through a debonded fracture feature of a HPLT 4 (Ti-10-2-3, 540°C HIP'd) tensile test sample: (a) showing the powder particles curvature (C) with no signs of microstructural debonding on the section plane (original magnification 1000x), and (b) an initial crack forming on prior particle boundary (original magnification 2000x).

Discussion

As-Compacted Material

As can be seen from the photomicrographs in Figures 4a-4d, all compacts reached full density during HIP'ing. No change in density or porosity in the microstructure was detected after the subsequent heat treatments (Figures 5a-5f). However, the tensile test results (Table IV) clearly show that the Ti-6-4 compacted at 595°C (HPLT 2) and the Ti-10-2-3 compacted at 540°C (HPLT 4) displayed almost no elongation in the as-compacted condition, the result of particle debonding (Figures 7b and 7d). The flat nature of some of the debonded particle boundaries is the result of SEP'ing the powder (Figure 2b). On the other hand, compacts HIP'd at higher temperature for each alloy (Ti-6-4 at 650°C and Ti-10-2-3 at 595°C) displayed adequate elongation. It should be noted that in the case of HPLT 1 and HPLT 3, a limited level of debonding is evident in the otherwise ductile fracture surfaces (arrow in Figures 7a and 7c, respectively). This limited debonding had no effect on tensile elongation but resulted in some degradation of the fatigue strength. A metallographic section [29] was made through a fracture region of a HPLT 4 sample which displayed an extensive interparticle debonding. The curvature of the debonded particle surfaces is very noticeable on the contour of the section plane (C in Figure 8a). However, no microstructural debonding, or other imperfections are seen in the underlying microstructure. The extensive particle debonding on the fracture surfaces of the lower temperature HIP conditions (HPLT 2 and HPLT 4) which is contrasted by the lack of voids detected in the microstructure, is an indication that low temperature/high pressure compaction can result in incomplete metallurgical bonding in a material that, otherwise, will look fully dense. On the section plane of Figure 8b, a side crack (arrow) is propagating from a debonded fracture zone into a prior particle region. The prior particle region does not show any detrimental features that can be related to this interparticle cracking.

The as-HIP'd microstructures of all four compacts are very fine due to the low compaction temperature which did not allow for much coarsening of the fine powder particle microstructure. The microstructure of HPLT 1 and HPLT 2 (Figures 4a and 4d) consists of a very fine alpha phase. Part of the fine alpha has a lenticular morphology, similar to the martensite needles of the as-produced particles (Figure 1a), and part is equiaxed (1-2μm) in a matrix of beta. The Ti-10-2-3 powder particle structure (Figure 1b) shows a columnar beta structure at the particle surface, the result of the high cooling rate. This microstructure then degenerates into a beta dendritic structure, the result of slower cooling rates inside the particle. In the as-compacted Ti-10-2-3 condition (HPLT 3 and HPLT 4), micron size alpha precipitation is visible after the compaction. In some regions (such as in the upper part of Figure 4c) the traces of the columnar structure are still visible in the as-HIP'd microstructure. The alpha precipitates are finer and more globular then previously observed in PM compacts HIP'd at higher temperatures [19, 20]. The good combination of high tensile strength and elongation of the higher temperature compacts of both alloys (HPLT 1 and HPLT 3, Table IV) is, probably, the result of this very fine alpha structure. Even the lower temperature extrusions display high tensile strength and it is only the lack of metallurgical bond between the powder particle which causes the very low ductility.

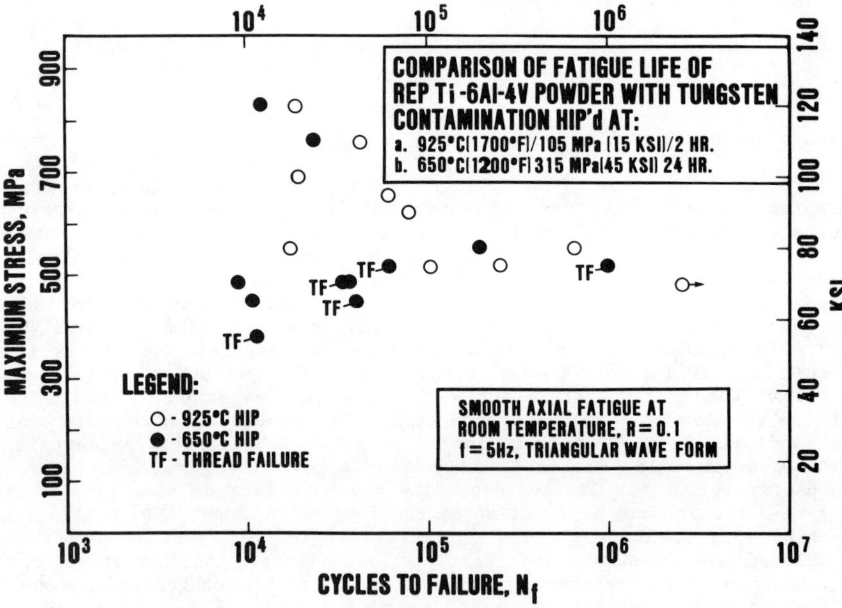

Figure 9 - S-N fatigue curve of HPLT 1 (Ti-6-4 HIP'd at 650°C) in the as-HIP'd condition.

Thus far only the as-HIP'd condition of HPLT 1 has been tested in high cycle fatigue. Since an old lot of REP powder which contained tungsten contaminants was used for this evaluation, these results were compared with the fatigue test data from powder of the same lot compacted at 925°C (1700°F)/105MPa (15ksi)/2 hr [30]. It can be seen that half of the data points of HPLT 1 are within the same population as the 925°C (1700°F) compacted material (Figure 9). However, a group of low lying data points and the indicated thread failures probably relates to the occasional debonding observed in the tensile fractures (Figures 7a and 7c), leading to notch sensitivity. An SEM fractographs of a fatigue sample shows a surface initiation (Figure 10a) at a debonded particle (Figure 10b).

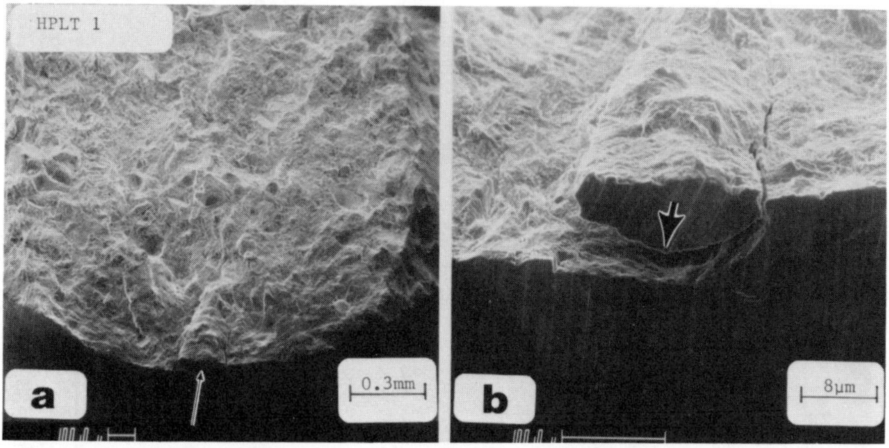

Figure 10 - SEM fractographs of fatigue crack initiation site in a sample failed prematurely: (a) low magnification (originally 50x), arrow marks initiation, and (b) higher magnification (originally 200x), arrow marks particle debonding zone at the crack initiation site.

Heat Treated Material

The level of recrystallization in the Ti-6-4 compacts was studied in detail for the lower temperature compact (HPLT 2). The amounts of recrystallization is shown in Figures 5a-5f in decreasing order. Full recrystallization is achieved both at 955°C (1700°F)/2 hr and 815°C (1500°F)/24 hr. Only in these two conditions was the tensile elongation of the compact restored (Table IV). It is probable that the temperature/time conditions were sufficient to result in good metallurgical bonding. The fact that the material recrystallized in the process may also have helped to improve this bonding. In a previous work on blended elemental Ti-6-4 powder compacts, it was demonstrated that recrystallization can partially close elongated voids [31]. In HPLT 1 compact elongation was increased from 8% to 22% after the recrystallization treatment. Lower temperatures (760°C and 705°C) or shorter annealing time (2 hr) resulted in a partially recrystallized structure (Figures 5c-5f) and almost no improvement in the tensile ductility (Table IV).

Heat treatment and aging response tests were conducted on Ti-10-2-3 HPLT 3 (Figure 11) and tensile tests were done on selected conditions (Table III). The microstructures of selected heat treatments are shown in Figure 6. In general, the solution treatments followed by 370°C (700°F) aging resulted in microstructure with almost no alpha precipitates, or with precipitates too small to be resolved in an optical level (Figures 6c and 6d). However, solution treatments followed by 510°C and 570°C (950°F and 1025°F) aging, resulted in microstructures with micron size globular and elongated alpha precipitates (Figures 6a and 6b) similar to those observed in previous work on similarly solution treated ingot metallurgy and PM products [20]. In the HPLT 3 compact these treatments resulted in a substantial increase in strength (from 990MPa to 1565MPa) with a substantial loss of tensile elongation (from 14% to 1%) (Table IV). In the HPLT 4 compact, where debonding was observed in the as-HIP'd condition, the relatively low solution treatment temperatures and short times did not result in any change from the debonding type of failure. The beta transus temperature of this alloy is approximately 760°C (1400°F), thus all solution treatments were done just below or in the beta phase field.

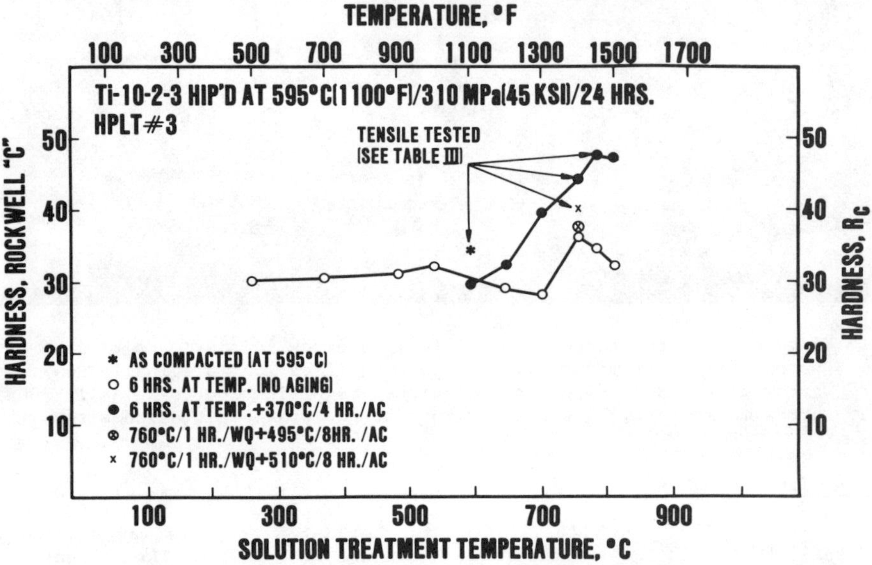

Figure 11 - Heat treatment and aging response (R_C vs. temperature) of Ti-10-2-3 HPLT 3. Conditions marked with an arrow were also tensile tested (Table III).

Summary and Conclusions

1. Compaction of rapidly solidified Ti-6-4 and Ti-10-2-3 titanium powders at low temperatures and high pressures by hot isostatic pressing (HIP'ing) was studied.

2. The compaction temperatures were 650°C (1200°F) and 595°C (1100°F) for Ti-6-4 and 595°C (1100°F) and 540°C (1000°F) for the Ti-10-2-3. The pressure was 315MPa (45ksi) in all cases for 24 hr.

3. All four compacts were found by metallography to be fully dense.

4. The relatively low compaction temperature retained the fine microstructure scale of the powder particles in all four compacts.

5. For each alloy, the compacts produced at the higher HIP temperature displayed good combination of tensile strength and elongation. However, the lower temperature compacts, displayed almost no elongation the result of powder particle debonding.

6. In the case of Ti-6-4, the elongation was restored after recrystallization treatment which also restored the particle debonding.

7. The solution treatment of the Ti-10-2-3 compacts resulted in higher tensile strength with a substantial loss of ductility.

8. While limited debonding in the higher temperature compacts had no effect on tensile ductility, a loss of some fatigue strength was observed.

Acknowledgments

The authors wish to thank T. F. Broderick and T. Egerer for their assistance in the early parts of this work. The assistance of N. G. Lovell, D. K. Patrick, C. M. Cooke, W. A. Houston, and R. D. Brodecki in the various stages of this work is highly appreciated. Finally the contribution of Ms. Karen A. Sitzman in typing the text is recognized. Parts of this work were done under U.S. Air Force Contract No. F33615-82-C-5078.

References

1. H. Jones, Rapid Solidification of Metals and Alloys, Monograph No. 8 (London: The Institute of Metallurgists, 1982).

2. S. J. Savage and F. H. Froes, "Production of Rapidly Solidified Metals and Alloys," Journal of Metals, 36(4)(1984), pp. 20-33.

3. F. H. Froes, D. Eylon, G. E. Eichelman, and H. M. Burte, "Advances in Titanium Powder Metallurgy, Journal of Metals, 32(2)(1980), pp. 47-54.

4. F. H. Froes and D. Eylon, "Powder Metallurgy of Titanium Alloys - A Review," in Titanium, Science and Technology, vol. 1, ed. G. Lutjering, U. Zwicker, and W. Bunk (Oberursel, West Germany: DGM, 1985), pp. 267-286; also in Powder Metallurgy International (PMI), 17(4)(1985), pp. 163-167 and continued in (5)(1985), pp. 235-238; also in Titanium Technology: Present Status and Future Trends, ed. F. H. Froes, D. Eylon, and H. B. Bomberger, (Dayton, OH: Titanium Development Association, 1985), pp. 49-59.

5. P. R. Smith, F. H. Froes, and J. T. Cammett, "Correlation of Fracture Characteristics and Mechanical Properties for Titanium-Matrix Composites," in Mechanical Behavior of Metal-Matrix Composites, ed. J. H. Hack and M. F. Amateau (Warrendale, PA: The Metallurgical Society, 1983), pp. 143-168.

6. P. R. Smith and F. H. Froes, "Developments in Titanium Metal Matrix Composites," Journal of Metals, (36)(3)(1984), pp. 19-26.

7. S. M. L. Sastry, T. C. Peng, and L. P. Beckerman, "Structure and Properties of Rapidly Solidified Dispersion Strengthened Titanium Alloys: Part II. Tensile and Creep Properties," Met. Trans. A, 15A (1984), pp. 1465-1474.

8. I. A. Martorell, Y. R. Mahajan, D. Eylon, and F. H. Froes, "Property Optimization of Ti-10V-2Fe-3Al by Microstructure Modification" (1985 TMS Fall Meeting, Abstract Book), p. 75.

9. D. Raybould, "Cold Dynamic Compaction of Pre-Alloyed Titanium and Activated Sintering," in Progress in Powder Metallurgy, vol. 38, ed. J. G. Bewley and S. W. McGee (Princeton, NJ: MPIF/APMI Publications, 1983), pp. 575-589.

10. N. C. Iyer, Westinghouse Research and Development Center (joint USAF/Westinghouse program), 1984-1986.

11. V. D. Linse, D. Eylon, T. F. Broderick, I. A. Martorell, and F. H. Froes, (an unpublished joint work: Battelle/Metcut/AFWAL/Westinghouse), 1983-1986.

12. D. Eylon, C. A. Kelto, A. Hayes, and F. H. Froes (work in progress), 1985-1986.

13. R. G. Vogt and D. Eylon (an unpublished work), 1983-1986.

14. T. F. Broderick, A. G. Jackson, H. Jones, and F. H. Froes, "The Effect of Cooling Conditions on the Microstructure of Rapidly Solidified Ti-6Al-4V," Met. Trans. A, 16A (1985), pp. 1951-1959.

15. F. H. Froes, D. Eylon, and H. B. Bomberger, eds., Titanium Technology: Present Status and Future Trends (Dayton, OH: Titanium Development Association, 1985).

16. R. R. Boyer, "Design Properties of a High Strength Alloy, Ti-10V-2Fe-3Al," Journal of Metals, 32(3)(1980), p. 61.

17. F. H. Froes and J. E. Smugeresky, eds., Powder Metallurgy of Titanium Alloys (Warrendale, PA: The Metallurgical Society, 1980).

18. J. E. Smugeresky and N. R. Moody, "Properties of High Strength, Blended Elemental Powder Metallurgy Titanium Alloys," in Titanium Net Shape Technology, ed. F. H. Froes and D. Eylon (Warrendale, PA: The Metallurgical Society, 1984), pp. 131-143.

19. R. R. Boyer, D. Eylon, C. F. Yolton, and F. H. Froes, "Powder Metallurgy of Ti-10V-2Fe-3Al," in Titanium Net Shape Technology, ed. F. H. Froes and D. Eylon (Warrendale, PA: The Metallurgical Society, 1984), pp. 63-78.

20. R. R. Boyer, D. Eylon, and F. H. Froes, "Comparative Evaluation of Ti-10V-2Fe-3Al Cast, P/M and Wrought Product Forms," in Titanium, Science and Technology, vol. 2, ed. G. Lutjering, U. Zwicker, and W. Bunk (Oberursel, West Germany: DGM, 1985), pp. 1307-1313; also in Powder Metallurgy International (PMI), 17(5)(1985), pp. 239-240.

21. C. F. Yolton and F. H. Froes (an unpublished work), 1978.

22. G. V. Scarich, G. R. Chanani, V. C. Petersen, and D. M. Weaver, "Relationships Between Mechanical Properties, Contaminants, and Fracture Topography of HIP Ti-6Al-4V Powder Metallurgy Products," Journal of Metals, 32(3)(1980), pp. 103-114.

23. D. Eylon, P. R. Smith, S. W. Schwenker, and F. H. Froes, "Status of Titanium Powder Metallurgy," in Industrial Applications of Titanium and Zirconium: Third Conference, ASTM STP 830, ed. R. T. Webster and C. S. Young, (Philadelphia, PA: ASTM Publications, 1984), pp. 48-65.

24. F. H. Froes, D. Eylon, G. Wirth, K-J. Grundhoff, and W. Smarsly, "Fatigue Properties of Hot Isostatically-Pressed Powders," Metal Powder Report, 38(1)(1983), pp. 36-41; also in Progress in Powder Metallurgy, vol. 38, ed. J. G. Bewley and S. W. McGee (Princeton, NJ: MPIF/APMI Publications, 1983), pp. 503-519.

25. P. R. Roberts and P. Loewenstein, "Titanium Alloy Powders Made by the Rotating Electrode Process," Journal of Metals, 32(3)(1980), pp. 21-35.

26. E. J. Kosinski, "The Mechanical Properties of Titanium PM Parts Produced from Superclean Powders," in Progress in Powder Metallurgy, vol. 38, ed. J. G. Bewley and S. W. McGee (Princeton, NJ: MPIF/APMI Publications, 1983), pp. 491-502.

27. D. Eylon and F. H. Froes, "High-Integrity Titanium Alloy Shapes by Powder Metallurgy," in Titanium Alloys in Surgical Implants, ASTM STP 796, ed. H. A. Luckey and F. Kubli (Philadelphia, PA: ASTM Publications, 1983), pp. 43-58.

28. Y. R. Mahajan, D. Eylon, C. A. Kelto, T. Egerer, and F. H. Froes, "Modification of Titanium Powder Metallurgy Alloy Microstructures by Strain Energizing and Rapid Omni-Directional Compaction," in Titanium, Science and Technology, vol. 1, ed. G. Lutjering, U. Zwicker, and W. Bunk (Oberursel, West Germany: DGM, 1985), pp. 339-346; also in Powder Metallurgy International (PMI), 17(2)(1985), pp. 75-78.

29. W. R. Kerr, D. Eylon, and J. A. Hall, "On the Correlation of Specific Fracture Surface and Metallographic Features by Precision Sectioning in Titanium Alloys," Met. Trans. A, 7A (1976), pp. 1477-1480.

30. P. R. Smith, D. Eylon, S. W. Schwenker, and F. H. Froes, "Effect of Heat Treatment on the Fatigue Behavior of PM Titanium Compacts," in Advanced Processing Methods for Titanium, ed. D. F. Hasson and C. H. Hamilton (Warrendale, PA: The Metallurgical Society, 1982), pp. 61-77.

31. I. Weiss, D. Eylon, M. W. Toaz, and F. H. Froes, "Effect of Isothermal Forging on Microstructure and Fatigue Behavior of Blended Elemental Ti-6Al-4V Powder Compacts," Met. Trans. A, (1986), in press.

TEM OF TITANIUM ALLOYS COMPACTED AT VERY HIGH PRESSURE AND LOW TEMPERATURE

I. E. Locci and G. E. Welsch

Department of Metallurgy and Materials Science
Case Institute of Technology
Case Western Reserve University
Cleveland, Ohio 44106

Abstract

Two alloys, Ti-6Al-4V and Ti-10V-2Fe-3Al, prepared from rotating electrode process powder (REP) and plasma rotating electrode powder (PREP) were investigated by transmission electron microscopy (TEM) and by energy dispersive X-ray analysis (EDX). The microstructural and microchemical analyses brought out several unusual results for the alloys which were HIP'ed at temperatures between 540 and 650°C and at pressure of 315 MPa. Despite the low temperatures, recrystallization took place in varying degrees, although grain growth was very limited. Partitioning of the alloy elements appears suppressed in the equiaxed microstructures, but has taken place in needle-like and domain-like microstructures. Here, discrete particles of beta phase have formed in the Ti-6Al-4V alloy, and domains of beta phase have formed in the Ti-10V-2Fe-3Al alloy. The equiaxed portions of the microstructures consisted of alpha phase only for both alloys, and no beta phase was detected.

One goal, the achievement of full densification without microstructural coarsening and without significant solute element redistribution, has been partially achieved. The micrographs in this paper are believed to be representative of the bulk of the samples. The boundary regions between former powder particles were not identified. Further analysis will be necessary to gain an understanding of the cohesiveness of the material in these locations.

INTRODUCTION

During compaction of rapidly solidified powders, it is usually desirable to maintain the fine microstructural and microchemical features obtained by the rapid freezing. To achieve this, it is often necessary to work at low compaction temperatures in order to prevent grain growth or precipitation of coarse second phase particles (1-4). This requires that compaction must be performed under higher pressure than normal or with other energy input.

Because of the limited availability of rapidly solidified titanium alloy powder, the present investigation was done on conventionally processed Ti-6Al-4V and Ti-10V-2Fe-3Al alloy powder, prepared by the rotating electrode process (REP) and by the plasma rotating electrode process (PREP). According to Ref. 5, the characteristics of REP and PREP alloy powders match those of rapidly solidified materials closely. The Ti-6Al-4V alloy is the most commonly used alpha + beta titanium alloy, while the Ti-10V-2Fe-3Al is a metastable beta alloy with high-strength potential (6). Cold deformation, also referred to as strain energizing process (SEP) (7), introduces additional internal energy and can aid in the compaction of powders at low temperature (8,9).

The objective of the present work was a microstructural and microchemical characterization of powder compacts which were hot isostatically pressed at 315 MPa pressure and at temperatures between 540 to 650°C using methods of transmission electron microscopy. The compacts had been prepared by Eylon and Froes (Ref. 10, in this volume). The processing procedure and the mechanical properties of the compacts are described in detail there.

Procedure

We followed the specimen designations given in Ref. 10. The alloys and their compaction conditions are summarized in Table 1. Table II contains the overall chemical compositions of the alloys.

Table I. Powder Treatment & HIP Conditions (Ref. 10)

Compact Designation	Alloy	Type of Powder	Powder Treatment	HIP Condition Temperature, °C (°F)	Pressure, MPa (ksi)	Time, h
HPLT 1	Ti-6-4	REP	--	650 (1200)	315 (45)	24
HPLT 2	Ti-6-4	REP	SEP	595 (1100)	315 (45)	24
HPLT 3	Ti-10-2-3	PREP	--	595 (1100)	315 (45)	24
HPLT 4	Ti-10-2-3	PREP	SEP	540 (1000)	315 (45)	24

Table II. Chemical Composition in Weight Percent (Ti = Balance)

Compact Designation/ Alloy/HIP Temperature	Al	V	Fe	C	N, ppm	O, ppm	H, ppm
HPLT 1/Ti-6-4/650°C	5.93	4.02	0.17	0.038	158	2050	157
HPLT 2/Ti-6-4/595°C	6.18	4.04	0.18	0.006	150	2088	71
HPLT 3/Ti-10-2-3/595°C	3.05	10.02	2.42	0.005	117	1234	75
HPLT 4/Ti-10-2-3/540°C	2.95	9.67	2.62	0.024	141	1643	150

Thin, electron-transparent foils were prepared from the compacts by standard techniques. The microstructural investigation was performed with a Philips EM400 transmission electron microscope. Microchemical analyses from various phases and precipitates were performed in the same microscope with energy dispersive X-ray analysis (EDX). For ease of comparison, the sequence of the alloys investigated was kept the same as in Ref. 10, and the Figures 1 through 4 in this report refer to the four specimens, respectively.

Results and Discussion

(1) Ti-6Al-4V Alloy HIP'ed at 650°C

The microstructure of this alloy is shown in Figure 1a. It shows a fine equiaxed alpha microstructure with the grain size being of the order of 1μm or less. The grains are generally free of dislocations, indicative of full recrystallization. Normally, heat treatment at this temperature for an extended period of time, such as 24 hours, would lead to a separation of vanadium-enriched beta phase and alpha phase. However, in the present microstructure no beta phase was found. Some scattered contrast within the alpha/alpha boundaries may be indicative of nucleating beta phase. Also, the EDX patterns taken from several regions within the equiaxed microstructure did not reveal any localized enrichment of vanadium or iron. An example is shown in Figure 1b. The conspicuous absence of any measurable quantity of beta phase leads to speculation that possibly the high pressure may have suppressed the formation of beta phase. At the 650°C HIP temperature, one would have expected some diffusional rearrangement of alloy elements and the formation of some distinct beta phase, see for example Ref. 11.

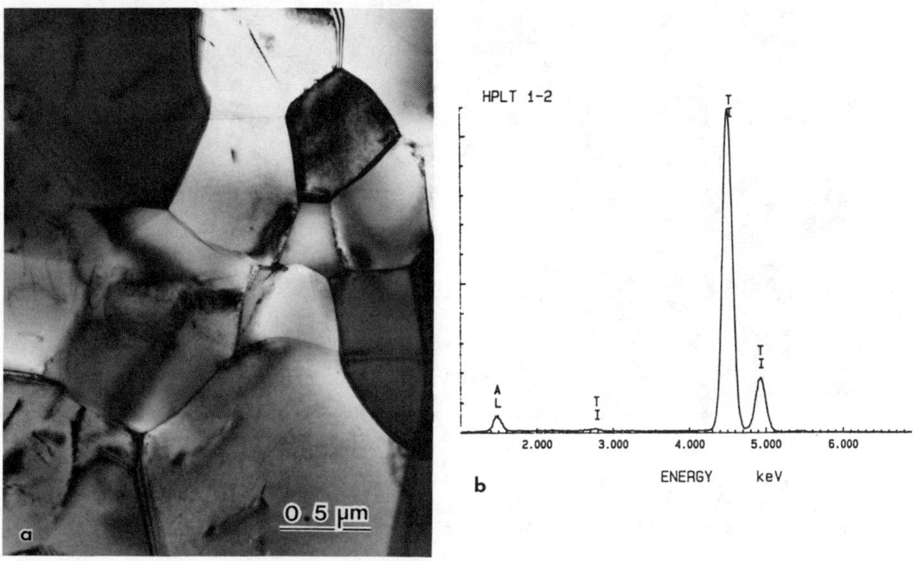

Figure 1. (a) TEM micrograph of Ti-6Al-4V, HIP'ed at 315 MPa and 650°C. The microstructure consists of fine, recrystallized alpha grains. No beta phase was found. (b) Typical EDX spectrum. Note, the vanadium concentration is insufficient to produce a K_β peak of V. The vanadium K_α peak is hidden under the titanium K_β peak.

(2) Ti-6Al-4V, HIP'ed at 595°C

In comparison to the alloy discussed before, the powder particles of this alloy had additionally been treated by the strain energizing process (SEP). The overall microstructure is shown in Figure 2a. It shows a mixed structure of submicron-width needles which are highly dislocated, and of recrystallized alpha grains with grain size ranging from 1-10μm. The recrystallized grains are easily distinguished by their characteristic absence of dislocations. A closeup of the equiaxed recrystallized microstructure in Figure 2b again shows absence of beta phase. The needle-like structure is shown in greater detail in Figures 2c,d (bright field and dark field, respectively). The needle-like structure contains beta precipitates which are imaged in the dark field micrograph with a 200 precipitate diffraction spot. The fact that only some precipitates are imaged with the diffracted beam indicates that there exist several orientation variances of the precipitates. The orientation relationship is analyzed in the diffraction pattern, Figure 2c. The $(0\bar{1}11)$ alpha plane matches the $(01\bar{1})$ beta plane. The beta phase nature of the precipitates is confirmed by the microchemical analyses in Figures 2e,f which show EDX patterns of the alpha matrix (2e) and of a beta precipitate (2f). Clearly, the beta precipitate is enriched with vanadium and iron. The EDX pattern of the alpha phase shows an absence of V and Fe peaks.

a b

Figure 2. (a) The microstructure of Ti-6Al-4V, HIP'ed at 315 MPa and 595°C is bimodal. It contains alpha needles and recrystallized alpha grains. (b) No beta phase was found in the recrystallized alpha structure.

(c-f cont'd. on next page.) (c) and (d) Bright field and dark field images of the needle-like structure. It contains precipitates of beta phase with the orientation relationship given by the overlapping diffraction patterns. (e) and (f) are EDX spectra of the alpha matrix and of a beta precipitate. Note the vanadium K_β peak and the iron peak in (f).

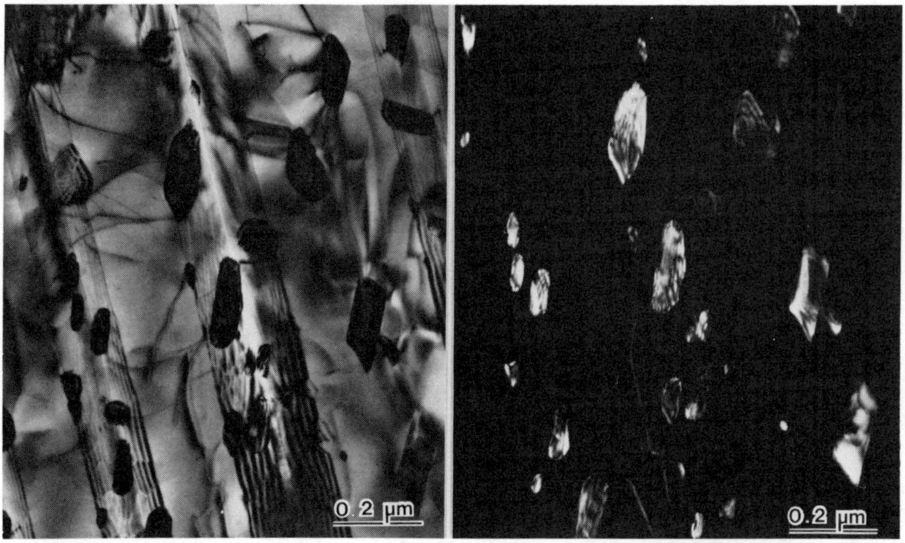

c

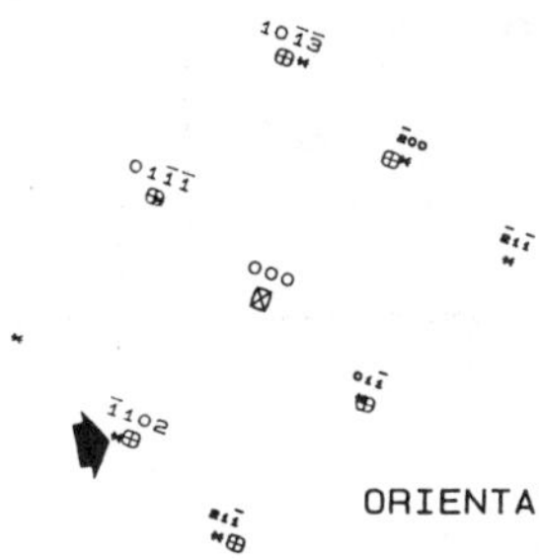

d

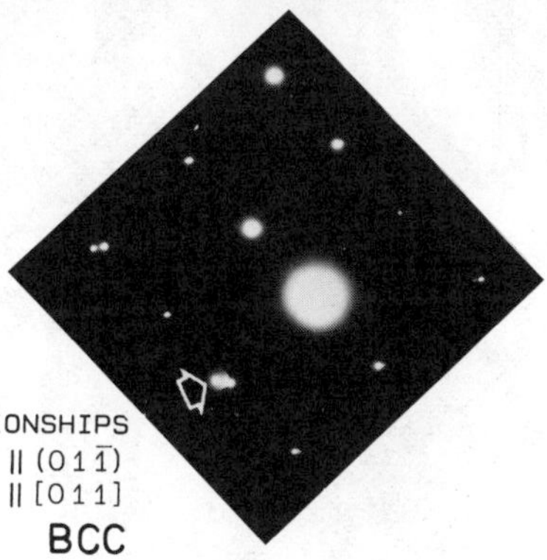

ORIENTATION RELATIONSHIPS
$(0\bar{1}11) \parallel (01\bar{1})$
$[5\bar{1}\bar{4}3] \parallel [011]$
HCP BCC

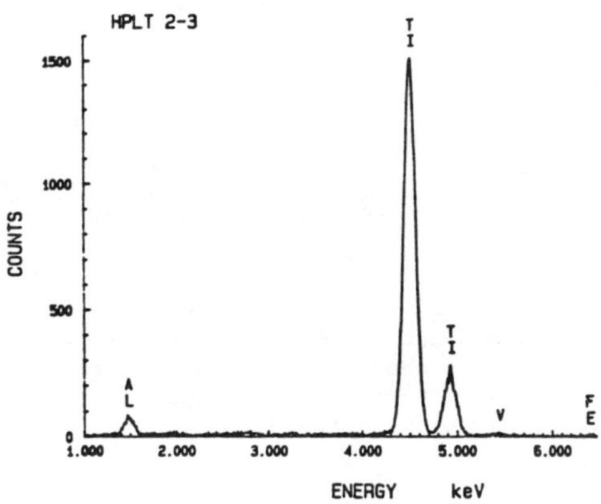

e

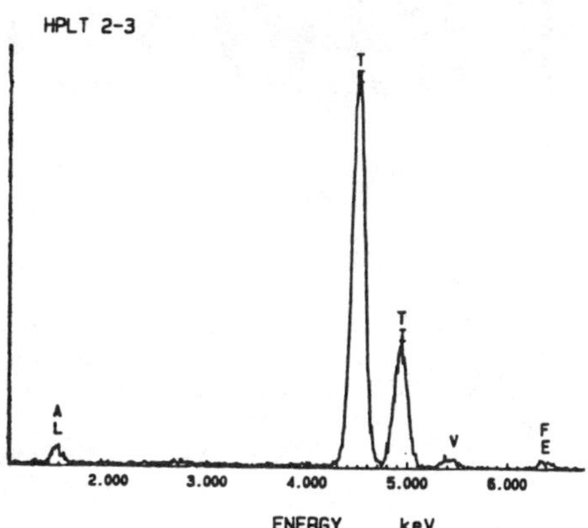

f

(3) Ti-10V-2Fe-3Al Alloy, HIP'ed at 595°C

The microstructure is a domain-like structure of alpha + beta phase with the majority alpha phase, Figure 3a. Here, the beta phase appears as the light contrast-free phase, but dislocations are visible under appropriate diffracting conditions. The alpha phase, on the other hand, contains a considerable amount of contrast from unidentified lattice defects which appear to be either dislocations clusters or fine precipitate clusters. EDX microchemical analyses of the two phases in Figures 3b, c show again absence of vanadium and iron peaks in the alpha phase and well-defined peaks of these two elements for the beta phase.

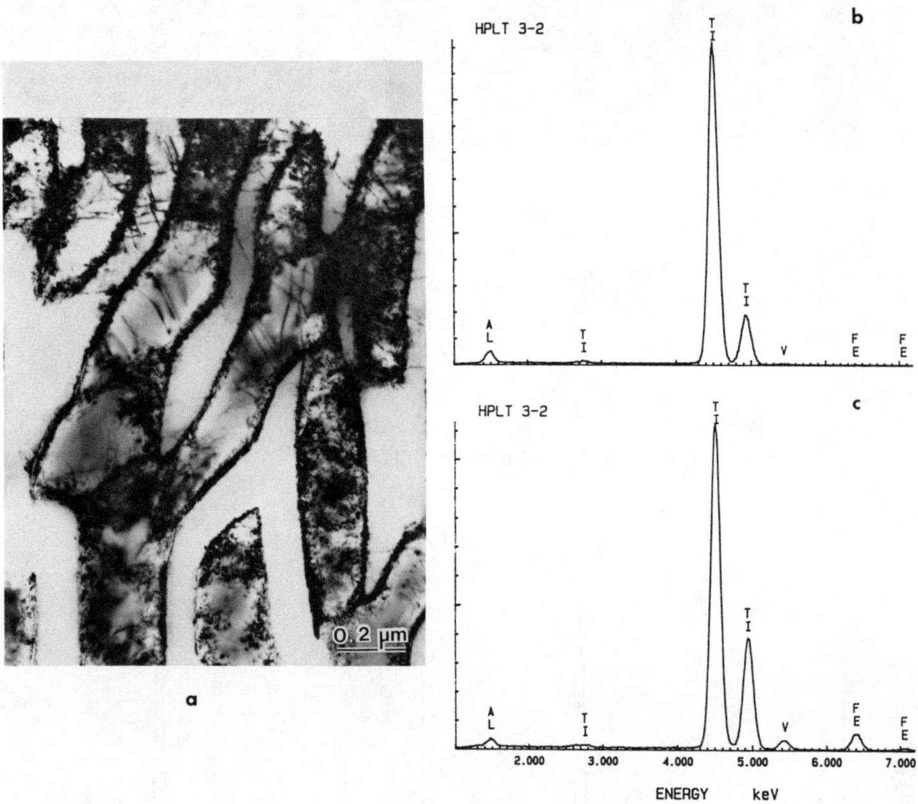

Figure 3. (a) The microstructure of Ti-10V-2Fe-3Al, HIP'ed at 595°C, is a domain-like mixture of alpha and beta phase. (b) and (c) EDX patterns of alpha and beta phase, respectively. The beta phase shows vanadium and iron peaks indicating their enrichment in beta. These peaks are absent in the alpha phase.

(4) Ti-10V-2Fe-3Al, HIP'ed at 540°C

In contrast to the allow discussed before, the powder particles of this alloy had additionally been subjected to the strain energizing process. The HIP'ed condition shows an equiaxed microstructure as evident in Figure 4a. Although equiaxed, the contrast in these grains is similar to the domain-like contrast in Figure 3a, namely some grains contain considerable quantity of strain contrast from dislocation clusters or unidentified precipitate clusters, and other grains are free of contrast. One would speculate that they are alpha and beta phase, however, microchemical analyses show absence of vanadium and iron peaks, which are characteristic of the enrichment of these elements in beta phase, for all of the equiaxed grains. This comes as a surprise that with the overall concentration of 10wt% vanadium there is no volume region with sufficiently large vanadium concentration to produce a vanadium K_β peak in the EDX spectrum. The K_α peak of V is hidden under the K_β peak of Ti. Clearly, the present microstructure does not represent equilibrium condition. On the observed scale, the microstructure appears fully dense and it contains a very fine grain structure, of the order of less then 1μm.

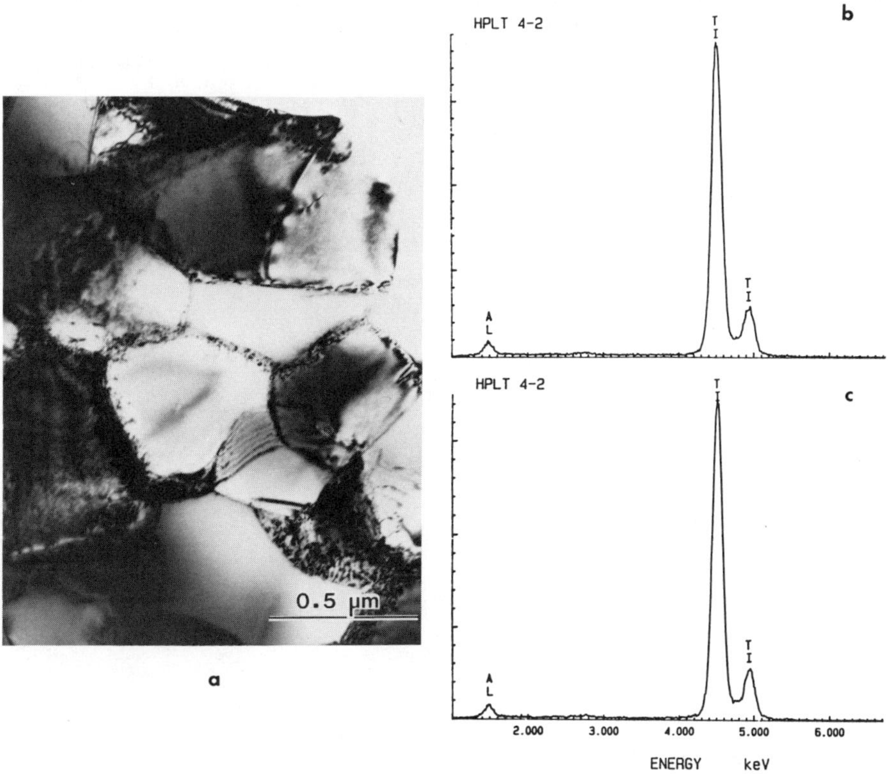

Figure 4. (a) the microstructure of Ti-10V-2Fe-3Al, HIP'ed at 540°C, is equiaxed and partially recrystallized. (b) and (c) EDX patterns taken from different grains are always devoid of vanadium and iron peaks, indicating lack of partitioning and local lenrichment of these elements.

SUMMARY

The microstructures and microchemistries of high pressure (315 MPa) HIP compacted titanium alloys at low temperatures (540 to 650°C) revealed the following characteristics.

Ti-6Al-4V alloy. Recrystallization of the rapidly solidified microstructure occurred despite the low compaction temperature. Two types of microstructures were observed: (a) equiaxed, consisting of alpha phase only, and (b) needle-like, consisting of alpha phase with beta precipitates. The beta precipitates were enriched in vanadium and iron.

Ti-10V-2Fe-3Al alloy: Recrystallization was also observed here. Again, two microstructural types occurred: (a) equiaxed, consisting of alpha phase only, and (b) domain-structure with alternating alpha and beta domains. Here, too, the beta phase was enriched in vanadium and iron.

The microstructures and microchemistries reported in this work are representative of the bulk of the HIP'ed samples. Interface regions between prior powder particles were not investigated.

References

1. F. H. Froes, D. Eylon, G. E. Eichelman, and H. M. Burte, "Advances in Titanium Powder Metallurgy, Journal of Metals, 32(2) (1980), pp. 47-54.

2. F. H. Froes and D. Eylon, "Powder Metallurgy of Titanium Alloys - A Review," in Titanium, Science and Technology, vol. 1, ed. G. Lutjering, U. Zwicker, and W. Bunk (Oberursel, West Germany: DGM, 1985), pp. 267-287.

3. P. R. Smith, F. H. Froes, and J. T. Cammett, "Correlation of Fracture Characteristics and Mechanical Properties for Titanium-Matrix Composites," in Mechanical Behavior of Metal-Matrix Composites, ed. J. H. Hack and M. F. Amateau (Warrendale, PA: The Metallurgical Society, 1983), pp. 143-168.

4. P. R. Smith and F. H. Froes, "Developments in Titanium Metal Matrix Composites," Journal of Metals, (36) (3) (1984), pp. 19-26.

5. T. F. Broderick, A. G. Jackson, H. Jones, and F. H. Froes, "The Effect of Cooling Conditions on the Microstructure of Rapidly Solidified Ti-6Al-4V," Met. Trans. A, 16A (1985), pp. 1951-1959.

6. R. R. Boyer, "Design Properties of a High Strength Alloy, Ti-10V-2Fe-3Al," Journal of Metals, 32 (3) (1980), p. 61.

7. G. V. Scarich, G. R. Chanani, V. C. Peterson, and D.M. Weaver, "Relationships Between Mechanical Properties, Contaminants, and Fracture Topography of HIP Ti-6Al-4V Powder Metallurgy Products," Journal of Metals, 32 (3) (1980), pp. 103-114.

8. C. F. Yolton and F. H. Froes, (Report 312-B, Colt-Crucible, June 1978).

9. D. Eylon, P. R. Smith, S. W. Schwenker, and F. H. Froes, "Status of Titanium Powder Metallurgy," in Industrial Applications of Titanium and Zirconium; Third Conference,, ASTM STP 830, ed. R. T. Webster and C. S. Young, Philadelphia, PA: ASTM Publications, (1984), pp. 48-65.

10. D. Eylon and F. H. Froes, "HIP Compaction of Ti Alloy Powders at Higher Pressure and Lower Temperature (HPLT)", in this volume.

11. G. Welsch, K. Gazioglu, G. Lutjering, and W. Bunk, "Deformation Characteristics of Age-Hardened Ti-6Al-4V", Met. Trans. A, 8A (1977) pp. 169-177

EXPLOSIVE COMPACTION OF TITANIUM AND TITANIUM ALLOY POWDERS

Vonne D. Linse

Battelle
505 King Avenue
Columbus, Ohio 43201

Summary

Explosive compaction is being investigated as an approach to fabricate high density, high integrity structures from titanium and titanium alloy powders. This approach is particularly attractive for the powders made by rapid solidification since the explosive compaction process allows consolidation at temperatures well below that which would destroy the unique microstructures and properties imparted by the rapid solidification process. Recent investigations have shown that with the appropriate particle shape, high integrity titanium alloy powder structures can be fabricated by explosive compaction.

Introduction

Advancements in rapid solidification techniques have created titanium alloy powders possessing unique microstructures and offering potential for improved properties over conventionally processed titanium alloy structures. The major problem with these powders has been the restriction of maximum allowable process temperatures for consolidating them into useful structures that realize the full potential of their rapidly solidified properties. The temperature must be sufficiently low that the microstructures of the powder will not be altered and the full property potential maintained. These temperatures are usually sufficiently low, however, that it is difficult to achieve the appropriate particle deformation in these higher strength powders with the full densification and interparticle bonding that is necessary to achieve the desired properties in the consolidated structure. As a result, nonstandard powder fabrication approaches that permit powder consolidation at lower temperatures are increasingly being investigated for their potential to solve these problems.

One of the nonstandard processes that has been receiving increasing attention for its potential to fabricate these powders is explosive powder compaction. It is a process that utilizes the high pressure, short duration pulse generated by a detonating explosive to consolidate powders at extremely high rates and at temperatures that are normally well below those used in conventional powder fabrication processes. Current research efforts are providing a significantly improved understanding of the fundamentals and controlling factors of the explosive powder compaction process that now make it a serious candidate for fabricating the titanium and titanium alloy powders. Also, recent studies in the scale up of the process and development of manufacturing concepts for it are providing strong indication of its economic competitiveness with other techniques for fabricating these powders. Advanced elevated temperature explosive compaction techniques also provide further potential for tailoring microstructures and properties of the compacted structures.

This presentation will give a general mechanistic overview of the explosive compaction process and the criteria for achieving high density powder compacts with accompanying high quality interparticle bonding. Examples of the particulate aspects associated with the deformation and densification of titanium alloy powder structures and how it relates to the fracture behavior of the final compacts will be shown. The particular importance of the powder particle shape and its effect on energy partition ing during compaction will be discussed. The effects of using elevated temperatures for compacting titanium alloy powders will also be discussed.

The Explosive - The Energy Source

In the explosive compaction process, the explosive is the self contained energy source which is capable of delivering very high energy densities on the basis of unit surface area or per unit weight of powder material being compacted. Each type of explosive produces a pressure and characteristic time duration of the pressure pulse based on its composition, density, and thickness. Typical detonation pressures generated in explosives can be selected to range from 1.0 to over 25.0 GPa (145,000 to over 3,625,000 psi) while the time duration of the pressure pulse can range from less than one to 10's of microseconds. When selecting the explosive for powder compaction, it is important to remember that the performance characteristics of the explosive must be matched to the characteristics of the powder and the structure being compacted in order to achieve the desired results.

The Compaction Zone

When the appropriate pressure pulse from the explosive is delivered through the powder container into the powder, a compaction zone is established and progresses through the powder structure as shown in Figure 1. It is in this rather narrow zone that the transition from the original powder to the fully consolidated state occurs. The front of this zone (shock front) travels through the powder at a velocity that depends on the input pressure of the explosive and the characteristics of the powder structure ([1]). It has been shown by a number of investigators ([2,3,4]) that this zone is divided into three distinct mechanistic regions of consolidation-particle rearrangement, localized flow, and bulk deformation. It is generally agreed that the first two regions, particle rearrangement and localized flow, are where the major portion of the compaction is to be accomplished.

Region I The particle rearrangement region is where the powder particles achieve their highest individual and relative velocities with significant densification occurring. This region is quite narrow (2 to 3 particle diameters in width) and almost no particle deformation and interparticle bonding is observed.

Region II The second region of localized flow is the most important in that it is within this region that the localized flow to complete most of the void elimination and achieve appropriate interparticle bonding must be accomplished. This region is usually considerably wider than the particle rearrangement region and it is here that the greatest

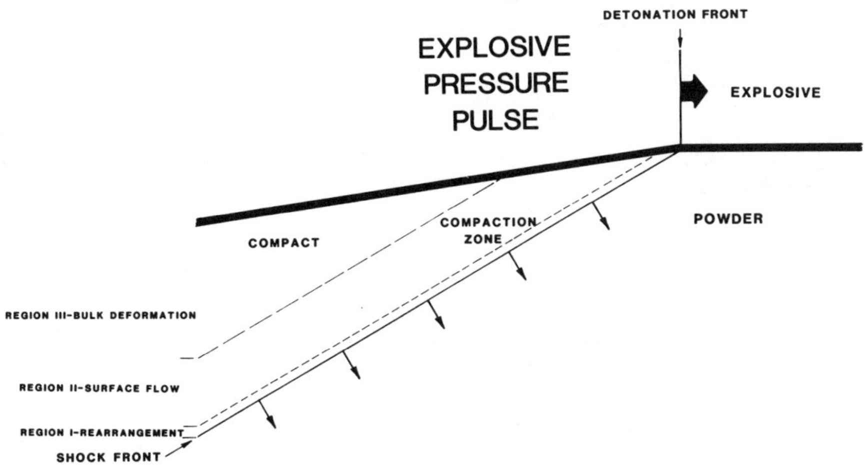

Figure 1 - Schematic representation of the compaction zone in the explosive compaction process showing the breakdown of the two major regions of void elimination.

amount of energy is deposited from the shock wave into the powder structure.

Region III If voids remain when this region is entered, further densification can only be achieved through bulk deformation at high pressures. It is desireable to minimize densification in this region, since most materials of interest to be compacted have limited ductility and will be susceptible to cracking if subjected to extensive bulk deformation in this region. Elevated temperature explosive compaction is beneficial in reducing the criticality of entering this region and allowing near theoretical densities to be achieved.

Void Elimination and Interparticle Bonding

In any powder compaction process, the basic requirement is to achieve theoretical density through material flow into voids or interstices with the accompanying break down of surface barrier layers to permit interparticle bonding that yields full structural integrity. The latter aspect of high quality interparticle bonding becomes increasingly critical as processing temperature is reduced in most powder consolidation processes. In the explosive compaction process, this problem is circumvented through the mechanisms of heating and thermal softening of the particle surfaces by friction and severe localized flow during the high velocity relative motion between particles. This allows the localized transport or displacement of the particle surface material from the areas of contact into the interstices as shown in Figure 2 ([5]). When densification is accomplished by this mechanism as opposed to bulk deformation of the particle, the severe surface flow

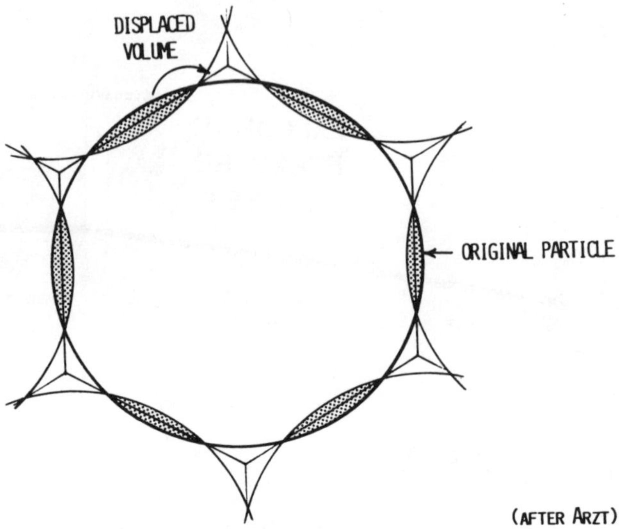

Figure 2 - Modified Arzt drawing of void elimination through localized flow.

breaks down the particle surface barriers and promotes interparticle bonding.

An extremely important aspect of achieving the appropriate amount localized surface flow and interparticle bonding during explosive compaction is the proper partitioning of the energy from the compaction wave between deposition in the particles and transmission to the neighboring powder particles. This partitioning has been noted to be optimum with irregular shaped particles where the localized surface deformation is maximized ([1]). The most unfavorable partitioning occurs with the two extremes of particle configuration, spherical and platelet (flake), where localized deformation and energy deposition is minimum. One of the major problems associated with earlier efforts to explosively compact rapidly solidified powders was found to be related to the spherical or platelet particle shape in which these powders typically result from the rapid solidification processing. This resulted in compacts having nonuniform compaction and structural integrity due to the inability to achieve the appropriate energy partitioning. The particle shape factor is a key element in achieving the desired results when explosively compacting rapidly solidified powders as will be discussed in the following section.

Titanium and Titanium Alloy Compaction

The use of explosive compaction to fabricate titanium alloy powder structures was investigated a number of years ago ([6]). These efforts were generally deemed successful in that Ti-6Al-4V alloy structures such as those shown in Figure 3 were compacted to densities of 97 percent of theoretical or greater with relatively good interparticle bonding using powders of irregularly shaped particles that were primarily available at that time.

Figure 3 - Solid cylindrical and tubular titanium alloy samples explosive compacted early with irregular shaped powders.

The major difficulty frequently experienced in these explosive compaction studies was the cracking that occurred in the compacted structures and the inability to predict or control it.

In the early 1970's, the first studies were conducted to explosively compact titanium alloy powders that had been made at increased solidification rates which were the predecessors to the present the rapidly solidified powders. These powders were the first truly spherical powders to which the explosive compaction process was applied to compact. It was quickly found that it was not possible to compact these powders at room temperatures as had been done previously with the irregular powders. Complete densification with appropriate interparticle bonding could not be achieved without excessive explosive energy inputs which in turn caused significant structural damage to the compacted structures.

Attempts to overcome this problem lead to the introduction of elevated temperature explosive compaction. In this work, it was found that compaction at temperatures of 870-955C (1600-1750F) would yield total densification and interparticle bonding with a greatly reduced tendency toward structural damage than had been experienced during room temperature compaction. At the lower end of the temperature range, compacted microstructures such as that shown in Figure 4 were observed where the basic features of the original powder particles were retained with some evidence of modification beginning to show at the particle boundaries where the deformation and thermal excursion are the greatest. At the upper end of the temperature range, the combination of temperature and energy deposited during the passage of the compaction wave lead to total recrystallization such as that shown in Figure 5. Samples up to 12.7-cm (5.0-in.) diameter by 15.2-cm (6.0-in.) long were successfully compacted in this study; however, no mechanical property data were ever generted from these compacted structures.

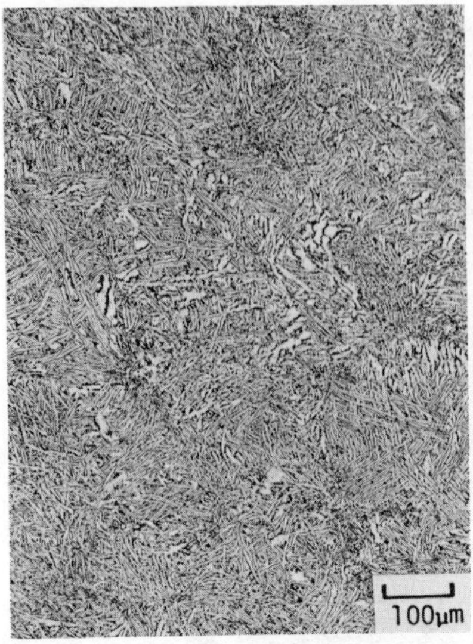

Figure 4 - Microstructure of sperical Ti-6Al-4V with identity of particles retained during compaction

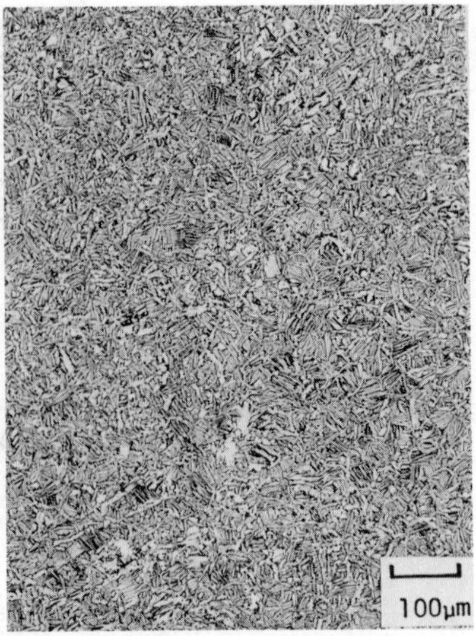

Figure 5 - Microstructure of spherical Ti6Al-4V with total recrystal lization during compaction

More recent studies with the explosive compaction of rapidly solidified spherical Ti-6Al-4V powders * have provided significant insight into understanding of the importance of particle shape. While these compaction studies were carried out at various elevated temperatures, the maximum temperatures were restricted to a maximum of 760C (1400F) to assure that the rapidly quenched microstructure and properties of the powder would be retained. Further, the 3.8-cm (1.5-in.) diameter sample size was much smaller than in the previous study.

It was found from these studies that achieving uniformity of densification and interparticle bonding from the exterior to the centerline of the sample became increasingly difficult with decreasing temperature. This was found through extensive analysis of the compacted microstructures to be attributable to the mode in which the energy from the compaction wave was partitioned between particle deformation and transmission through the powder structure. Particularly revealing were Scanning Electron Microscopy (SEM) examinations of the surfaces of fractured sections from the compacted samples. When the energy deposition within the compaction zone was insufficient, poor interparticle bonding resulted and the fractures were interparticle in nature as shown in Figure 6. When the proper amount of energy was partitioned, however, the fracture mode changed to ductile transparticle as shown in Figure 7 with little evidence of the original particles remaining.

---

*. A portion of this work was conducted under AF Contract No. F33601 83-MJ057 with powders provide by AFWAL

Figure 6 - Photomicrograph showing interparticle fracture in compact explosively compacted with insufficient energy.

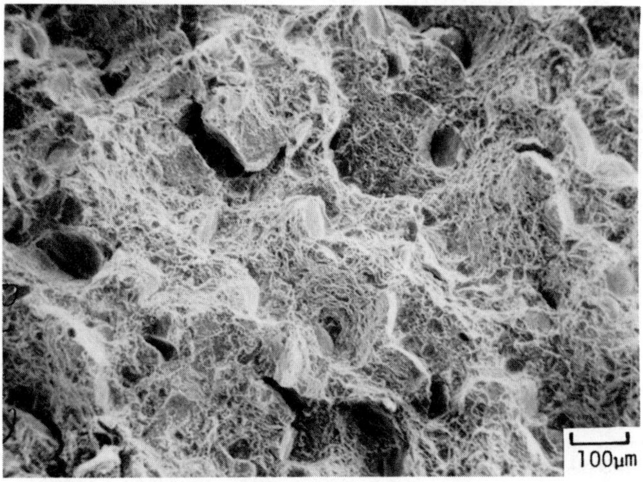

Figure 7.Photomicrograph showing ductile transparticle fracture in compact with the proper amount of energy partitioned into localized deformation of the particles.

Another technique used in the early stages of these studies that served as a reliable indicator of the interparticle bond integrity was the machining behavior of the compacted titanium alloy during container removal and cleanup. When complete, high quality bonding had been achieved, the surface of the titanium would machine to an almost polished finish while producing a continuous machine chip as shown in Figure 8. If the interparticle bonding was incomplete, the surface finish would quickly begin to appear dull and have a sandpaper like texture. Furthermore, the machine chips would break into short segments.

Figure 8 - Machined titanium alloy compact showing high quality surface finish and continuous machine chip due to high integrity interparticle bonding.

In general, it has been concluded that it is impractical to explosive compact spherical powders due to the inability to properly partition the appropriate amount of energy to the localized deformation that is required for interparticle bonding. While elevated temperature explosive compaction enhances the potential for fabricating these spherical powders, it is still questionable whether proper energy partitioning can be achieved on a practical scale at the restricted fabrication temperatures that are placed on the rapidly solidified titanium alloys.

The more appropriate course for explosively compacting most metal powders including titanium and its alloys has been to alter the particle shape to a more irregular configuration thus forcing a larger amount of deformation and heating to take place near the particle surfaces. The high strain rate and temperature sensitivity of titanium alloys and its effectiveness in working them is well known (7). The results of studies currently in progress with extremely irregularly shaped particles of low alloy titanium powders are demonstrating that uniform, full densification and interparticle bonding can be obtained throughout the compact at 600C (1110F) which is well below that required for spherical powders. Similar favorable compaction results with uniform energy partitioning were recently observed with stainless steel powders (1). A typical microstructure from one of these titanium compacts with the alpha structure (white) surrounding the particles as a result of the extreme thermal cycle at the particle surfaces is shown in Figure 9. This combined with the excellent structural integrity of the compact is a strong indication that near optimum compaction is being achieved for this alloy. Mechanical property tests are currently being

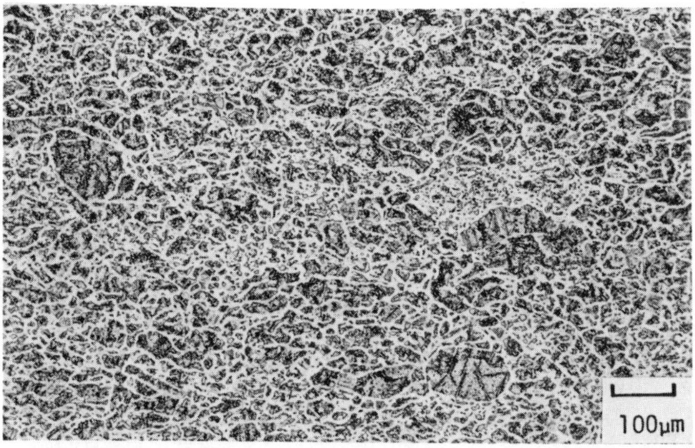

Figure 9 - Microstructure of irregular shaped titanium alloy powder explosive compacted at 600C.

conducted on these compacts. Favorable mechanical properties have been reported on similarly made compacts (8)

Conclusions

Recent studies involving the explosive compaction of titanium alloy powders have shown that optimum compaction results are achieved when powders having irregularly shaped particle are compacted. This permits the best control of the energy partitioning and deposition into the powder structure that yield the most uniform densification and interparticle bonding. Also these irregular shaped powders can be successfully compacted at temperatures well below the critical temperatures at which rapidly solidified titanium alloys loose their unique microstrucures and property potential. The major problem remains to tailor the appropriate irregular particle shape into the rapidly solidified powders.

References

1. V. D. Linse, "The Particulate Nature of Explosive Compaction", Presented at Explomet '85, Portland, Oregon (July, 1985), to be published.

2. R. A. Pruemmer and G. Ziegler, "Structure and Annealing Behavior of Explosively Compacted Alumina", in Proc. of Fifth International Conference on High Energy Rate Fabrication, Denver, Colorado (1975).

3. O. V. Roman and V. G. Gorobtsov, "Fundamentals of Explosive Compaction of Powders, Shock Waves and High-Strain-Rate Phenomena in Metals, M. A. Meyers and L. E. Murr, editors (Plenum Press, 1980).

4. V. D. Linse, "Dynamic Compaction of Metal and Ceramic Powders", Proc. of the Sagamore Army Materials Research Conference on Innovations in Materials Processing, Volume 30 (Plenum Press, New York and London, 1985), 381-404.

5. E. Arzt, "The Influence of An Increasing Particle Coordination on the Densification of Spherical Powders", Acta Metall., 30 (1982), 1883-1890.

6. R. W. Leonard, D. Laber, and V. D. Linse, "Advances in Explosive Powder Compaction", in Proc. of the Second International Conference of the Center for High Energy Forming, Estes Park, Colorado, (1969).

7. A. U. Sulijoadikusumo and O. W. Dillon, Jr., "Work Softening of Ti-6Al-4V Due To Adiabatic Heating", Metallurgical Effects of High Strain Rates, Edited by R. W. Rhode, et al, (Plenum Press, 1973).

8. V. G. Gorobtsov and O. V. Roman, "Hot Explosive Pressing of Powders", The International Journal of Powder Metallurgy & Powder Technology, 11, 1, (January, 1975).

PRODUCTION OF METAL MATRIX COMPOSITES

FROM RAPIDLY SOLIDIFIED TITANIUM ALLOY FOILS

D. Eylon[a], C. M. Cooke[b], and F. H. Froes[c]

[a]Metcut-Materials Research Group, now with the University of Dayton
Graduate Materials Engineering, Dayton, OH 45469

[b]Metcut-Materials Research Group, P.O. Box 33511
Wright-Patterson Air Force Base, OH 45433

[c]Air Force Wright Aeronautical Laboratories, Materials Laboratory
AFWAL/MLLS, Wright-Patterson Air Force Base, OH 45433

Abstract

Potential for use of rapidly solidified (RS) titanium alloy foils for the production of metal matrix composites (MMC) has been explored. The foils used were 2mm wide thin ribbons (40-100μm) produced by the pendant drop melt extraction (PDME) method. The alloys investigated were Ti-6Al-4V (Ti-6-4), Ti-11.5Mo-6Zr-4.5Sn (Beta III), Ti-15V-3Cr-3Sn-3Al (Ti-15-3), and Ti_3Al+Nb. Vacuum hot press (VHP) compaction to full density of foils and fibers was accomplished for all alloys at temperatures lower then reported for similar MMC compacts produced from conventionally rolled titanium alloy foils. The lowest temperature at which full bonding was achieved was 730°C (1350°F) for Ti-6-4, Beta III, and the Ti-15-3 matrices and 925°C (1700°F) for Ti_3Al+Nb. Due to the lower compaction temperature, the matrix/fiber reaction zones were narrower than those observed in conventional MMC products. The reaction zone thickness with Borsic fibers at the lowest successful compaction temperature was 0.1μm for the Beta III and 0.3μm for the Ti_3Al+Nb.

Introduction

Two of the major reasons for the slow acceptance of titanium alloy base metal matrix composites (MMC) into the aerospace market are the relatively high cost of titanium foil and the magnitude of the matrix/fiber reaction zone [1]; the latter leading to mechanical property degradation. The rolling of titanium alloy foils requires numerous cycles of vacuum anneal cold rolling using a low reduction ratio to prevent cracking. Recently a relatively lower cost Ti-15V-3Cr-3Sn-3Al (Ti-15-3) foil has been introduced commercially. This alloy can undergo larger amounts of cold deformation between anneals which cuts down the cost of processing [2]. However, other alloy classes like alpha+beta, near-alpha or Ti_3Al base materials are very difficult to roll into foil. Despite advances in control of grain size in titanium alloy rolled foils [3], allowing reduced compaction temperatures, even lower fabrication temperatures are desirable to reach high mechanical property levels due to reduction in reaction zone size.

In recent years, the development of rapid solidification (RS) technology [4, 5] has made alloy foils available with a relatively small grain size, the result of the high cooling rates. The objective of the present work was to explore the possibilities of using RS titanium alloy foils for producing MMC. The fine grain size of the RS foils should promote flow at lower temperatures leading to a smaller reaction zone. At the same time the RS technology can produce lower cost foils thus allowing both of the problems with current foil to be circumvented simultaneously.

A representative cross-section of titanium alloys was selected for this exploratory study. Ti-6Al-4V (Ti-6-4) is the most commonly used alloy in MMC and many other aerospace applications. Beta III (Ti-11.5Mo-6Zr-4.5Sn), a relatively rich metastable beta alloy [6] is an example of an alloy which is very difficult to process into a fine grain size by conventional rolling practices [7, 8]. The Ti-15-3 is an emerging cold formable alloy [9, 10] with a strong potential for titanium MMC [2]. The two beta alloys need to be compacted below the beta transus temperature (760°C [1400°F]) to produce optimum matrix properties upon subsequent aging. Ti_3Al+Nb alloy is a representative of the Ti-aluminides class which can be used as a base material for high temperature components.

Experimental Procedures and Results

The foils selected for this work were produced as 2mm wide ribbons by the pendant drop melt extraction (PDME) process [5]. A schematic of this unit is shown in Figure 1. The foils nominal composition, thickness, and average beta grain size in the as-solidified condition are listed in Table I.

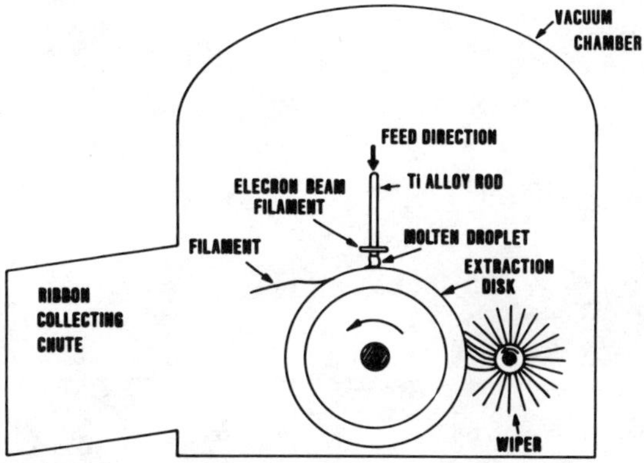

Figure 1 - A schematic of the pendant drop melt extraction (PDME) unit.

Table I. Characterization of Ribbons Used in This Study

Matrix Alloy	Average Ribbon Thickness, μm[a]	Average Ribbon Beta Grain Size[b], μm
Ti-6Al-4V (Ti-6-4)	86	5.8
Ti-11.5Mo-6Zr-4.5Sn (Beta III)	63	2.5
Ti-15V-3Cr-3Sn-3Al (Ti-15-3)	90	14.3
Ti_3Al+Nb	72	2.8

[a]Based on measurements of 4 segments per alloy, 20 readings per segment.

[b]Based on 3 location linear intercept measurements along the middle of the longitudinal cross-section.

The microstructure of the four ribbons on a longitudinal section plane is shown in Figures 2a-2d.

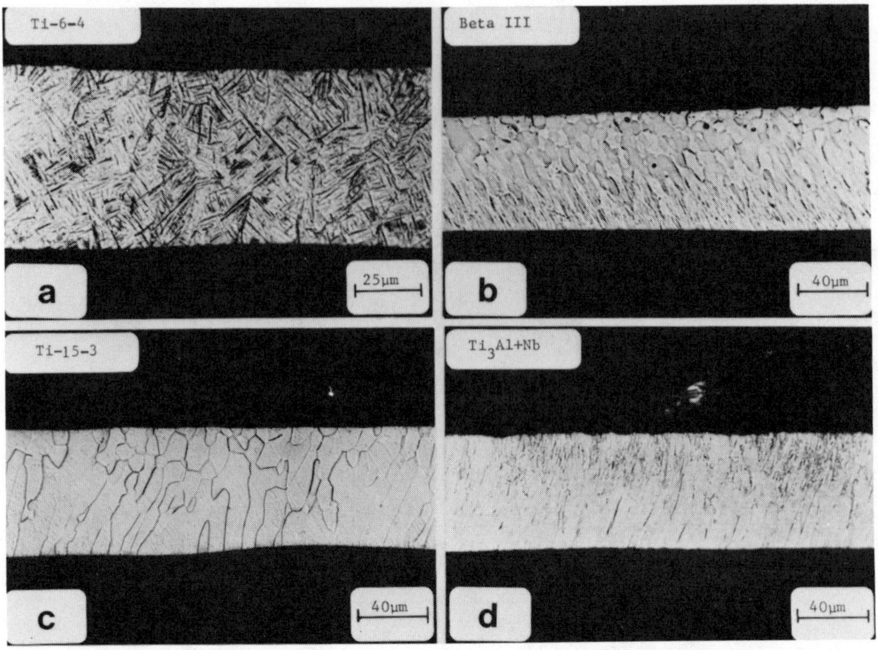

Figure 2 - Longitudinal cross-section of as-produced ribbons. Wheel side at the bottom. Original magnifications are noted: (a) Ti-6-4, 600x; (b) Beta III, 400x; (c) Ti-15-3, 400x; and (d) Ti_3Al+Nb, 400x.

The ribbons were compacted with B_4C, Borsic, and SCS-6 fibers by the vacuum hot press (VHP) process. Prior to compaction ribbons were cut into less than 25mm (1 in.) segments and placed into a carburized steel cup (Figure 3) which was lined with CP titanium foil. Fibers were placed on top of the ribbons followed by an additional layer of ribbons and an additional CP titanium foil cover. A 25mm (1 in.) plug of carburized steel was fitted into the cup and the entire assembly was fitted into a die for hot pressing.

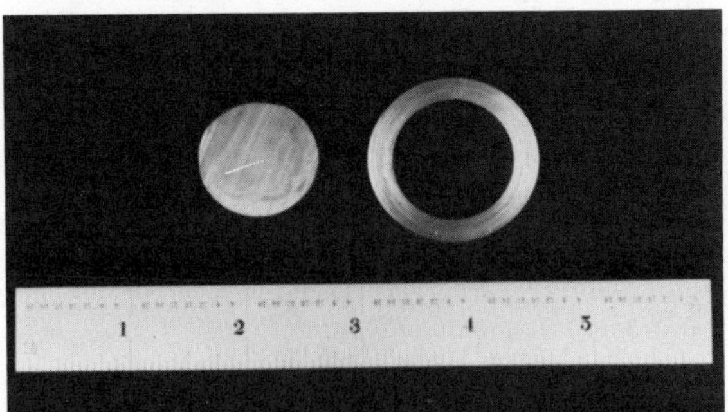

Figure 3 - VHP tool set (cup and plug) used for compaction.

The compaction conditions are listed in Table II which also lists the bonding quality, the fiber condition (based on metallography), and the matrix/fiber reaction zone thickness.

Table II. Compact Conditions and Characteristics

Matrix Alloy	Compaction Temperature, °C (°F)	Pressure, MPa (ksi)	Time, hr	Compact Bonding	Fiber Condition	Reaction Zone Size, μm
Ti-6-4	730 (1350)	70 (10)	24	complete	cracked	0.5-1.0
	730 (1350)	70 (10)	2	very poor	cracked	N/A*
Beta III	730 (1350)	70 (10)	24	complete	cracked	0.1
	730 (1350)	70 (10)	2	very poor	cracked	N/A
Ti-15-3	730 (1350)	70 (10)	24	complete	cracked	1.0-2.0
Ti_3Al+Nb	980 (1800)	100 (15)	2	complete	no cracks	0.5-1.0
	925 (1700)	240 (35)	2	complete	cracked	1.0
	925 (1700)	100 (15)	24	complete	no cracks	1.0-2.0
	925 (1700)	100 (15)	2	partial	cracked	1.0
	870 (1600)	100 (15)	24	complete	no cracks	1.0
	815 (1500)	240 (35)	2	complete	cracked	0.3
	815 (1500)	100 (15)	2	poor	cracked	0.5
	730 (1350)	240 (35)	2	very poor	cracked	N/A
Ti_3Al+Nb Thinned IM	925 (1700)	240 (35)	2	very poor	fibers lost (Fig. 4a)	N/A

*Not available: no fibers in an acceptable condition for reaction zone measurements.

An example of the ease of MMC fabrication using RS ribbon compared to use of foil produced by grinding ingot material to the required thickness is shown (Figures 4a and 4b) for the Ti_3Al+Nb alloy compacted at 925°C/240MPa/2 hr. While the coarse grain ingot material foil did not bond under this condition (Figure 4a), full bonding was achieved using RS foil (Figure 4b).

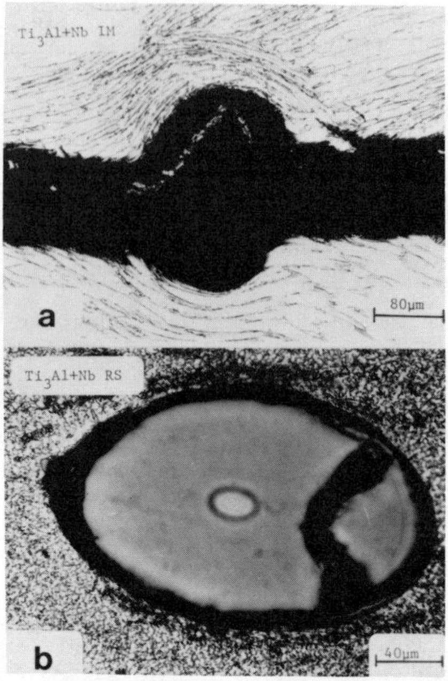

Figure 4 - Comparison of sections through MMC compacted (925°C/240MPa/2 hr) Ti_3Al+Nb material (a) thinned ingot; and (b) RS ribbon. Original magnifications: 200x and 400x, respectively.

Metallographic sections of ribbons/fibers compacted under various conditions are shown in Figures 5a-5d for all four matrix alloys.

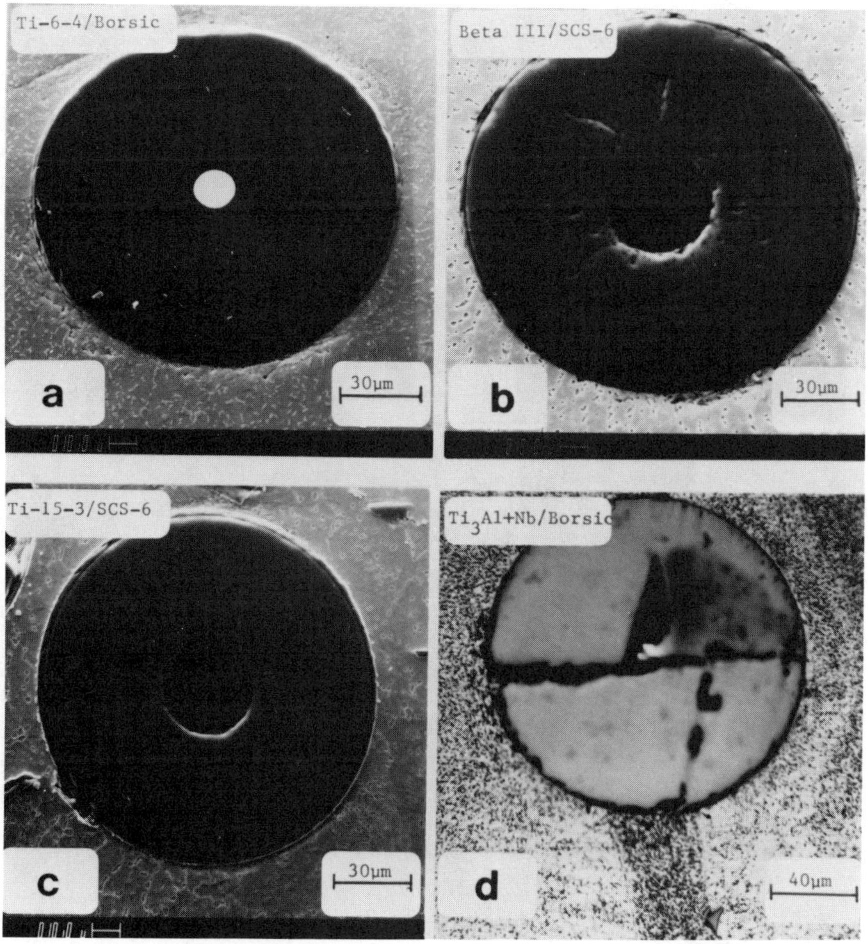

Figure 5 - Section through bonded matrix/fiber zones of (compaction conditions and original magnifications are noted): (a) Ti-6-4/Borsic (730°C/70MPa/24 hr, 500x); (b) Beta III/SCS-6 (730°C/70MPa/24 hr, 500x); (c) Ti-15-3/SCS-6 (730°C/70MPa/24 hr, 500x); and (d) Ti_3Al+Nb/Borsic (815°C/240MPa/2 hr, 400x).

As-compacted matrix/fiber reaction zones of all four matrix alloys are shown in Figures 6a-6d.

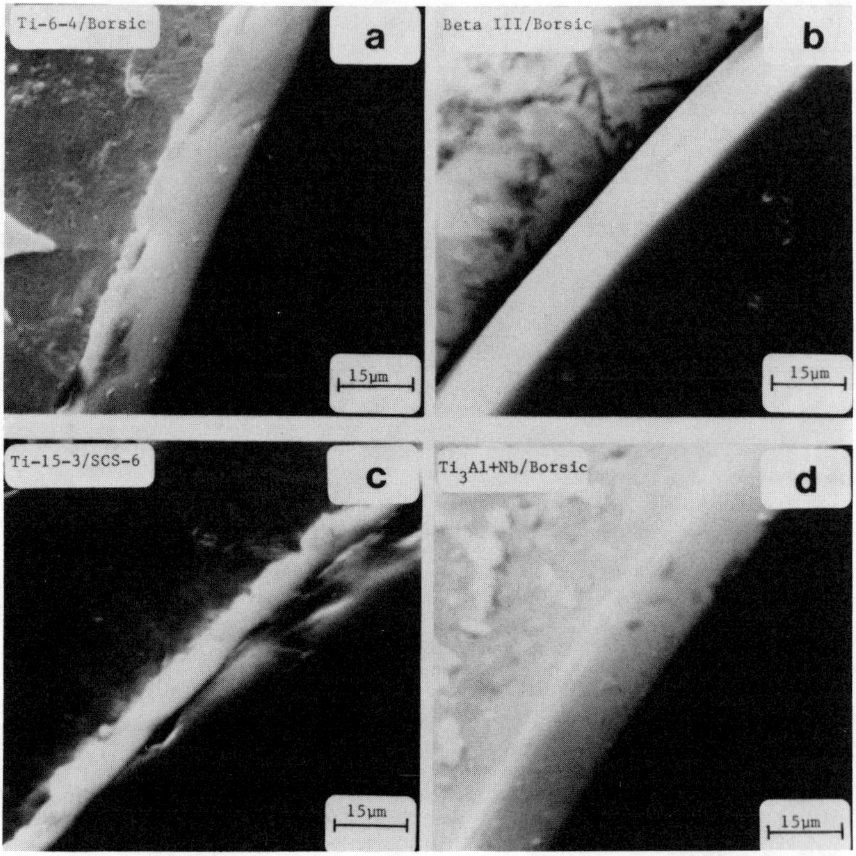

Figure 6 - Matrix/fiber reaction zone of (compaction conditions and original magnifications are noted): (a) Ti-6-4/Borsic (730°C/70MPa/24 hr, 1000x); (b) Beta III/Borsic (730°C/70MPa/24 hr, 1000x); (c) Ti-15-3/SCS-6 (730°C/70MPa/24 hr, 1000x); and (d) Ti_3Al+Nb/Borsic (815°C/240MPa/2 hr, 1000x).

The as-compacted ribbon microstructures of all four alloys are shown in Figures 7a-7d. Average grain size measured from similar micrographs and based on 3 location linear intercept measurements are listed in Table III.

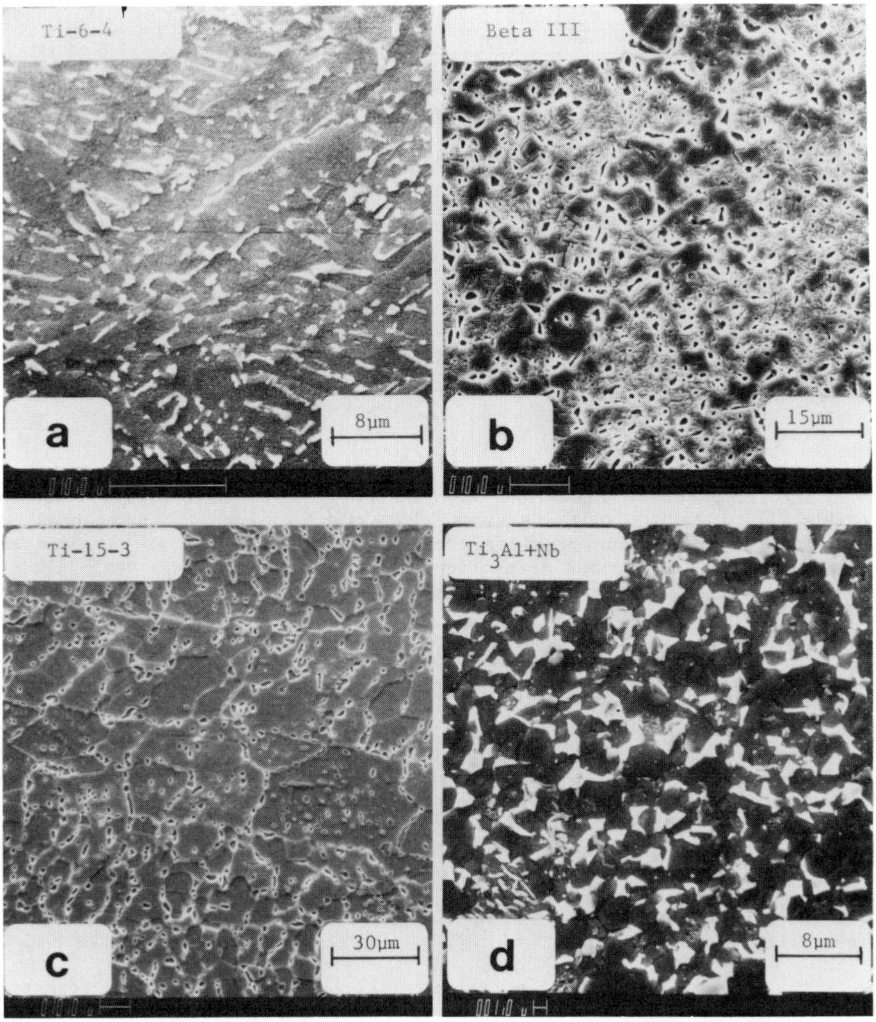

Figure 7 - The as-compacted microstructure of RS ribbon matrix of (original magnifications are noted): (a) Ti-6-4, 2000x; (b) Beta III, 1000x; (c) Ti-15-3, 1500x; and (d) Ti_3Al+Nb, 2000x.

Table III. Matrix Grain Size

Matrix Alloy	Average Grain Size[a], μm	Compaction Conditions, °C/MPa/hr
Ti-6-4	3.6	730/70/24
Beta III	5.2	730/70/24
Ti-15-3	4.9	730/70/24
Ti_3Al+Nb	1.7	925/100/24

[a] Based on 3 location linear intercept measurement.

Discussion

The compaction conditions for achieving full matrix/fiber bonding (Table II) clearly indicate that RS ribbons can be compacted at temperatures lower than previously reported for conventional titanium MMC products [1]. The minimum compaction temperature for Ti-6-4, Beta III, and Ti-15-3 was 730°C (1350°F). Ti_3Al+Nb was compacted at as low as 815°C (1500°F).

It should, however, be noted that to achieve full compaction at these low temperatures, long exposure time (24 hr), higher compaction pressures (250MPa) or combination of both were often used (Table II). The long exposure time is economically undesired and the high compaction pressure may cause fiber cracking or fiber deformation (Figures 4b and 5d). The choice of such extreme compaction conditions was aimed at establishing the lowest possible temperature for the process.

On the other hand, the lower compaction temperatures resulted in very thin Borsic fiber/matrix reaction zones even when the exposure time was 24 hr. The reaction zone of Beta III (730°C compaction) and Ti_3Al+Nb (815°C compaction) was only 0.1 and 0.3μm, respectively (Figures 6b and 6d, Table II).

The lower compaction temperature is probably the combined result of the fine grain size inherent in the RS ribbons (Table I and Figure 2) and the high compaction pressure and long cycle time. The ribbon beta grain sizes of all alloys examined is smaller than the grain size typically obtained by sheet rolling practices. In addition, the RS grains undergo morphological changes during the low strain rate/low temperature deformation into a lower aspect ratio configuration while maintaining the small grain size levels obtained by the RS process (compare Figure 2 with Figure 7 and Table I with Table III). Maintaining fine grain size and equiaxed morphology during the compaction process allows superplastic deformation processes to continue throughout the whole ribbon/fiber bonding cycle.

Summary and Conclusions

1. The use of foils from the titanium alloys Ti-6Al-4V (Ti-6-4), Ti-11.5Mo-6Zr-4.5Sn (Beta III), Ti-15V-3Cr-3Sn-3Al (Ti-15-3), and Ti_3Al+Nb,

produced by rapid solidification (RS), for fabrication of metal matrix composites (MMC) was studied in this work.

2. In all alloys, full compaction of ribbons/fibers was achieved at temperatures lower than reported for conventional MMC products made from conventionally rolled foils of the same composition. The lowest satisfactory compaction temperatures were 730°C (1350°F) for the Ti-6-4, Beta III, and Ti-15-3, and 815°C (1500°F) for the Ti_3Al+Nb.

3. The fiber/matrix reaction zone was narrower than in conventional rolled foil compacts due to the lower compaction temperature. Reaction zones with Borsic fibers of 0.1 and 0.3μm were measured for Beta III and Ti_3Al+Nb, respectively, at the lowest satisfactory compaction temperature conditions.

4. The fine grain size of the RS ribbons was maintained in the compact matrices during the bonding cycle promoting superplastic deformation behavior which is possibly responsible for the relatively low temperature fiber/matrix compaction.

Acknowledgments

The authors wish to thank Mark S. Dodd, N. Glenn Lovell, and Robert D. Brodecki for their assistance in the various stages of this work. Also the contribution of Ms. Karen A. Sitzman in typing the text is recognized. Parts of this work were done under Air Force Contract No. F33615-82-C-5078.

References

1. P. R. Smith and F. H. Froes, "Developments in Titanium Metal Matrix Composites," Journal of Metals, 36(3)(1984), 19-26.

2. A. G. Hicks and H. W. Rosenberg, "Ti-15-3 Foil Properties and Applications," Beta Titanium Alloys in the 1980's, ed. R. R. Boyer and H. W. Rosenberg (Warrendale, PA: The Metallurgical Society, 1984), 231-237.

3. P. R. Smith and F. H. Froes, "Titanium Metal-Matrix Composites," Patent No. 4,499,156, February 12, 1985.

4. H. Jones, Rapid Solidification of Metals and Alloys (London: The Institute of Metallurgists, 1982).

5. S. J. Savage and F. H. Froes, "Production of Rapidly Solidified Metals and Alloys," Journal of Metals, 36(4)(1984), 20-33.

6. F. H. Froes and H. B. Bomberger, "The Beta Titanium Alloys," Journal of Metals, 37(7)(1985), 28-37.

7. F. H. Froes, C. F. Yolton, and J. P. Hirth, "Grain Growth in Beta III Titanium," Beta Titanium Alloys in the 1980's, ed. R. R. Boyer and H. W. Rosenberg (Warrendale, PA: The Metallurgical Society, 1984), 185-207.

8. J. P. Beckman and C. F. Yolton, "Beta III (Ti-11.5Mo-6Zr-4.5Sn)," Beta Titanium Alloys in the 1980's, ed. R. R. Boyer and H. W. Rosenberg (Warrendale, PA: The Metallurgical Society, 1984), 401-408.

9. P. J. Bania, G. A. Lenning, and J. A. Hall, "Development and Properties of Ti-15V-3Cr-3Sn-3Al," Beta Titanium Alloys in the 1980's, ed. R. R. Boyer and H. W. Rosenberg (Warrendale, PA: The Metallurgical Society, 1984), 209-228.

10. H. W. Rosenberg, "Ti-15-3 Property Data," Beta Titanium Alloys in the 1980's, ed. R. R. Boyer and H. W. Rosenberg (Warrendale, PA: The Metallurgical Society, 1984), 409-425.

Subject Index

THE UNDERLINED PAGE NUMBER(S) DENOTE LOCATIONS IN WHICH THE SUBJECT IS DESCRIBED OR DEFINED IN GREATER DETAIL.

Author Index